MPJ's Ultimate Math Lessons

Over 100 classroom-tested projects and ideas for Algebra, Pre-Algebra, and Geometry

by Chris Shore

with contributions from the readers of *The Math Projects Journal*

Design by Greg Rhodes

The Math Projects Journal
Murrieta, CA 92562

Dedicated to my children, Rylie and Preston,
whose own math education was on my mind with the creation of each lesson.

Second Edition

Library of Congress Control Number: 2003111322

Printed in the United States of America

ISBN 0-9724057-0-4

The Math Projects Journal
23983 Crowned Partridge Lane
Murrieta, CA 92562
1-877-MATH-123
http://www.mathprojects.com

Contents

Contents *(continued)*

ALGEBRA LESSONS (continued)

GEOMETRY LESSONS

ARTICLES

Contents *(continued)*

ARTICLES (continued)

ADDITIONAL ACTIVITIES

Introduction

The greatest influence on student success is the decisions that the teacher makes in the classroom on a daily basis. These decisions have more of an impact than any demographic factor the child bears, or any instructional material the school purchases. Yet, the American math teacher is trained to follow the textbook, demonstrate examples and administer homework. We have not been trained to make informed decisions on effective ways of teaching mathematics. Overcoming that hurdle is what this book is all about.

For six years, *The Math Projects Journal* has offered innovative tools for teachers who are looking to step out of the traditional model of instruction, without leaving the traditional curricula, and to make decisions on behalf of the students. (see *The MPJ Story*) This book is a compilation of *MPJ*'s best lessons, articles and activities that have been published in our newsletter. To truly understand both the intent and the content of this book, you need to understand the following premises:

1. **The primary purpose of these lessons is the mathematical competency of our students.**

   Yes, the activities in this book appeal to the interest of students and, henceforth, are considered to be fun and engaging. However, we want the students to be intellectually engaged, not just physically or socially. Therefore, while some of these lessons involve such educational fads as cooperative learning, active learning, and manipulatives, these methods are implemented only when they take students beyond superficial memorization, and assist them in gaining a deep and rich understanding of conventional mathematical content. Our students need to be able to solve, factor, and graph equations as well calculate probability, surface area and volume, and ultimately, prove mathematical conjectures. In the current climate of high stakes testing, without exceptions or excuses, our students must know their stuff.

2. **There is no one best way to teach.**

   There are several effective ways to teach, and an innumerable amount of ineffective ways. From personal experience and from the feedback of *MPJ*'s readership, I can say with confidence that the lessons offered here are very effective. I believe all the lessons in this book are good; some are great; and several are absolutely awesome. That does not make them a panacea. The very nature of the book implies that there always exists another way to approach a topic, which may help the teacher reach students that previously were unreachable.

3. **It is the teacher that teaches, not *MPJ*.**

   If you simply disseminate our student handouts as worksheets to be collected and graded as homework, then you will still be following the conventional, textbook-driven model of instruction. If you choose to do this, you will have wasted your time and money with this book. These lessons are only a vehicle for you, the instructor, to impart mathematical knowledge, understanding and skill to your students. You are the teacher. *MPJ* is only a tool.

4. ***MPJ* very loosely defines the term projects.**

   I have been called to task a few times for my use of that word. Many people view projects as long-term, activities in which the student creates some product, much like a model of the solar system at a science fair. Some of our projects are like that. Some are not. In fact, the nature of the projects vary as widely as the topics they address. They range in length from ten minutes to an entire week. Some are student-driven, while others are more teacher-directed. *MPJ* has never been formulaic, so the term *projects* refers to any non-traditional lesson that helps the students understand mathematics. Semantics aside, anything that works is good by us.

Introduction *(continued)*

With that said, here is what you will find in *MPJ's Ultimate Math Lessons*:

Projects, projects and more projects
The best of *MPJ*'s previously published lessons have been compiled by course (Pre-Algebra, Algebra and Geometry) in a sequence that corresponds to traditional curricula. The majority of the projects have both lesson plans as well as student handouts. This book also contains some new student handouts for some of our most popular lessons that did not have them previously.

Articles
Since the teachers' decisions are so crucial to successful learning, our most informative and thought provoking articles are included here. As a whole, these articles embody the general philosophy of *The Math Projects Journal*, and are intended to radically change the way you think about education and the way you approach your teaching of mathematics. There are also articles of a more pragmatic nature, addressing issues of testing, grading, portfolios, etc. This section also includes various samples of Greg Rhodes' cover art, which has been such a special part of *MPJ*'s history and spirit.

Activities
Several issues of the newsletter included a section titled "Whatever Its Worth." As the subtitle indicated, these were activities that qualified as "really cool stuff that you don't know what to do with, but is worth doing anyway." These activities are relatively brief, but offer substantial mathematical content.

If you have read this far, then I know you share in the goals and purpose of *MPJ*, and that you desire to take your students mathematically deeper and higher than any drill-n-kill, textbook regime can ever attempt to go. On behalf of your students, I applaud your courage. On behalf of *The Math Projects Journal*, I thank you for your support.

Chris Shore
Editor in Chief, *The Math Projects Journal*

The MPJ Story

It seems like *The Math Projects Journal* was born just yesterday, and yet, it is hard to remember what teaching was like for me before the days of nation-wide collaboration and the myriad of lessons that have been shared from teachers around the world through this publication. For those of you who are new to *The Math Projects Journal*, here is *MPJ*'s story.

Six years ago, I had an idea that would allow me to share math project lessons and philosophies with other math project enthusiasts. I recruited Greg Rhodes as a partner in this endeavor because he was a techno-guru and a very talented young teacher. He was also a former student teacher of mine, so we were kindred spirits when it came to unconventional math lessons. Our school had recently been wired for Internet access, so we thought we would create a simple web page that could reach out into the world of cyberspace and form a network of teachers that could share ideas and lessons. Before that plan was launched, the single web page grew to a multi-page, interactive web site, with an accompanying hardcopy, bimonthly newsletter, *The Math Projects Journal*.

Initially, lesson ideas for this infant newsletter were fueled by a rather unique collaborative experience. At that time, Greg and I, along with some of the other teachers in our math department, were involved in a training program that encouraged teachers to investigate alternate means of mathematics instruction. That training spurred the creation of a department-wide instructional program that led to extraordinary student success on the state achievement tests. The first year of *MPJ*, was, in essence, a documentary of the lessons that we used in that program.

The journal grew quickly. After that first year of publishing, we expanded the newsletter from eight pages to twelve in order to offer student handouts that would augment the teacher lesson plans. The expanded format also allowed us the luxury of displaying cover art, which showcased Greg's talents as a graphic designer, and promoted the fun-loving spirit of *MPJ*.

MPJ matured in substance as well as style. As our readership expanded, we began to receive submissions from teachers outside of our own instructional program. *The Math Projects Journal* was becoming the network of math teachers that it was intended to be. Since the journal was quickly becoming an influential voice for this network, my interest in education research was now stirred by a responsibility to learn more about the art and science of teaching. In my quest for this knowledge, I was amazingly blessed to be associated with many intelligent people who were directly involved in the current research, and their teaching powerfully impacted the content and the philosophy of *MPJ*.

One of those impactful people was Dr. William Schmidt, of the Third International Mathematics and Science Study. We heard Dr. Schmidt speak on the need for a more focused curriculum in this country, but we wanted information on how teachers could influence change from within the walls of their own classrooms. By his own graciousness and commitment to the message, Dr. Schmidt granted us an hour-long interview and shared the basic findings of TIMSS. I later dubbed these findings "The Four and Half Principles of Quality Math Instruction." Those basic principles (standards, concepts, substance, accountability, and rapport) became the overarching philosophy that, since then, shaped my teaching and the lessons within *MPJ*. You will see that *MPJ*'s lessons and articles are focused more on student understanding and learning, and less on just offering something fun and different in the classroom.

As *MPJ* began to pick up steam, my family and I moved to another part of the state. I left a math department that had achieved so much, and began anew at another high school. I was blessed again with a very enthusiastic group of teachers who were willing to experiment. We began a Lesson Study Program in which teachers collaborated to create, implement and observe innovative math lessons. Many of the lessons developed in that program were published in *MPJ*. This collaborative effort also involved other teachers from our district including

THE MPJ Story *(continued)*

many of its middle school teachers. Concurrently, I was asked to share the research findings with our district's elementary school teachers. Working with math teachers from across the grade levels gave me invaluable new insights into the broader issues challenging math educators, which further influenced the content of the journal.

Due to the reputation of the newsletter, I started to receive invitations to speak at conferences and other events. Eventually, through Prime Presentations, I began conducting training workshops around the country, while maintaining my position as a high school math teacher. The message of teaching for the conceptual understanding of a limited number of topics by holding students accountable to rigorous mathematical substance was very well received. Sharing with teachers both locally and nationally further extended the *MPJ* network and established many collaborative relationships. Interestingly, these three aspects of my career - teaching, training, writing - so mutually influenced each other that the lines between them blurred, to the point at which they have all, some how, left their mark on the pages of *MPJ*.

That is how *The Math Projects Journal* got to this point, ready to move into a new phase of its life. Currently, *MPJ* is going through several big changes. First of all, Greg will be leaving to focus on his new career as a programmer and web developer; his influence on the journal will be sorely missed. Also, *MPJ* will no longer be publishing the newsletter in hardcopy form. Instead, it will now offer new lesson plans and student handouts online. These lesson pages will be accessible on our web site to anyone who purchases this book (see back page). With the new flexible format, we can now post some of the elementary school lessons that we have developed as well. With the creation of this book, *MPJ* will now have access to catalogs, bookstores and other avenues that are closed to periodicals. In its new form, *MPJ* will be able to reach more people with more material, and thus, have a greater impact.

While its function has changed, the initial intent of *The Math Projects Journal* still remains. *MPJ* will continue to create innovative lessons, challenge teachers' perceptions of effective mathematics instruction, and foster its network of committed math educators.

Acknowledgements

Anyone who has impacted me as a teacher, a trainer or an author has in some way affected the creation of this book. Thank you to:

First and foremost, the readers of *The Math Projects Journal*. The journal would not have survived the six years that it did without you who supported it so enthusiastically. Your students are blessed to have teachers who have both the willingness to think outside the box, and the courage to step outside it as well. Special thanks go to those who submitted lessons to *MPJ*, especially our international contributors, from whose continual correspondence I have learned so much. I am also grateful to College Preparatory Mathematics (CPM) for allowing us to publish modified versions of their lessons.

Dr. Harris Shultz of California State University, Fullerton and Ed Rodevich of the Orange County Department of Education. My participation in your C-Cubed program radically changed my teaching practices. The first lessons published by *MPJ* were born in that teacher-training program. You also funded the initial publications of *MPJ*. Ed and Harris, this book is just as much the fruit of your labor as it is mine. Through the C-Cubed training, I met Dr. Uri Treisman of the University Texas, Austin. You inspired the cooperative learning model espoused in these pages (though you claim that you have no memory of that), and your work with remedial students has given me the faith that, indeed, we can reach these students by challenging them with serious mathematics.

Dr. William Schmidt of the University of Michigan. You granted us the interview on the TIMSS report when *MPJ* was a fledgling newsletter. Your candid and simple message on the results of the international studies spawned the "Four and a Half Principles of Quality Math Instruction" which shaped all future lessons in the journal, and forever changed the way I personally view education. I pray that this book will perpetuate in others the radical paradigm shift that you affected in me.

My master teacher, Don Stoll. From my first day of student teaching you challenged me to place myself in the student's seat. I was to keep them engaged, by keeping them in mind. You encouraged me to be myself in the classroom and to experiment freely; I have the bruises and scars to show that I have continued to do both.

Dr. Michael Elliot and Dr. Dennis Rader, formerly of United States International University. I originally enrolled in your classes simply for salary credit and you introduced me to a whole new world of intellectual thought. Mike, you also taught me to look at the big picture and to avoid the reductionist mentality which so plagues the American education system. Dennis, you also gave me my storytelling license; as you can see from the context of these lessons, I am wielding that license regularly. You both gave me my first speaking engagements, which consequently began my career as a teacher trainer.

Dr. Tom Bennett of California State University, San Marcos. You have taught me so much about how students learn mathematics, and have given me great opportunities to share that knowledge with others. The seed of the Lesson Study Program that you planted in our district has already grown tall and born great fruit.

My colleagues and dear friends at Trabuco Hills High School. Those of you who, out of either curiosity or friendship, joined me in the C-Cubed program — our success on the state exams gave credibility to what otherwise would have been just another faithful but failed experiment in youthful idealism. And to the ITC team, I consider teaching with you for that year in the Freshman Academy to be one of the highlights of my career. It was from that experience of working with teachers from other content areas that I gleaned much of my insights on the accountability and assessment issues in mathematics.

My many colleagues and new friends at Temecula Valley High School and throughout the district, with whom and from whom I have learned so much. Many of your lessons created in the Lesson Studies Program made their way into this book. You all are a big part of the reason why I enjoy going to work each day. You were all so receptive of these strange new ideas from the new kid in town. I cannot tell you how grateful I am for that.

Acknowledgements *(continued)*

Carol and Larry McGehe, of Prime Presentations, for giving me a national audience as a workshop presenter. Your faith in me was a great gamble at that time. You have had a great impact on my life and career. You run a top-notch organization, and I consider it a great pleasure to call you colleagues and friends.

Denny Bethke, you offered critical guidance regarding printing and publishing when *MPJ* was in its infancy. Donna Beecham of Maurice Printers, and Rod and Mary Ann Davenport of Postal Connections, you offered discounts at a time when *MPJ* really needed it. I sense that your generosity stemmed from a belief in the cause, as well as a kind heart.

Chris Grant, for helping us take our first steps in desktop publishing, web design, and for the host of advice on starting an online business. Your enthusiasm for this project is always felt and appreciated.

My mother and first teacher, Jo Lancaster, who, by overcoming so much with such grace, taught me lessons in sacrifice, perseverance, passion, love and wisdom. My sister and brother, Denise and Keith, your own successes in coaching and business leadership have inspired your older brother, and our childhood stories have made their way into more of my lessons than you really want to know. I love you!

My children, Rylie and Preston. So many times you happily volunteered as guinea pigs for new lessons, which gave me so many insights into how children think and learn. Your enthusiasm for Daddy's job is evident in your smiles. I intend to take advantage of these times when you still want me to teach in your classrooms and still want to come visit mine. You are true joys!

My wife, Casie, who taught me to ask the troubled and struggling students "What's up?" You have been instrumental in my quest for humility, which is the only way I ever would have believed it was my duty to reach out to those students. On this project, you were my sounding board, my editor, and my cheerleader, who always challenged me to stick to the spirit of *MPJ*. I treasure your belief in my work. Endeavors like this have their fair share of risks. As always...thanks for the adventure. I adore you!

Greg Rhodes, *MPJ* 's graphic artist and web designer, my partner, colleague and great friend. You jumped at this crazy idea without even flinching, and provided a sense of style that will forever make *The Math Projects Journal* unique. You also provided wisdom at critical times. As a man of integrity, you stuck with this project out of both duty and friendship, which I will never forget. Our partnership possesses a very special synergy. You are awesome, my friend. Pang, Greg's wife, I owe you! Louise Cryer, Greg's mother, you advised us to publish the newsletter quarterly rather than monthly. We eventually published bimonthly and quickly learned the extent of your wisdom.

God. You blessed me with inspiration, talent, creativity, endurance, a multitude of opportunities, mentors, and resources, and most importantly...faith. I ask that this book impact the young people who are touched by its lessons. As always, grant us your blessings so that we may continue to do your work.

Lastly, but most significantly, I wish to thank my students. I pray that I serve you well.

Pre-Algebra Lessons

1in = 5 mi
3.5 in = 17.5 mi

Ms Liters

Me	63"	height	Doll	15"	Rat: 4.2
	21	head		5	
	~~32~~	~~chest~~		~~6~~	
	26	waist		6.2	
	32	ins		7.6	
	9	foot		2.1	

Barbie?

Hills High School, CA

Concepts
Ratios, proportions, similarity, measurement, Fundamental Theory of Similarity (optional)

Time: 1-2 days

Materials
An assortment of Barbie dolls and action figures (Batman™, G.I. Joe™, Power Rangers™, etc.) Student Handout

Preparation
Tell the students in advance to bring dolls and action figures on the day of the lab. It helps to offer an incentive for the students (extra credit, etc.) Each group will need one male and one female doll. It will be helpful to have extras on hand.

Have students work with a partner or in groups. Instruct them to take the following measurements of both Barbie™ and one male action figure: height, head (circumference), chest, waist, inseam, and foot (length). The measurements are written in the first blank column of each of the charts. The numbers already given in the chart are the average measurements for females and males. The students will use these numbers to calculate their size change magnitudes.

Once students have completed the measurements, walk them through the first conversion, using height to be the first standard. For example, if Barbie is 12 inches tall and the average woman is 65 inches tall, what would the other enlarged measurements be? Divide 65 by 12, to get a ratio of 5.6, then multiply each of the measurements by this ratio. Emphasize that this represents one "possible" body shape. The next conversion will use 21 inches as the standard for the head. Calculate this new ratio and repeat the process. This represents another "possible" body shape.

After the walk through, allow the students to complete the chart by calculating the conversions for the other standards. Reinforce that each column is one possible enlargement of the doll and that the body shape enlargements will not always come out looking the same.

To conclude the activity, the students should summarize their findings, and make conjectures in regards to why manufacturers chose the particular proportions of the dolls. There are four questions on the handout to guide the students in their analysis.

TEACHER COMMENTS

- Stress that each column in the chart is considered a different enlargement. Emphasize that for #1, you might be enlarging the doll to an average height, but for #2 you are enlarging the doll to an average head.
- It is helpful to model an example of the lab before the students begin. With a doll and tape measure, show them how to measure the doll's attributes, fill in the chart, calculate the magnitude, and enlarge all the measurements.
- The primary concept here is ratio and proportion, which can be dramatically demonstrated by comparing the bodies represented in each column. For example, at average height, Barbie has a large head with small feet, which means she cannot stand. (She could not at doll size, either.) Given an average foot size, her height is gargantuan, as is her head, yet her feet still are not big enough to support her. This is because each attribute of her body was multiplied by the same ratio; in other words, each body enlargement is proportional to the others.

NOT

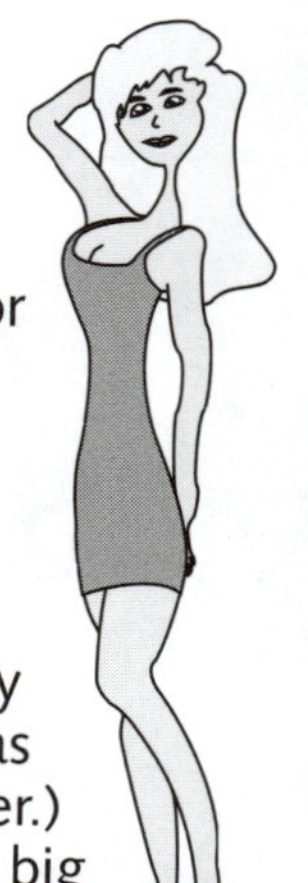

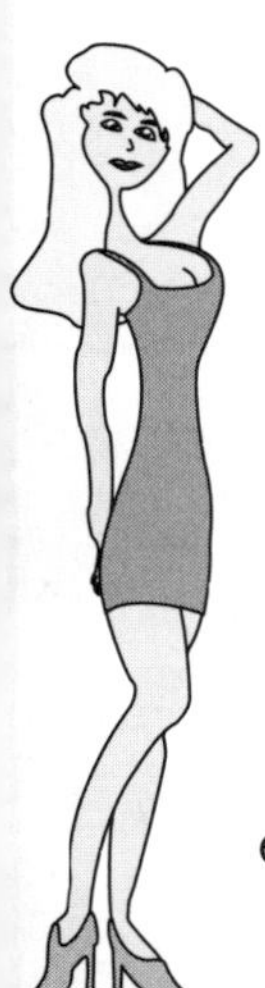

How Big is Barbie?

If Barbie™ was the size of an average woman and Batman™ an average man, what would they look like?

First, find two dolls or action figures (one male, one female) and measure various attributes of their bodies (see charts below). Second, calculate an appropriate scale factor to enlarge a certain dimension to a certain size. For example, in the chart below, the height of the average male is listed as 72 inches. After you measure your male doll, multiply the figure's measurements by some factor in order to enlarge it to 72 inches. Then use that scale factor to determine the other enlarged measurements. You will perform this process several times for each figure. The numbers provided are the hypothetical average measurements for females and males.

DATA CHARTS

MALE	Doll	Height	Head	Chest	Waist	Inseam	Foot
Ratio							
Height		72"					
Head			23"				
Chest				40"			
Waist					32"		
Inseam						32"	
Foot							11"

Divide avg / doll measurements

FEMALE	Doll	Height	Head	Chest	Waist	Inseam	Foot
Ratio		6.5	7	7.2	4	6	9
Height	10	65"					
Head	3		21"				
Chest	5			36"			
Waist	6				24"		
Inseam	5					30"	
Foot	1						9"

Remember: Each column represents a different possible enlargement of the doll. Look for their distinct characteristics. What makes one different from another?

ASSIGNMENT

Once you have the chart completed, answer the following questions about your data:

1. What are your initial impressions as you look at your results?
2. What would the figures look like if they were real people? Describe each possibility in detail.
3. How do the female's measurements in the height column compare to those in the foot column?
4. Why do you think the figures are designed with such measurements? Do these designs have social impact?

Extra Credit you + a partner

Option #1: Take measurements of yourself and scale them backwards (reduce them) to determine the dimensions of a doll or action figure modeled after you. Write a paragraph describing what you would look like as a doll. You may also draw a picture to illustrate.

Option #2: Using the given measurements in one of the charts as the "average" male or female, scale the measurements backwards (reduce them) to determine the dimensions of a 12-inch doll or action figure with proportional measurements. Then compare the doll with your original doll's measurements and write a paragraph explaining the similarities and differences.

m&m™ Count & Crunch

Submitted by Debbie Osborne, Trabuco Hills H.S. Mission Viejo, CA

PART ONE

Give students an overview of the entire project. Inform them that they will apply their probability and statistics skills to make a prediction on the number of candies for each color in the one-pound bag. They are not to eat the candies from the individual bags; they will be saving these at the conclusion of each day. There will be a chance to eat the candies from the larger bag at the conclusion of the project. (The numbers here correspond to those on the student handouts.)

1. Have the students estimate the number of candies in the bag and the quantity of each color. In the chart provided, they are to record their guesses as raw numbers as well as percentages.
2. Each group is to open its bag and the members are to count the actual quantity of each color and the total number of candies. These are to be recorded, and the percentages calculated.
3. The students create a bar graph to represent their data.
4. The next step involves creating pie charts. Be sure the students understand that the pie charts are to accurately represent the percentage of each color for both the estimated and actual candy counts. For instance, if the blue candy count is 13%, then the blue pie sector should measure 13% of 360 which is approximately 37 degrees.

PART TWO

5. Have the students determine the weight and diameter of a single candy. The weight of an individual candy can be determined by dividing the number of actual candies by the total weight which is printed on the bag. The instructor may tell the students this simple strategy or leave them to explore a variety of strategies with weights and scales.
6. Step six of the project stresses unit conversion. Make sure that students know that, although there are 12 inches in a foot, there are 144 square inches in a square foot. In part 6d, assign each group one color that they are to use to answer the question.

Concepts
Probability & statistics, ratio & proportion, percentage, unit conversion, area, bar graphs and pie charts

Time: 3-4 hours

Materials
1 individual package of m&m's per group of 3-5 students. 1 one-pound bag of m&m's per class, protractors, rulers, coloring pencils/markers and student handout.

Preparation
Since this project incorporates such a variety of skills, it serves best as reinforcement or review. Students should be familiar with some of the basic principles of probability and statistics. You will need to save each group's bag of candy at the conclusion of each day. A piece of tape with a name on it can be used to seal and identify each bag.

PART THREE

7. Step seven is the statistical study. Have a chart on the board in which each group can present its data. The class can then compute the mean, median, mode and range. This is a good time to discuss which of the central tendencies best represents the data.
8. Students now shift their attention to the one-pound bag by predicting the quantity of each color and the total number of candies. Most teachers choose to demonstrate this with a proportion, yet when left to their own invention, many students will convert all ratios to decimal form and simply multiply it to all the appropriate quantities. Counting the candies in the large bag is best accomplished by distributing the m&m's among all the groups and compiling the results.

PART FOUR

9. For the student-derived probability experiment, assign each group a different situation to test. For instance, one group can test the simple probability of drawing a red candy from the bag, while another group can test the probability of drawing a green then a blue candy. Each group will calculate the probability of their assigned situation for their small bag, then make 50 trials and record the relative frequency. Once the experiment is complete, have the students compare the results of the trials with the calculated probability.

Count & Crunch

1. Estimate the quantity of each color and the total number of candies in your bag. Also guess the percentage of the total that each color comprises. Record your results in the left chart below.

2. Open your bag and count the actual number of each color and calculate the percentage.

ESTIMATIONS

Color	Number	%
Red		
Brown		
Yellow		
Green		
Orange		
Blue		
TOTALS		

ACTUAL DATA

Color	Number	%
Red		
Brown		
Yellow		
Green		
Orange		
Blue		
TOTALS		

3. Make a bar graph showing your estimations and the actual count of each color.

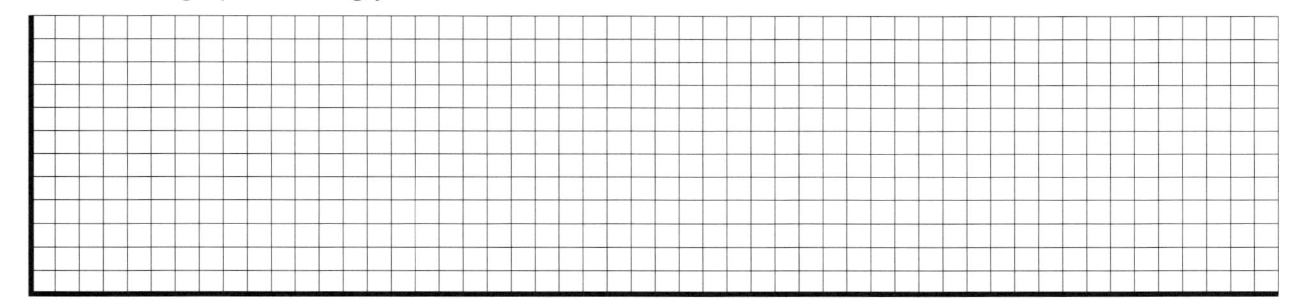

4. Make two pie charts, each representing the percentage of each color in the bag. The first pie chart will represent your estimations, the second is to represent the actual counts. Each sector of the pie chart should be proportional to the percentage it represents. For instance, if you are graphing 13% for yellow, then your yellow sector should measure 13% of 360 degrees.

ESTIMATIONS

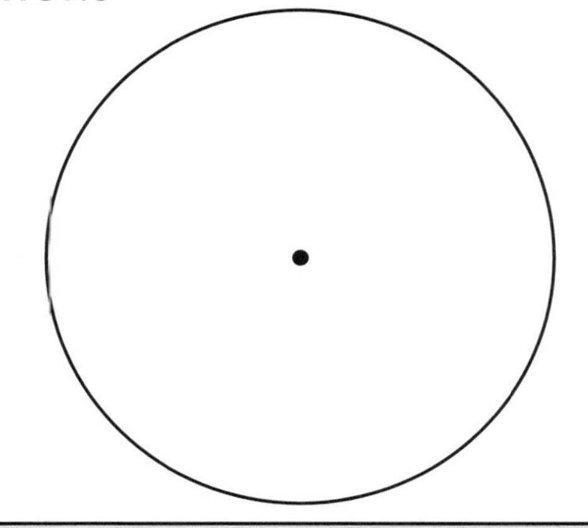

ACTUAL DATA

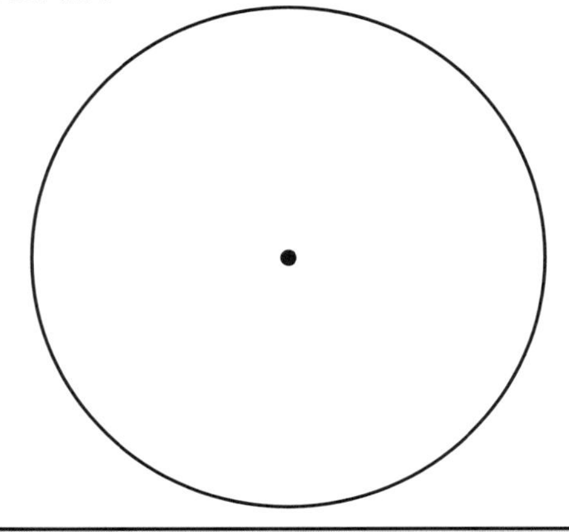

m&m™ Count & Crunch

5. Characteristics of one m&m candy (show all calculations)
   a. Weight:

   b. Diameter:

6. Unit Conversion & Area
   a. How many m&m's fit in an inch?
      ... in a foot?
      ... in a mile?

   b. How many fit in a square inch?
      ... in a square foot?
      ... in a square mile?

   c. How many would cover a football field?
      (The dimensions of a football field, not including the end zones, are 100 yards by 55 yards.)

   d. How many bags of m&m's would it take to cover the field in one color?
      Group's assigned color:

7. Statistical Study
   a. Record the m&m data from all the groups in the class:

Group	Red	Brown	Yellow	Green	Orange	Blue
#1						
#2						
#3						
#4						
#5						
#6						
#7						
#8						
#9						
TOTALS						

   b. Calculate the mean, median, mode and range for each color.

	Red	Brown	Yellow	Green	Orange	Blue
Mean						
Median						
Mode						
Range						

m&m™ Count & Crunch

8. Proportions: Small Bag vs. Large Bag
   a. Using the figures from your group's small bag, estimate the quantity of each color and the total number of m&m's in the one-pound bag.

   b. Using the figures from the class statistics, estimate the quantity of each color and the total number of m&m's in the one-pound bag. Be sure to designate which statistics your group chose to use.

   c. As a class, open the one-pound bag and count the actual quantity of each color and the total.

	a) Est. from Group's Bag		b) Est. from Class's Stats		Actual Bag
Color	**Small Bag**	**Large Bag**	**Class Stats**	**Large Bag**	**Large Bag**
Red					
Brown					
Yellow					
Green					
Orange					
Blues					
TOTALS					

   d. Which provided a more accurate prediction, the data from the individual bag or the class statistics?

9. Probability
   a. Compute the probability of each of the following occurrences, if the first candy is replaced.

   b. Compute the probability of each of the following occurrences, if the first candy is not replaced.

First Draw	Second Draw	a) Probability w/ Replacement	a) Probability w/o Replacement
Blue	Red		
Red	Yellow		
Brown	Brown		
Blue	Orange		
Yellow	Red		
Green	Green		

   c. Do your odds improve or diminish by not replacing the candy?

   d. Conduct an experiment to test one of the above. On another sheet of paper, describe your experiment and record the data, analyze and display the results.

PROJECT

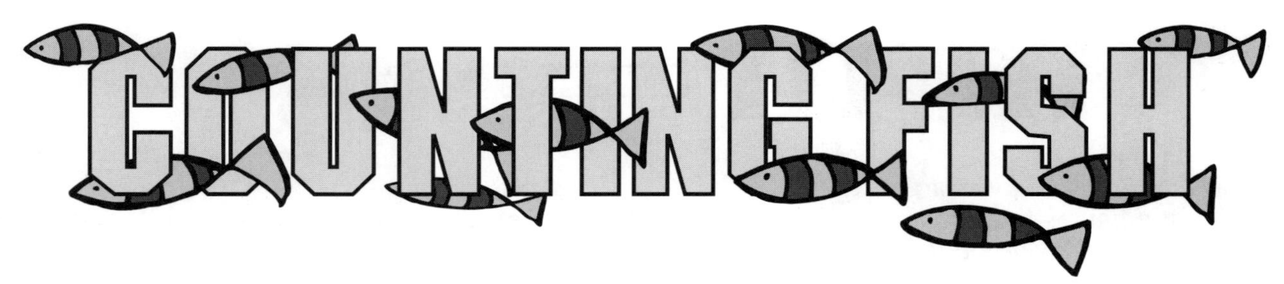

LESSON PLAN

The students will use ratios and proportions, as well as sampling to predict the number of beans in a plastic sandwich bag. The process mimics one method wildlife professionals use to estimate fish and game populations. In this case, the activity models a fish population in a lake.

Begin by having the students read the front side of the handout. The students have probably seen some kind of television documentary in which scientist are capturing and tagging large game to be released back into the wild. The steps for the procedure are fairly explicit; however, it is suggested that you model the first sampling.

Concepts
Ratio and proportion, sampling

Time: 1-2 hours

Materials
Student handout (optional); Each group needs: a ziplock sandwich bag half-full of brown beans (approximately 1,000), a smaller ziplock bag half-full of white beans, and a small paper cup. One large store-bought bag of brown and a small bag of white pinto beans will usually suffice for an entire class.

Preparation
Place the smaller bag of white beans and the paper cup inside the large bag with the brown beans and zip it shut. This is one self-contained package for a single group.

SAMPLING PROCEDURE

1. Scoop a portion of the brown beans from the large bag using the paper cup. Suggest that for the first sampling they fill the cup about halfway. For the model, you do not need to actually take the time to count the beans. Simply use an example, "If I have 50 brown beans, I will replace them with 50 white beans." Set aside the brown beans, and dump a bunch of white ones back into the bag.

2. Shake and knead the bag very well. Stress that we want the beans evenly distributed. It is easy to mix when the rest of the bag is puffed with air. Re-scoop any quantity of beans.

3. Pretend that you count the total number of beans — "One, two, three ... one hundred," — as well as the number of white versus brown beans that are in the cup. Remind the students of our assumption that the ratio of white beans to total beans in this cup will be the same as the ratio of white beans to total beans in the bag.

4. It is up to you as to how much of the actual solution of the proportion you wish to model. Emphasize that the calculations should be recorded as well as the solution. The margin of error is the ratio of their answer to the true count, so the students will have to complete step six, before they calculate this for the first sampling.

5. Tell them that they will then separate the different colored beans and count all the brown ones. They love that thought! It actually does not take as long as it might sound, especially if the group divides the beans and counts in "little heaps."

6. The students will repeat steps 1-5 and record their results. They should decide whether to sample a larger or smaller quantity than the first in order to get a more accurate estimate. Larger usually proves best. Assure them that they only have to count the entire bag of beans once, even though they will be sampling twice. Leave time for them to pick out the white beans at the end of the activity.

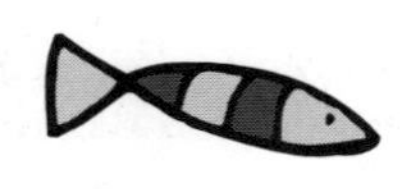

SCENARIO

You are a fish and game warden for Lake Holattawatta. One of your jobs is to estimate the number of fish in the lake. The last warden, Larry Lamo, lost his job, because he couldn't do it right. At first, he tried to swim among the fish with scuba gear and count them all. He kept losing track and would have to start over. In his frustration, he gave up and said that Holattawatta had a holattafish. Needless to say, the Fish & Game department fired him and hired you.

You were granted the job because of your mathematical skills in sampling. You intend to net several samples of fish, tag the fish and release them back into the lake. After an adequate amount of time, you will net several samples again. The ratio of tagged fish to untagged in your sample will give you an idea of how many fish are in the lake.

For example, assume you capture, tag and release 100 fish. You then recapture 200 fish, of which 4 are tagged. You will assume that the ratio of fish originally-tagged to total fish in the lake is equivalent to the ratio of recaptured-tagged fish to total recaptured fish. In other words ...

$$\frac{\text{(tagged original)}}{\text{(total lake)}} \quad \frac{100}{x} = \frac{4}{200} \quad \frac{\text{(tagged recaptured)}}{\text{(total recaptured)}}$$

Solving the proportion yields a total lake population of 5,000 fish.

PROCEDURE

Since Larry Lamo lost his job for inaccurately counting the fish, you want to get this right. Therefore, you are going to conduct an experiment to validate your procedure.

1. In the large baggy (lake) are brown beans (untagged fish). Scoop a portion of them in the paper cup (net) and replace the brown beans in the cup with white beans (tag the fish). COUNT the white beans and pour them into the big baggy (release the tagged fish back into the lake).
2. Shake the baggy very well so that the beans are evenly distributed (wait an adequate amount of time). Again, scoop some beans (recapture some fish).
3. Count the number of white versus brown beans that are in the cup (tagged versus untagged).
4. Use the data you have gathered in a proportion to approximate the number of beans in the baggy (fish in the lake). Calculate the margin of error after you complete step six.
5. Once you are done, count the total number of brown beans that you started with, and judge the accuracy of your procedure.
6. Repeat steps 1-4 and record your results. Decide what you can do differently this second time to improve your accuracy.
7. Determine which sampling was more accurate and explain why.

FIRST SAMPLING

1. Total brown beans scooped from the baggy and replaced with white beans: 1) ____________

2. Total number of beans (brown and white) re-scooped: 2) ____________

3. Number of white beans only in the re-scooped sample: 3) ____________

4. Proportion to be used: Solve:

   ________ = ________

   a) Number of total brown beans estimated in the original baggy: 4a) ____________

   b) Margin of error: 4b) ____________ %

THE TRUE COUNT

5. Total number of brown beans COUNTED to be in the baggy: 5) ____________

SECOND SAMPLING

1. Total brown beans scooped from the baggy and replaced with white beans: 1) ____________

2. Total number of beans (brown and white) re-scooped: 2) ____________

3. Number of white beans only in the re-scooped sample: 3) ____________

4. Proportion to be used: Solve:

   ________ = ________

   a) Number of total brown beans estimated in the original baggy: 4a) ____________

   b) Margin of error: 4b) ____________ %

Which sampling gave you a more accurate estimate and why?

____

____

____

Don't Break My Stride

How fast does the average high school student walk in miles per hour? How long would it take that person to walk from school to the nearest shopping mall?

In this activity, students will calculate their walking rate and the length of their strides. They will then use basic statistics to approximate a given distance, such as a school parking lot or gymnasium, as well as use unit conversion to approximate the time it will take to walk a considerably longer distance. Finally, the students will graph and compare their data.

Concepts
Rate, mean, median, mode, range, plotting points, formulas

Time: 3-4 days

Materials
1 stopwatch per group of students, student handout, Spreadsheet software (optional)

Preparation
Measure and mark a 100 foot distance somewhere on campus. Choose a moderately long distance (Given Distance) like the school parking lot or athletic field and know the approximate length.

LESSON PLAN

Day 1: The Data Collection
Each student in the group walks the 100 foot distance three times. For each of the three trial walks, the students' steps are counted while the walk is timed by a partner. The time in seconds and the number of steps are recorded for each trial on #1 of the student handout. All group members are responsible for recording the data of the other members. Students should write the names of the group members next to the corresponding walker number. Once this is completed, the group has one student make two trial walks of the "Given Distance," timing the walk and counting the steps as with the 100 foot trials and recording the data in #2. Both the walker's number (1-4) and the name of the "Given Distance" (e.g. parking lot) are recorded on the student handout.

Day 2: Statistics and Rate
The students calculate the mean, median, mode, and range for times and steps and record them in #3 on the student handout. Encourage students to do all computations individually, and compare answers, rather than divide the labor and copy answers. Next, in #4 the students calculate the average rate in feet per second by dividing the average time in #3 by the distance (100 feet). They then convert to miles per hour in order to see their rates in a recognizable form. Lastly, the students calculate the length of their strides by dividing the average number of steps by the distance and recording their answers to the nearest hundredth of a foot in #5. To complete #5, students convert their answers from decimal feet into feet and inches.

Day 3: Graphing
On graph paper, students create a one-quadrant graph, labeling the domain as "time" and the range as "number of steps." Students then graph each group member's average time and average steps as an ordered pair. Then students accurately predict the number of steps they would walk in ten seconds and also in forty seconds and plot these points. If done correctly, the three points for any given student, should be collinear.
The students draw lines through each set of three data points and answer the questions: "According to the graph, which member of your group takes the most steps in one second?" and "What characteristic of the graphs brings you to that conclusion?"

TEACHER COMMENTS
1) If you have the equipment, this project is an excellent opportunity to incorporate computer technology. The worksheet can be used to teach students how to create a spreadsheet and input formulas. For instance, the average rate can be calculated by dividing 100 (feet) by value in the cell containing the average time.
2) The data from this activity may be saved and revisited when studying linear equations and slopes.
3) Although we use this project very early in the year to explore new topics, some teachers might find it more appropriate to use it later in the school year for reinforcement or enrichment.

Don't Break My Stride

1. You and each member of the group are to make three trial walks. Count your steps and time for each trial.

	Time			Number of Steps		
	Trial 1	Trial 2	Trial 3	Trial 1	Trial 2	Trial 3
Walker 1						
Walker 2						
Walker 3						
Walker 4						

2. Choose ONE person from your group to walk the distance designated by the instructor. Record both the steps and times for two trials.

	Trial 1	Trial 2	Length
Time			
Steps			

Location ____________________

Walker ____________________

3. Calculate the mean, median, mode, and range of both times and steps for all members.

	Statistics for TIME				Statistics for STEPS			
	Mean	Median	Mode	Range	Mean	Median	Mode	Range
Walker 1								
Walker 2								
Walker 3								
Walker 4								

4. Use the average time to calculate the rate per second.

	Average Rate	
	ft/sec	MPH
Walker 1		
Walker 2		
Walker 3		
Walker 4		

5. Use the average number of steps to calculate the average stride length (in feet).

	Stride Length	
	Feet	Inches
Walker 1		
Walker 2		
Walker 3		
Walker 4		

6. It is __________miles to______________________. The approximate time to walk there is _______________.

4-DIGIT PROBLEM

BY COLLEGE PREPARATORY MATHEMATICS; SACRAMENTO, CALIFORNIA

THE ASSIGNMENT

Arrange four 8's to make each integer from 1 through 5. You are permitted to use any of the four operations (addition, subtraction, multiplication and division) as well as parentheses, roots and exponents. You may also use any number you wish for an exponent or root, but other than that, only 8's are allowed. For instance, 8^2 is permissible, but 2 • 8 is not. The number 88 may count as two 8's, but the number 18 is not permissible. You must use exactly four 8's for each solution. For example: 88/8 + 8 = 19.

Concepts
Order of operations, roots and exponents

Time: One hour

Materials
Butcher block paper (optional)

Preparation
Students need a general understanding of the order of operations. For the optional activity, on several sheets of butcher block paper place the numbers 1-100 in a column down the left margin. Usually, it is best to place 1-25 on the first sheet, 26-50 on the next, etc. Then hang these sheets on a wall in the room.

LESSON PLAN

Begin the class with the following warm-up activity. Show them how 8 + 8 + 8 + 8 = 32 and 8 + 8 • 8 + 8 = 80. Then ask them to arrange four 8's to produce 19, using any or all of the four operations. You may get several like the example given above. Then write the problem 8 + 88/8 on the board and ask your students to simplify it. Many students may erroneously give the answer as 12 (the real answer is 19). Use this example to illustrate the importance of the Order of Operations. Then, challenge them to modify this example without changing the order of the numbers to produce 12: (8 + 88)/8. You can then use this example to illustrate the use of parentheses.

For each integer, there are several solutions. Keep the class discussions open to include them all, and point out the merits of each. Then, have the students create another number, like arranging four 8's to produce 15: 8 + 8 - 8 ÷ 8. Students often like to get tricky and see who can come up with the most clever solution. After they practice, discuss with your students useful tools such as square roots, cube roots, and the exponent of zero.

Helpful Hints:

8/8 = 1	8 - 8 = 0	8 + 8 = 16	88/8 = 11
$8^2 = 64$	$\sqrt[3]{8} = 2$	$8^0 = 1$	$\sqrt{8 + 8} = 4$

Once they have the idea of how the four digit problem works, challenge the students to find each of the integers 1-5. Some possible solutions are listed below.

1: 8 + 8/8 - 8 or 88/88 or $(8 + 8 + 8 + 8)^0$
2: 8/8 + 8/8 or $8^0 + 8^0 + 8 - 8$
3: 88/8 - 8
4: $(8 \cdot 8)^{1/3} + 8 - 8$ or $8^0 + 8^0 + 8^0 + 8^0$
5: $\sqrt{(8 + 8)} + 8/8$

FURTHER INVESTIGATION

For homework you may have students find solutions for the integers 6-10. For future problems, you may use any 4 digits to produce any type of answer. For instance, arrange four 5's to produce 21. Arrange four 3's to produce a negative number, or zero, or even a fraction.

For an ultimate challenge, invite the students to find solutions to the integers 1-100. Place the butcher block paper on the wall, and write the students' solutions next to the corresponding number as they come up with them. Leave a marker handy, and challenge the students to find the rest. Allow the sheets to remain on the wall for several weeks and watch them go for it!

Wallflowers

adding & multiplying integers

THE NEUTRAL FIELD

As the story on the student handout shares, a girl is represented by a "positive" sign and a boy by a "negative" sign. A dancing couple is then represented by a positive paired with a negative, which equates to a "zero." Numerous couples on the dance floor are several of these zeros, which we will call "the neutral field." All the examples in this lesson start with a neutral field of five zeros, but more may be added by the students.

ADDING & SUBTRACTING INTEGERS

1. After colorfully describing the wallflower scenario (see top of first student handout), show the students the first representation: "6 - 8." In the context of the scenario: 6 girls are waiting for dance partners, then 8 girls leave. Model the diagram on the board or overhead. The initial 6 positives are the wallflowers. To represent eight girls leaving, cross-out 8 positive signs (or remove eight positive tile separators from the overhead). To do this, you remove the six female wallflowers, then two more girls from the dance floor, leaving two boys without a dance partner. The students can see the two remaining male wallflowers (negative two). Have them record their answer (-2) in arithmetic form on the handout.

2. For the second scenario, have students publicly offer possible wallflower scenarios. The correct scenario for -3 + 4 is: There are three boys waiting for partners, then four girls enter the dance. How many wallflowers are left? To draw an accurate diagram, the students must show the initial -3 as three negative signs. The four girls are drawn as four positive signs, three of which correspond to (dance with) the three negatives. This leaves one girl (positive one) remaining. Circle this remaining one (+1), then have students record their result arithmetically. (i.e. -3 + 4 = 1).

3. For #3-6, have the students make their attempts at the diagrams and written scenarios. Then have volunteers offer these on the board for discussion and correction. As usual, prod students to look for the underlying patterns and rules that will allow us to do these problems without the diagrams and/or manipulatives. Problem #7 addresses the same topic in a different context, thus it assists in checking for understanding.

Concepts
Adding, subtracting, multiplying and dividing integers

Time: 2 hours. First hour, Adding/Subtracting; second hour Multiplying/Dividing.

Materials
Tile Separators, Student Handout

Preparation
You can effectively conduct this lesson by drawing positive and negative signs on the board, but tile separators make for great manipulative displays on an overhead. Purchase a small bag of tile separators at a hardware store. They look like small rubber "plus" signs. Cut the crossbars off half of them to make "minus" signs.

SOLUTIONS

1) 6 - 8 = **-2**

6 girls are waiting and 8 girls leave.

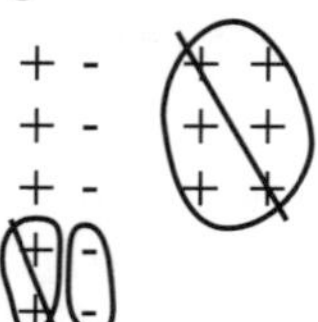

2) -3 + 4 = **1**

3 boys are waiting and 4 girls arrive.

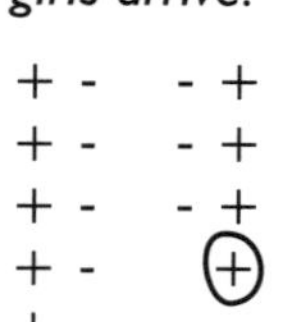

3) -3 - 4 = **-7**

3 boys are waiting and 4 girls leave.

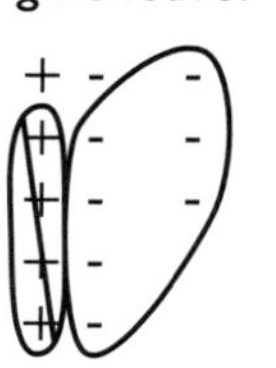

4) 5 - (-1) = **6**

5 girls are waiting and 1 boy leaves.

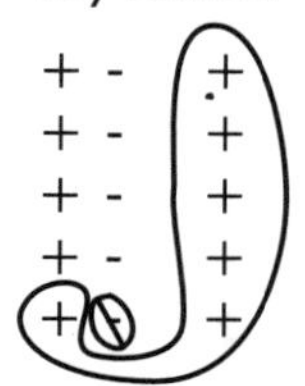

5) -2 - (-5) = **3**

2 boys are waiting and 5 boys leave.

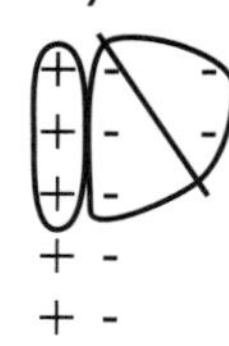

6) -8 + -4 = **-12**

8 boys are waiting and 4 more boys arrive.

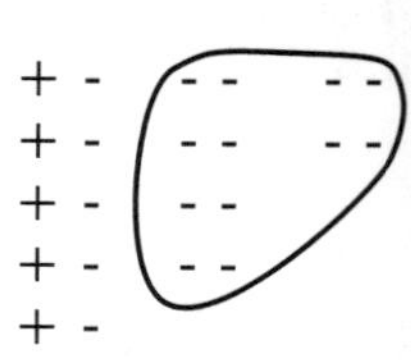

Wallflowers

MULTIPLYING & DIVIDING INTEGERS

Multiplication is based on grouping; therefore, grouping is emphasized in this phase of the wallflower lesson. The multiplier (first value in the expression) designates the number of groups, and the multiplicand (second value) designates the number of people in each group. The sign of the multiplier designates whether the groups are arriving at (+) or leaving from (-) the dance. The sign of the multiplicand designates the gender of the group, girls (+) or boys (-). For example, 4(-2) is translated into a wallflower scenario as "4 groups of two boys enter the dance." In contrast, -2(4) is translated as "2 groups of 4 girls leave the dance."

4. In #8, use the examples above to explain the idea of grouping and the correlations between the expressions and the wallflower scenarios. Challenge the students to offer the scenario for 3(-5). In the wallflower scenario, this translates to "3 groups of 5 boys arrive at the dance." Then have the students draw the three groups of five negative signs. It helps to have the students circle each of these groups.

5. In #9, again challenge the students to generate the scenario for -2(3): "2 groups of 3 girls leave the dance." This is the critical point in the lesson where the meaning of the leading negative sign comes forth for the students. Arithmetically speaking, the negative two implies that we are subtracting two groups of something (in this, case subtracting 2 groups of 3 from zero). In the context of the wallflowers, emphasize that the "-2" implies that two groups are leaving the dance. Have the students draw an extra pair of dancers (+ -) on the dance floor, circle two groups of three positives, then cross out these groups. Six negatives remain.

6. In #10, the students then play with the idea of "a negative times a negative," which means to subtract a certain number of groups of a negative value. In the wallflower context, it means that a certain number of groups of boys are leaving the dance. In this particular case, it is "3 groups of 2 boys leaving." Again, have the students draw an extra pair of dancers (+ -) on the dance floor, circle three groups of two negatives, then cross out these groups. Six positives remain.

7. In #11-13 have the students make their attempts at the diagrams and written scenarios. Then have volunteers offer these on the board for discussion and correction. As with all concrete contexts, we want the students to make some abstract generalizations. After all, trying to represent "16.2 - 3.2(-4.5)" with tile separators would be a cumbersome task, if not an impossible one. Therefore, prod the students to see the general rules for multiplying integers ("two negatives make a positive" etc.), and record these in #14. Then have the students investigate the two scenarios offered in #15 a & b: $^{-12}/_{3}$ = -4 and $^{-15}/_{-5}$ = 3, respectively. The students should see that the rules for multiplication, also hold true for division.

SOLUTIONS

8) 3(-5) = **-15**

3 groups of 5 boys arrive.

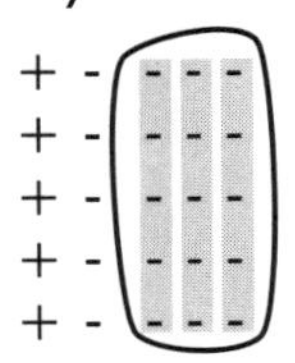

9) -2(3) = **-6**

2 groups of 3 girls leave.

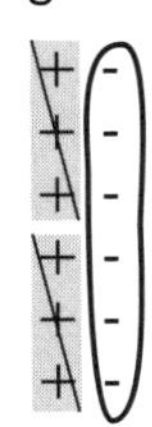

10) -3(-2) = **6**

3 groups of 2 boys leave.

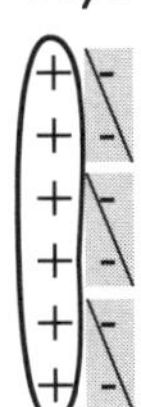

11) 4 - 2(-3) = **10**

4 girls are waiting, 2 groups of 3 boys leave.

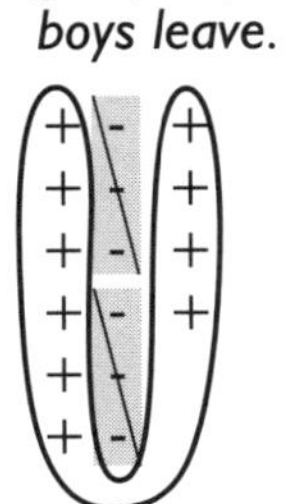

12) 3 + 4(-2) = **-5**

3 girls are waiting, 4 groups of 2 boys arrive.

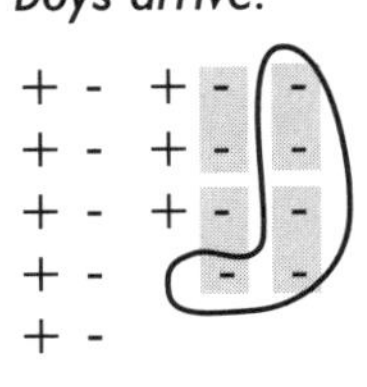

13) -3(-2) - 1 = **5**

3 groups of 2 boys leave, then a girl leaves.

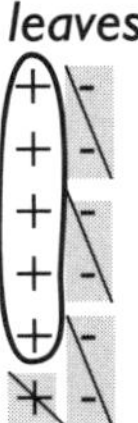

Wallflowers
adding & subtracting integers

Algebra High School is holding its annual Integer Dance in the gymnasium. As with all dances, there is not an equal number of girls and boys, so there are always wallflowers sitting at the dance. Wallflowers are people on the perimeter of the dance floor waiting for a dance partner. The number and gender of the wallflowers can be represented by a mathematical model. (After all, this is Algebra High School.) Girls are represented by positive signs (+), and boys by negative signs (-). A dancing couple, one positive matched with one negative (+ -), equates to zero. Several dance partners make up the dance floor, mathematically termed a "neutral field." Therefore, the given diagram in #1 represents five couples dancing and six girls waiting to dance (female wallflowers).

For #1-6, a) write a scenario that may be represented by the expression, b) represent it with a "wallflower" diagram, c) evaluate the expression.

1. 6 - 8 = _____

   Scenario: ______________________

   ____

   ____

   \+ -   + +
+ -   + +
+ -   + +
+ -
+ -

2. -3 + 4 = _____

   Scenario: ______________________

   ____

   ____

   \+ -
+ -
+ -
+ -
+ -

3. -3 - 4 = _____

   Scenario: ______________________

   ____

   ____

   \+ -
+ -
+ -
+ -
+ -

4. 5 - (-1) = _____

   Scenario: ______________________

   ____

   ____

5. -2 - (-5) = _____

   Scenario: ______________________

   ____

   ____

6. -8 + -4 = _____

   Scenario: ______________________

   ____

   ____

7. You start with $10 in your wallet, lose a bet for $16, then another bet for $4. How much money do you now have/owe?

Wallflowers
multiplying & dividing integers

As the night progresses at the Annual Integer Dance at Algebra High School, students begin to greet friends and hang out in groups. For example, the expression in #8 represents 3 groups of 5 boys coming into the dance, while #9 represents 2 groups of 3 girls leaving the dance.

For #8-13, a) write a scenario that may be represented by the expression, b) represent it with a "wallflower" diagram, c) evaluate the expression.

8. 3(-5) = _____
   Scenario:____________________
   ____
   ____

   \+ -
   \+ -
   \+ -
   \+ -
   \+ -

9. -2(3) = _____
   Scenario:____________________
   ____
   ____

   \+ -
   \+ -
   \+ -
   \+ -
   \+ -

10. -3(-2) = _____
    Scenario:____________________
    ____
    ____

    \+ -
    \+ -
    \+ -
    \+ -
    \+ -

11. 4 - 2(-3) = _____
    Scenario:____________________
    ____
    ____

12. 3 + 4(-2) = _____
    Scenario:____________________
    ____
    ____

13. -3(-2) - 1 = _____
    Scenario:____________________
    ____
    ____

14. Are there any rules or patterns for multiplying integers?

15. Do these rules for multiplying integers apply to division as well?
    a) You spend $12 dollars over the course of three days. You spend the same amount each day. How much do you spend each day? (Write an expression to represent this scenario)

    b) You spend $15 dollars total by spending $5 a day. How many days does it take you to spend the entire $15?

THE POSTMAN ALWAYS RINGS TWICE

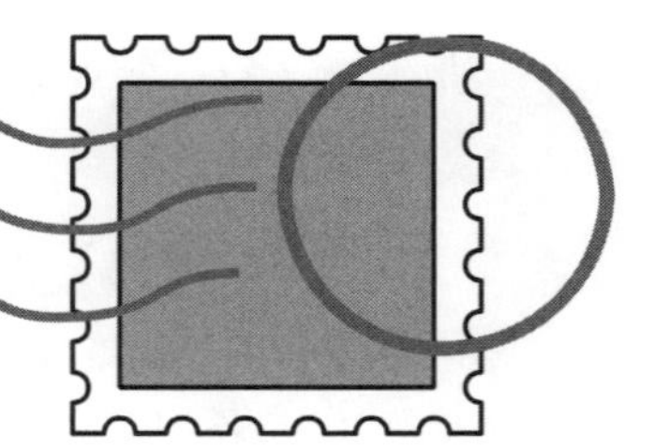

Inspired by the Math Centre, New Brunswick University

LESSON PLAN

1. For each of the scenarios #1-3, have the students write the expression for the morning delivery of the postman (all addition). Then have them extend the expression with the afternoon delivery (all subtraction). In #1, for example, the students should first write: 10 + -7. Then they should continue writing so that the expression looks like:10 + -7 - 10. After each scenario emphasize the following points with the students:

#1.	10 + -7 - 10	Adding a negative has the same result as subtracting a positive.
#2.	10 + -7 + -3 - (-7)	Subtracting a negative has the same result as adding a positive. Also adding two negative integers yields another negative integer.
#3.	9 + 12 + -5 + 3 - (-5) - 12	This is a summary of the four possible combinations of adding and subtracting positive and negative integers.

Concepts
Adding, subtracting, multiplying and dividing integers. Simplifying expressions using multiple operations.

Time: 1 day

Materials
Student handout

Preparation
Practice the scenarios on your own before presenting to students.

2. The scenarios in #4 & 5 have students explore the idea of multiplying and adding integers within the same expression. Again, have the students write the expression for the morning delivery of the postman (all addition). Then have them extend the expression with the afternoon delivery (all subtraction).

#4.	3(10) + 2(-7) - 2(10) - (-7)	A negative times a negative is a positive.
#5.	5(-20) + (-7) + 4(3) - 3(20) - 2(3)	A negative times a positive is a negative.

3. In order to check for true understanding, have the students then write postman scenarios for the expressions in #6. Have students share these publicly. Exercises #7 & 8 offer traditional practice.

#6. a) The postman delivers a check for $5, and another check for $3. He then takes back the $5 check.
b) The postman delivers a check for $8, a bill for $3, and another bill for $5. He then takes back the bill for $3, and the check for $8.
c) The postman delivers two bills for $5 each, and four checks for $3 each. He then takes back one bill for $5.

THE POSTMAN ALWAYS RINGS TWICE

Your neighborhood has an absent-minded postman who delivers mail to the wrong addresses in the morning, then has to go back and correct the mistakes in the afternoon. The mail is predominantly one of two types: checks and bills. The checks represent money that you receive and bills represent money that you owe. For each scenario below, write an expression that represents the amount of money you have/owe after the postman's second trip through your neighborhood. Then evaluate the expression to determine that amount of money.

1. The postman delivers a check for 10 dollars and a bill for 7. He then takes back the check for 10.

2. The postman delivers a check for $10, a bill for $7, and another bill for $3. He then takes back the bill for $7.

3. The postman delivers a bill for $9, a check for $12, another bill for $5, and another check for $3. He then takes back the bill for $5 and the check for $12.

4. The postman delivers three checks for $10 each and two bills for $7 each. Then, he takes back two of the checks and one of the bills.

5. The postman delivers five bills for $20 each, another bill for $7, and four checks for $3 each. Then, he takes back three of the bills for $20, and two of the checks.

6. Write a postman scenario of the following expressions, then evaluate the expression:
   a) $5 + 3 - 5$

   b) $8 + (-3) + (-5) - (-3) - 8$

   c) $2(-5) + 4(3) - (-5)$

7. Simplify: a) $6 - (-3) + -5 - 7$ b) $-7 - 8 + -9 - (10)$ c) $3(-6) - 4(-3) + -5(2)$ d) $-4(7) - 8(2) - (10)$

8. Evaluate $x - y + z$: a) if $x = -5$, $y = 6$, and $z = -4$ b) if $x = 7$, $y = -5$, and $z = -3$

   Evaluate $xy - z$: c) if $x = -3$, $y = 5$, and $z = -4$ d) if $x = -2$, $y = -6$, and $z = -4$

LESSON PLAN

1. Have a container (coffee mug or soup can) filled with pennies and a pile of extra pennies prominently on display in class. Have the students read the scenario of Duffy and the piggy bank, and state that the penny mug will serve as our model for the piggy bank. Ask the students how many pennies are in the bank. Of course, no one knows for sure, so they may use "x" to represent the number of pennies in the bank.

2. Simulate Duffy adding and removing pennies to and from the bank. Invite a student to choose a random number of pennies from the pile and add them to the bank. The student should count the added pennies aloud so that everyone may record this on the handout. (Assume 5 pennies are added.) Ask the students how many pennies now are in the bank. We do not know exactly, but we know that there are 5 more than we started with. Have the students record this as $x + 5$. Then have another student remove a random number of pennies from the bank. (Assume 7). How many pennies are now in the bank? The students record $x + 5 - 7$. Once more, have a student add a random number of pennies from the pile to the bank. (Assume only one penny). The students record $x + 5 - 7 + 1$. Simplify this expression to get $x - 1$, which means that we have one less penny in the bank than we started with.

3. Run a similar simulation for the scenario in #2. Then have students write an algebraic expression and a written response for #3. Then have students write their own scenarios to #4 and share publicly.

   Solutions to #4

a) $x - 5 + 3 - 4$	Duffy removes 5 pennies, adds back 3, then removes 4.
b) $x + 2 - 10 + 15 - 6$	Duffy adds 2 pennies, removes 10, adds 15, then removes 6.
c) $8 - 3 + x - 5 + 2$	Duffy has 8 pennies in the bank, removes 3, adds an unknown amount, removes 5, then adds 2.

4. Now display THREE containers filled with pennies and a pile of extra pennies. Have the students read the scenario of Duffy and the three piggy banks. Remind the students that each bank has the same number of pennies, and then ask how many pennies total are there in all the banks. Again, no one knows for sure, so "3x" may be used to represent the number of pennies in all the banks.

5. Run simulations similar to those for the single piggy bank for #6 & 7. This time, though, the same number of pennies is added to each of the three piggy banks. Assuming for #6 that 8 pennies are added and 5 removed, the students can write, a) $3x + 3(8)$, b) $3x + 3(8) - 3(5)$. This simplifies to $3x + 24 - 15$ (We have added a total of 24 pennies and removed a total of 15). The expression further simplifies to $3x + 9$, which means that we have 9 more pennies than we started with. (Three more in each bank). This expression can be factored to $3(x + 8 - 5)$ which means that we can think about what is being done to one bank, then multiple that by 3.

6. Have students write a piggy bank scenario for both parts of #9. The first focuses on what is being added to or removed from each of 4 banks. The second focuses on what is being done to one bank, then multiplying that result by 5 banks.

Concepts
Simplifying and evaluating expressions with integers

Time: 2 days

Materials
Three containers and several dollars worth of pennies, student handout

Preparation
Practice the scenarios on your own before presenting to students.

Duffy has an unknown number of pennies in his piggy bank. He periodically adds pennies to the bank, and on occasion, takes pennies from the bank. For each scenario below, write an expression that represents the amount of pennies that Duffy has in the bank.

1. a) Duffy adds ___ pennies b) then Duffy removes ___ pennies c) then adds ___ more pennies

2. a) Duffy removes ___ pennies b) then Duffy removes ___ more pennies c) then adds ___ pennies

3. Duffy removes 6, then removes 8 more, adds 20, and then removes 9. Write an expression for this situation. What does your expression say about the amount of pennies in the piggy bank?

4. Write a piggy bank scenario of the following expressions, then evaluate the expression:

   a) $x - 5 + 3 - 4$

   b) $x + 2 - 10 + 15 - 6$

   c) $8 - 3 + x - 5 + 2$

5. Evaluate $x - y + z$: a) if $x = -3$, $y = 5$, and $z = -4$ b) if $x = 8$, $y = -6$, and $z = -4$

   Evaluate $xy - z$: c) if $x = -3$, $y = 5$, and $z = -4$ d) if $x = -2$, $y = -6$, and $z = -4$

Duffy has the same unknown number of pennies in each of his three piggy banks. He periodically adds the same number of pennies to each bank, and on occasion, takes the same number of pennies from each bank. For each scenario below, write an expression that represents the amount of pennies that Duffy has in the bank.

6. a) Duffy adds ___ pennies to each b) then removes ___ pennies from each

7. a) Duffy removes ___ from each b) then removes ___ more from each c) then adds ___ to each

8. Factor the above expressions: a) for #6: b) for #7:

9. Write a piggy bank scenario of the following expressions, then evaluate the expression:

   a) $4x - 4(3) + 4(5)$ b) $5(x + 3 - 11)$

10. a) Factor & simplify #9a) above: b) Distribute & simplify #9b) above:

11. Evaluate: a) $3(-4) =$ b) $-2(-5) =$ c) $-7(3) =$

    d) $8 - 3 + 7 =$ e) $-7 - 3 + 4 =$ f) $-15 \div -5 =$

    g) $8 - (-2) =$ h) $12 \div -6 =$ i) $-12 - (9) + 4 =$

12. Simplify & Verify: a) $3x + 5y - 8x + 6y$ b) $-4x - (-5x)$ c) $-3(4x - 2y)$

13. Evaluate: $5x - 7y$; if $x = -9$, and $y = -3$

CANDY BARS: ADDING FRACTIONS

LESSON PLAN

For the sake of discussion, we will assume that the students are using green and red multi-link cubes (blocks).

1. Students should first represent the half of a candy bar. This would be a 2-block bar with one green and one red (shaded) block. Then they need to represent the one-third. The students will most likely create a 3-block bar with one block being red. However, the candy bars need to be the same size, so challenge the students to create two bars that have the same number of blocks, but are still one-half and one-third red, respectively. The students will easily create bars of six blocks each, 3 red and 2 red, respectively.

   When we then add the fractions, we need to combine the red blocks into one (or more) bars that are the same size as the original (6 blocks). We create one 6-block bar, with five of the six blocks being red.

   We then represent the problem arithmetically: $\frac{1}{2} + \frac{1}{3} = \frac{3}{6} + \frac{2}{6} = \frac{5}{6}$
   What is the rule for getting $\frac{3}{6} + \frac{2}{6} = \frac{5}{6}$?

2. The students will test out this rule on the next three problems in #2. Stress the importance of keeping all the candy bars within each problem the same size. This forces the students to find common denominators. They should then represent each problem arithmetically, and test their arithmetic rule to see if it generates the same solution as their geometric representation did.

Concepts
Adding fractions. Geometric and arithmetic representations.

Time: 1-2 days

Materials
Multi-link cubes or colored blocks

Preparation
Each pair of students needs 20 cubes of one color, and 20 of another color.

SOLUTION DIAGRAMS (cubes & student sketches)

#1 #2a

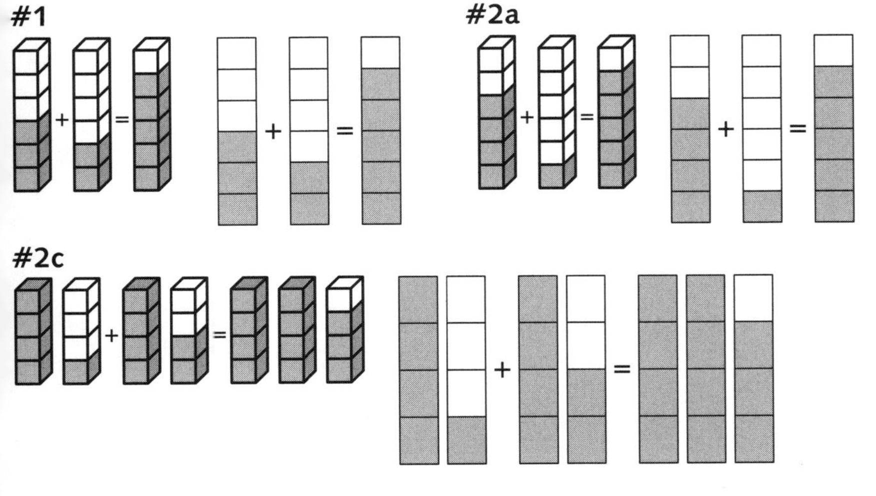

#2b

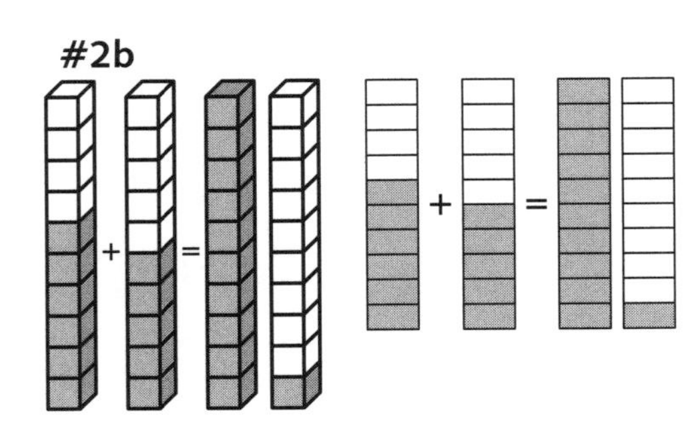

#2c

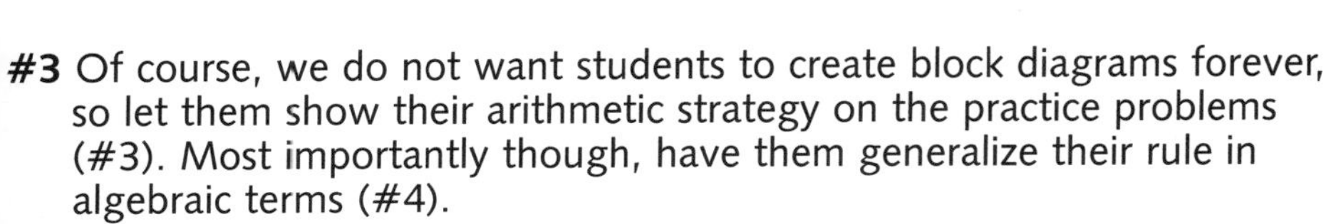

#3 Of course, we do not want students to create block diagrams forever, so let them show their arithmetic strategy on the practice problems (#3). Most importantly though, have them generalize their rule in algebraic terms (#4).

CANDY BARS: ADDING FRACTIONS

Your friend shares a candy bar equally with you. Another friend shares the same kind of candy bar among three of you.

1. a) Show in the first bar below your portion of the first candy bar.
   b) Show in the second bar your portion.
   c) Show in the third bar, how much you have total.
   d) How much more do you need in order to eat a whole candy bar?

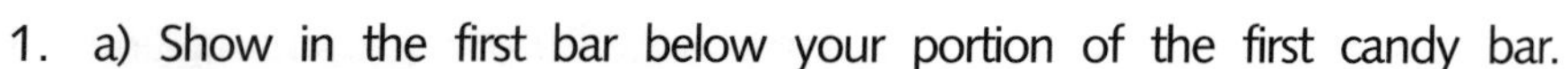

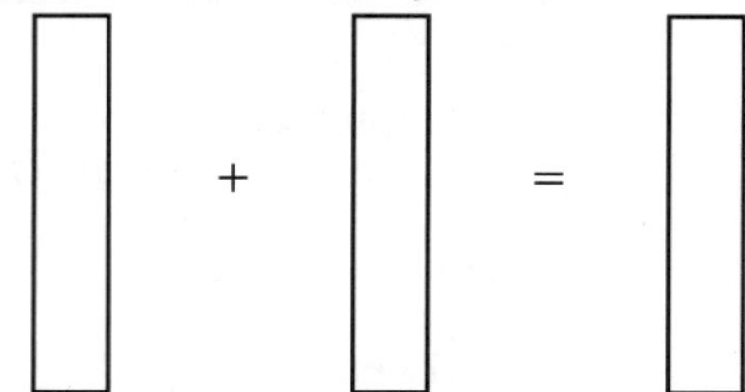

   e) Is there an easier way to figure this out, without drawing the diagrams? Show it.

2. Show how both methods above can be used to solve the following problems.

   a) $\frac{2}{3} + \frac{1}{6} =$

   $\square + \square = \square$

   b) $\frac{3}{5} + \frac{1}{2} =$

   $\square + \square = \square$

   c) $1\frac{1}{4} + 1\frac{1}{2} =$

   $\square + \square = \square$

3. Practice:
   a) $\frac{2}{3} + \frac{1}{2} =$

   b) $\frac{3}{2} + \frac{1}{4} =$

   c) $2\frac{1}{4} + 1\frac{1}{3} =$

4. Generalization:
   $\frac{a}{b} + \frac{c}{d} =$

BROWNIES: MULTIPLYING FRACTIONS

LESSON PLAN

For the sake of discussion, we will assume that the students are using green and red multi-link cubes (blocks). The multi-link cubes may be used to represent multiplying fractions in two different ways. The first is in the context of the candy bars:

1. Have students represent one-half of a candy bar. Again they will most likely show 2 blocks with one block being red. Then on a second candy bar, show one-half of that one-half. Since the students cannot show half of a block, encourage them to go back and reconfigure their original candy bar. The students will easily create a 4-block bar with 2 red blocks. On the second bar they will also have 4 blocks, of which only 1 is red ($^1/_2$ of the $^1/_2$). Have them go through the same process for the next problem $^1/_2$ of $^3/_4$. (Note: They should create the $^3/_4$ candy bar first, then the one-half of the three-fourths.)

2. Once they have completed these exercises and recorded their diagrams, have the students generate an arithmetic rule for multiplying fractions, and show it in #2. They instinctively multiply the denominators to determine the number of blocks in the stick. They are then to test this rule both geometrically and arithmetically in #3. Of course, they must generalize their rule algebraically as well in #4.

Concepts
Multiplying fractions.
Geometric and arithmetic representations

Time: 1-2 days

Materials
Multi-link cubes or colored blocks, student handout

Preparation
Each pair of students needs 20 cubes for each of 2 colors.

SOLUTION DIAGRAMS (Candy Bar Method - cube models)

#1a #1b

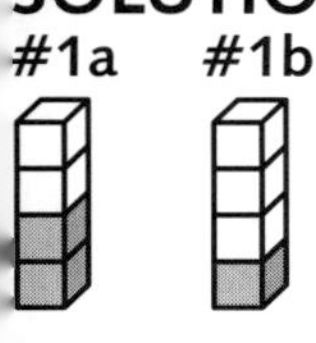

#1c #1d

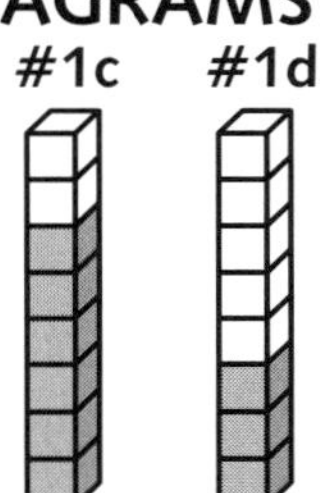

#3a

#3b

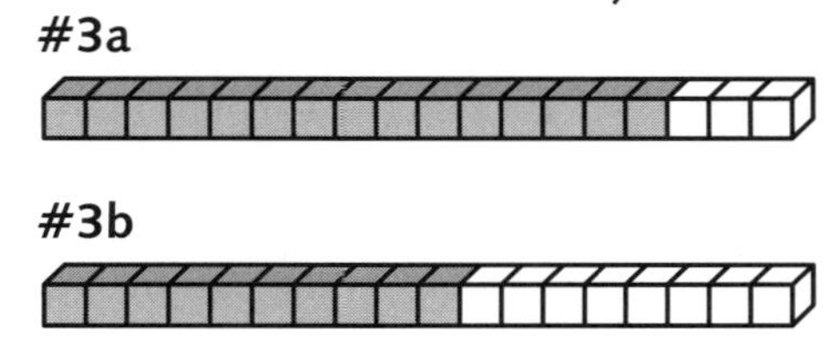

The second representation is the area model — in the context of sharing brownies.

3. In #5, first ask the students what multiplication is represented by the given scenario ($^2/_3$ of $^1/_2$). Challenge them then to make a brownie (rectangle) that can be portioned into halves horizontally, and thirds vertically. How many cubes does it take to make that brownie? Partition the diagram on the handout into the six parts. Make half of that brownie red and lightly shade half of the brownie (3 blocks). Go back to the original brownie and make two-thirds of it red and lightly shade two-thirds on the diagram. Darken the overlapping shaded region in the diagram; then represent that overlapping region with red blocks on the brownie. How much of the brownie is this? Does this model still support your arithmetic rule that you created earlier?

4. Have students practice both the geometric representation and the arithmetic rule for #6.

SOLUTION DIAGRAMS (Brownie Method - cubes & student sketches)

#5

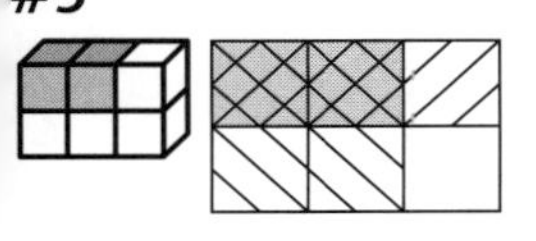

#6a

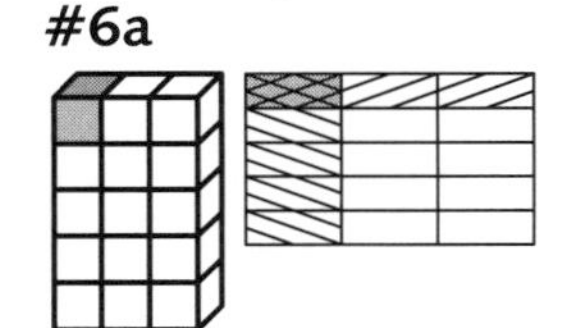

#6b

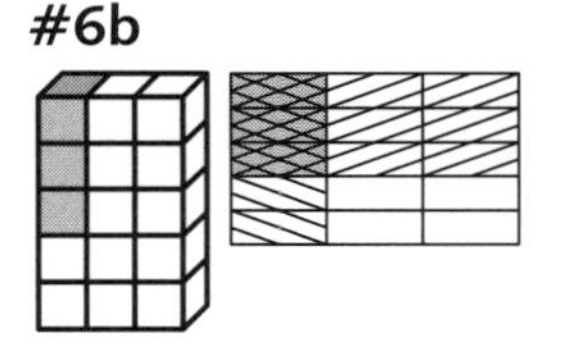

#6c

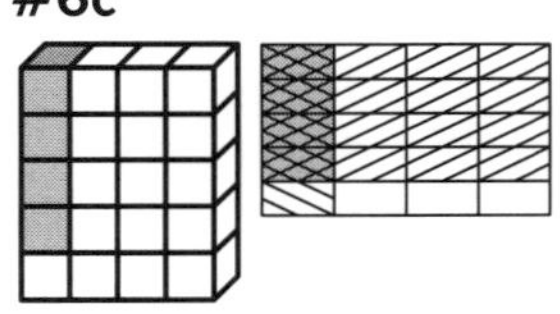

#6d

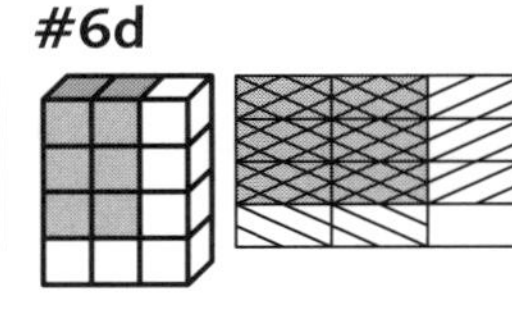

BROWNIES: MULTIPLYING FRACTIONS

1. With multi-link cubes...
   a) Show $\frac{1}{2}$ b) Show $\frac{1}{2}$ of $\frac{1}{2}$ c) Show $\frac{3}{4}$ d) Show $\frac{1}{2}$ of $\frac{3}{4}$

2. Is there an easier way to represent this, without drawing the diagrams? Show this symbolic method.
   a) $\frac{1}{2} \bullet \frac{1}{2}$ b) $\frac{1}{2} \bullet \frac{3}{4}$

3. Show how both methods above can be used to solve the following problem:

   a) Show $\frac{5}{6}$ b) Show $\frac{2}{3}$ times $\frac{5}{6}$

4. What is the outcome of $\frac{a}{b} \bullet \frac{c}{d}$?

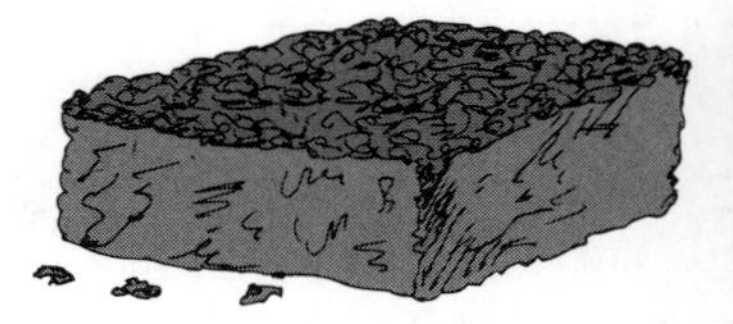

5. Your friend shares a brownie equally with you. You share your portion of the brownie among you and two friends (three portions). One of the friends says "No thank you," so you keep his portion.

   a) Show, horizontally, the first portion that you receive.
   b) Show, vertically, the portion of it that you keep.
   c) What operation with fractions does this represent.

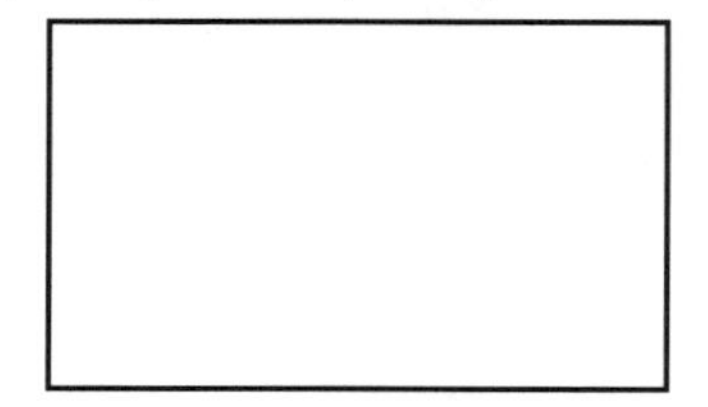

6. Represent the solution to the following problems both geometrically and symbolically.

   a) $\frac{1}{3} \bullet \frac{1}{5}$ b) $\frac{1}{3} \bullet \frac{3}{5}$ c) $\frac{1}{4} \bullet \frac{4}{5}$ d) $\frac{2}{3} \bullet \frac{3}{4}$

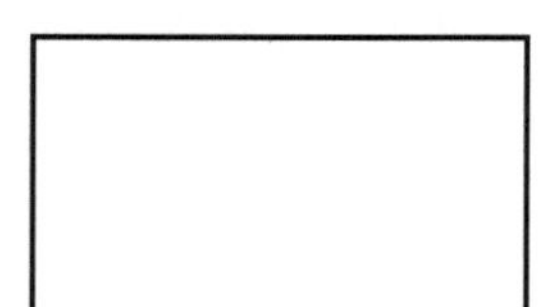

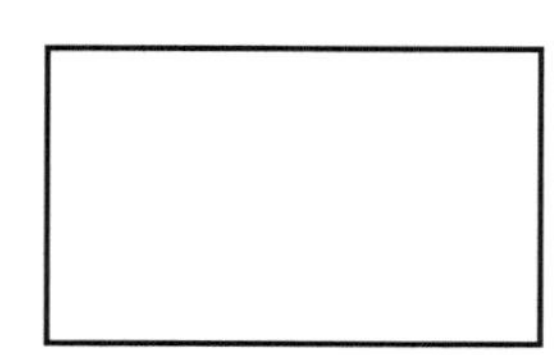

BEEF JERKY: DIVIDING FRACTIONS

LESSON PLAN

For the sake of discussion, we will assume that the students are using yellow, green and red multi-link cubes (blocks).

> ***Concepts***
> Dividing fractions. Geometric and arithmetic representations
>
> ***Time:*** 2 days
>
> ***Materials***
> Multi-link cubes or colored blocks
>
> ***Preparation***
> Each pair of students needs 10 cubes of one color, 10 of a second color and 10 of a third.

1. Use the question in #1 to hook the students. Line up 5 single cubes and ask them if these are an appropriate model for the beef jerky. They likely will say no, because they cannot represent half of a cube. The students will probably offer 5 stacks, each having two cubes. Have them draw this model on the handout, and shade half of just one stick of jerky. (Make only one cube of one stack red.) This shading of the diagram helps students see the concept of "How many halves go into five wholes?" The students should then circle as many halves as possible (ten halves). Having them separate the 5 sticks into the 10 halves helps them better understand also.

   Show students the traditional model of inverting and multiplying; and ask them to analyze it. The first 10 represents the number of single cubes in our model. The 1 represents the number of single cubes in each portion that we wish to share.

2. Have the students then represent the problem in #2: $6 \div \frac{2}{3}$. They should have 6 stacks of 3 cubes. The first stack only should have 2 red cubes. When they circle as many two-thirds on their diagram as possible, also have them separate the six sticks of jerky into as many groups of 2 as possible. (There should be nine.) Parallel this with the traditional model of inverting and multiplying. The eighteen represents the number of single cubes in our model. The two represents the number of cubes in each portion that we want to share. The question of "How many times does two-thirds go into six?" has been changed to "How many times does two go into eighteen?" The students can SEE this!

3. Have the students practice with modeling division of fractions with the four problems provided. Parts c & d are a challenge because they will be dividing a fraction by another fraction. For part c, have the students form a stick with two colors. The students start getting savvy and anticipate the division by a third so they make a stick of 15 cubes, 12 of which are green ($\frac{4}{5}$) and the remainder are yellow. Now they are to make one-third of the total jerky stick (5 cubes) red. This will be five cubes. How many sets of five cubes go into the 12 cubes? (Two with two-fifths left over!) For part d, the students will have a stick of six cubes, three of which are green ($\frac{1}{2}$). They then need to shade two-thirds of the six (4 cubes), but they can't because they only have 3 cubes. So the answer must be less than one. Circling the four cubes in the diagram shows that they have $\frac{3}{4}$ of a single portion!

SOLUTION DIAGRAMS (Brownie Method - cubes & student sketches)

#1 = 10

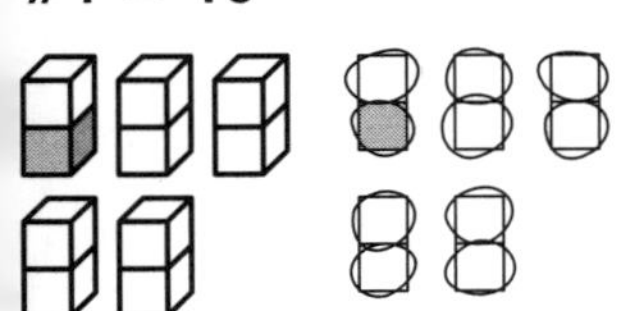

#2 = 9

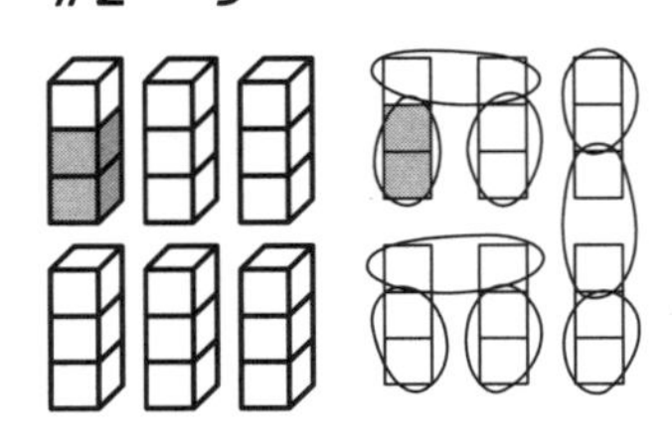

#3a = 10

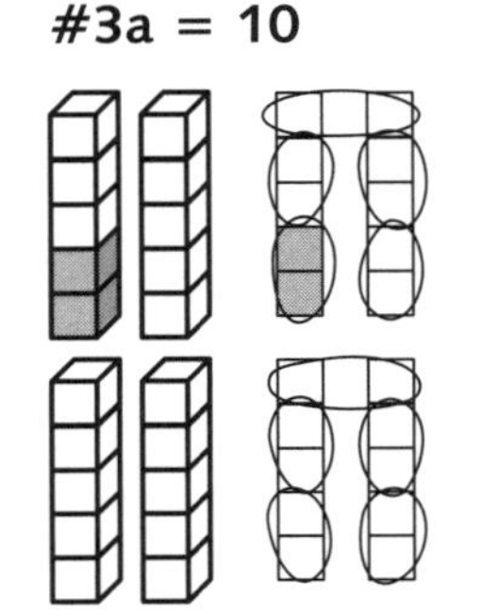

#3b = $7\frac{1}{2}$

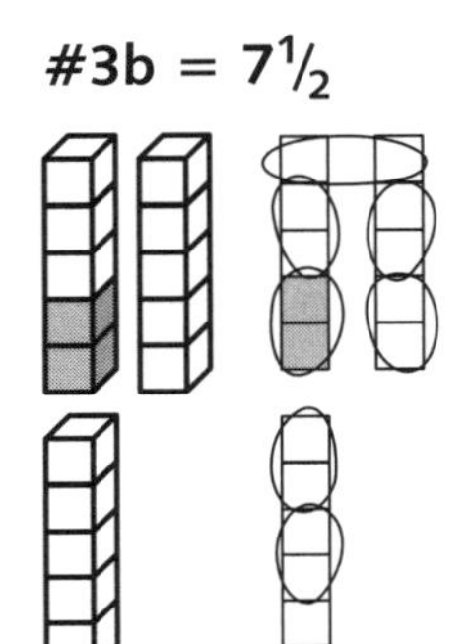

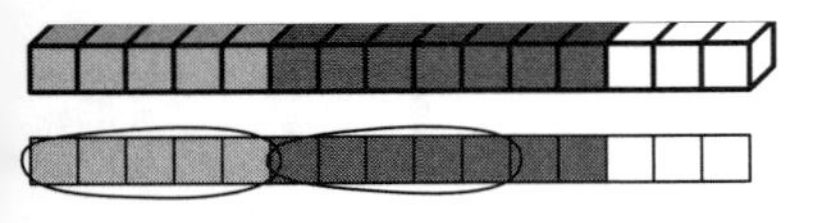

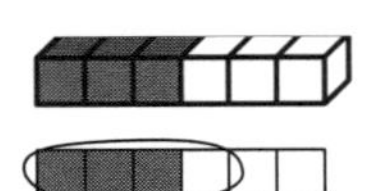

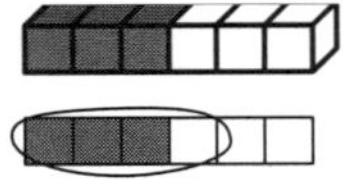

BEEF JERKY: DIVIDING FRACTIONS

1. You have 5 sticks of beef jerky. To how many friends can you give half of a stick of jerky?

   a) Use multi-link cubes to model the 5 beef jerky sticks. Record your multi-link models below.

   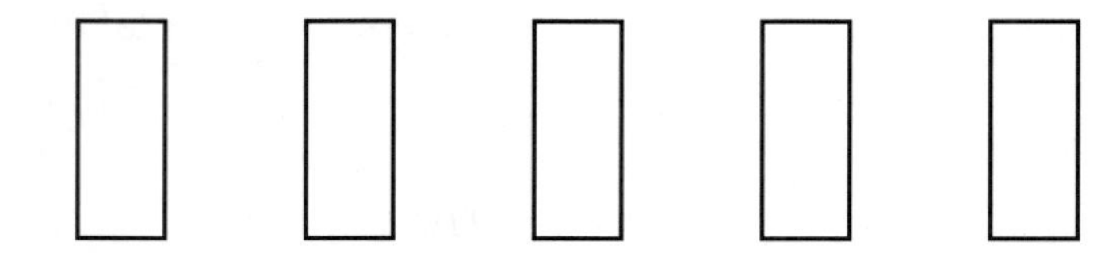

   b) On only one stick of jerky in the diagram, shade the portion that you wish to share with each person.

   c) Circle as many shares of that portion that you think the 5 sticks will yield. How many is that?

   d) This model is traditionally represented and solved in the following manner:

   $5 \div \frac{1}{2} = 5 \bullet \frac{2}{1} = \frac{10}{1} = 10$

   What does the first 10 represent? What does the 1 represent?

2. a) Represent the following division problem with the cubes and record that model below: $6 \div \frac{2}{3}$
   Be sure to shade the $\frac{2}{3}$ then circle as many of those $\frac{2}{3}$ portions as possible.

    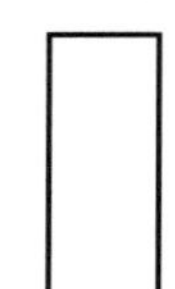 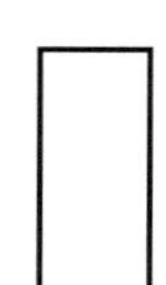 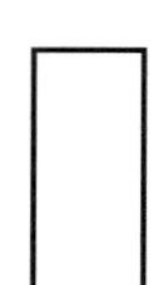  

   b) This model is traditionally represented and solved in the following manner:

   $6 \div \frac{2}{3} = 6 \bullet \frac{3}{2} = \frac{18}{2} = 9$

   What does the 18 represent? What does the 2 represent?

3. Represent the following division problems with the cubes and record those models below. Be sure to shade the appropriate portions and circle as many of those portions as possible.

   a) $4 \div \frac{2}{5}$ b) $3 \div \frac{2}{5}$ c) $\frac{4}{5} \div \frac{1}{3}$ d) $\frac{1}{2} \div \frac{2}{3}$

♣ ♦ ♥ ♠ Playing with a Full Deck ♠ ♥ ♦ ♣

Submitted by Randy Hoffman, Trabuco Hills High School, Mission Viejo, CA

LESSON PLAN

Explain the rules of the game to the students as described on the top of the handout. The students will be choosing one of the seven outcomes for each of the six cards in a round. Strongly encourage the students to experiment with different strategies as they play successive rounds.

To facilitate the game, have each student commit to their six guesses before the round begins. Then draw one card at a time from a deck in front of the class for all to see. After each card is drawn and recorded, return it to the deck and shuffle. A few days prior to the assignment, two or three rounds can be played at the beginning or end of class so that the students get very familiar with the game.

Once the data from these rounds is collected, the students complete the calculations and graphs on the handout. The questions on the handout and their solutions below are appropriate for a beginning Algebra lesson on probability with the following exceptions. On #4, most Algebra students will be able plot the points, but only Algebra 2 students will be able to write the equation for the exponential regression. To help students with writing this equation, have them think of it as a compound interest (decay) problem: $1 = 35r^{75}$. The objective for #7 is that students recognize that choosing a face card is the best bet. When the expected values are calculated, we see that 35 points for an event that has a probability of $^1/_{52}$ is not an adequate reward. In contrast, a face card should come up nearly one out of every 5 cards, but a correct choice is awarded three times that value, namely 15 points.

Concepts
Simple probability (Pre-Algebra), multiple probability, (Algebra), exponential Probability (Algebra 2), graphing

Time: 2-3 hours

Materials
Student handout, graph paper, calculators, one deck of cards

Preparation
Students should already have a strong understanding of simple probabilities.

SOLUTIONS

1. Results of individual rounds will vary
2.

Any Non-Face Card:	$^{10}/_{13}$	77%
A Specific Color:	$^1/_2$	50%
A Specific Rank:	$^6/_{13}$	46%
A Specific Suit:	$^1/_4$	25%
Any Face Card:	$^3/_{13}$	23%
A Specific Number:	$^1/_{13}$	8%
A Specific Number & Suit:	$^1/_{52}$	2%

3. See graph to the right
4. $1 = 35r^{75}$
   $\sqrt[75]{^1/_{35}} = r$
   $r = .9537$
   $y = 35(.9537)^x$
5. Answers will vary, see example on student handout
6. Most likely, the high scores occur at the .02% range, while lower scores occur at both higher and lower probabilities.

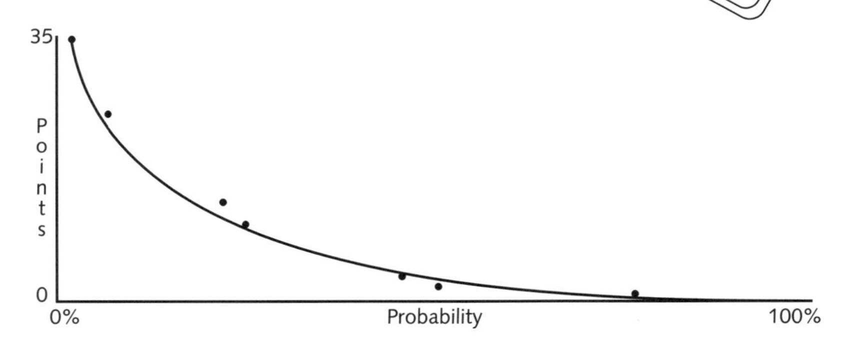

7. / 8.

Any Non-Face Card:	1.3	1
Any Face Card:	4.3	5
A Specific Color:	2	2
A Specific Rank:	2.2	3
A Specific Suit:	4	4
A Specific Number:	13	13
A Specific Number & Suit:	52	52

9. Guessing a face card correctly awards over 3 times the expected value. Since no other choice offers much of a return, guessing a face card is "the best bet."

♣♦♥♠ Playing with a Full Deck ♠♥♦♣

Submitted by Randy Hoffman, Trabuco Hills High School, Mission Viejo, CA

The object of this game is to acquire as many points as possible by guessing the outcome of cards drawn from a deck. Six cards will be selected one at a time and replaced to the deck to be reshuffled each time. For each draw, you are to chose one of the options listed below regarding the nature of the card. The point values for a correct guess are also shown. For instance, if you guess a card is going to be a LOW card and the 2 of hearts (2♥) is drawn, you earn 3 points. If you are wrong you are awarded zero. For each round played, guess the six cards in advance, record the actual draws and points earned, and then tally your points. Experiment with a variety of strategies as your eventual goal will be to choose the optimal guessing strategy for this game. A sample round is demonstrated below. Good luck and have fun!

Options	Points
Any Non-Face Card	1
A Specific Color: (Red or Black)	2
A Specific Rank: Low (Ace through 6) or High (8 through King)	3
A Specific Suit: Club, Heart, Diamond, Spade	10
Any Face Card: (Jack, Queen, or King)	15
A Specific Number (like a 5)	25
A Specific Number & Suit (like the nine of clubs)	35

1) Complete the following rounds

Round 1: Sample

Trial	Guess	Card	Pts earned
1	Red	3♣	2
2	High	10♥	0
3	Club	2♦	10
4	Ace	A♥	0
5	Face	K♦	15
6	4 Hearts	4♥	0
		Total	27

Round 2

Trial	Guess	Card	Pts earned
1	____	____	____
2	____	____	____
3	____	____	____
4	____	____	____
5	____	____	____
6	____	____	____
		Total	____

Round 3

Trial	Guess	Card	Pts earned
1	____	____	____
2	____	____	____
3	____	____	____
4	____	____	____
5	____	____	____
6	____	____	____
		Total	____

Round 4

Trial	Guess	Card	Pts earned
1	____	____	____
2	____	____	____
3	____	____	____
4	____	____	____
5	____	____	____
6	____	____	____
		Total	____

Round 5

Trial	Guess	Card	Pts earned
1	____	____	____
2	____	____	____
3	____	____	____
4	____	____	____
5	____	____	____
6	____	____	____
		Total	____

Round 6

Trial	Guess	Card	Pts earned
1	____	____	____
2	____	____	____
3	____	____	____
4	____	____	____
5	____	____	____
6	____	____	____
		Total	____

Playing with a Full Deck

2. Determine the probability of each outcome of the game (write as a fraction and a percentage).

Options	Probability
Any Non-Face Card	________
A Specific Color: (Red or Black)	________
A Specific Rank: Low (Ace through 6) or High (8 through King)	________
A Specific Suit: Club, Heart, Diamond, Spade	________

Options	Points
Any Face Card: (Jack, Queen, or King)	________
A Specific Number (like a 5)	________
A Specific Number & Suit (like the nine of clubs)	________

3. Graph the probability and the points earned for each category. Establish the probability of the outcomes as the domain (0% to 100%), and the points earned as the range (0 to 60).

4. The graph looks like a curve of exponential decay. Draw a best fit curve for this data, and write the corresponding equation.

5. Calculate the probability (fraction and percent) for drawing the exact combination of cards that you guessed. Write your answer in scientific notation and as a percentage.

		Scientific Notation	Percentage	Total Points
Round 1: $^1/_2 \cdot {}^6/_{13} \cdot {}^1/_4 \cdot {}^1/_{13} \cdot {}^3/_{13} \cdot {}^1/_{52}$	=	1.9×10^{-5}	.002%	27
Round 2: ________________		________	________	________
Round 3: ________________		________	________	________
Round 4: ________________		________	________	________
Round 5: ________________		________	________	________
Round 6: ________________		________	________	________

6. In which rounds do you have the highest probability? In which rounds did you win the most points? Do you see any connections?

7. An expected value is what an outcome is expected to be worth according to its probability of occurring. For example, choosing a specific card correctly has a probability of occurring once every 52 times. Therefore, it should receive 52 points. Calculate the expected values of the other outcomes. Write your answer as if you are completing the following statement "the outcome is expected to occur one out of every ____ times."

Options	Expected Value
Any Non-Face Card	________
A Specific Color: (Red or Black)	________
A Specific Rank: Low (Ace through 6) or High (8 through King)	________
A Specific Suit: Club, Heart, Diamond, Spade	________

Options	Expected Value
Any Face Card: (Jack, Queen, or King)	________
A Specific Number (like a 5)	________
A Specific Number & Suit (like the nine of clubs)	________

8. According to the expected values, what should the whole-number point values be for each of the outcomes?

9. According to the graph, probabilities and expected values, what are the "best bets" in the game and why?

Algebra Lessons

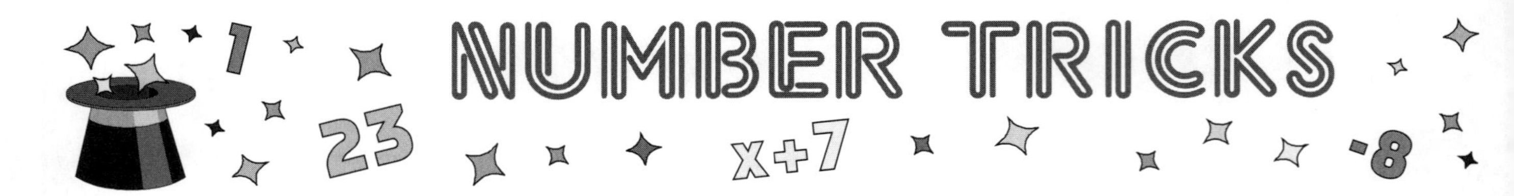

NUMBER TRICKS

LESSON PLAN

Before giving the handout to the students, have them complete the first number trick mentally. Have several students place their solutions on the board showing each of their steps. Once the numeric solutions are completed, show the students how to write the algebraic expression for the number trick. The simplified expression shows that they can predict the result without choosing any numbers at all.

Concepts
Simplifying algebraic expressions

Time: One hour

Materials
Student handout (optional)

Preparation
None

Distribute the handout and have the students fill in the columns for Trick #1. They should write their conjecture, "the result will always be three," as well as their simplified expression, "3." Next, lead them through Trick #2 just as you did in #1. The emphasis here is the use of the distributive property, as students will need to use parenthesis because they are multiplying the intermediate result, not the original number or the constant.

Trick #3 catches them off guard, because they all get a different number. Again have the students put the solutions on the board, and the pattern becomes obvious. They all get the original number that they started with. The algebraic solution should then render "x." In a similar fashion, Trick #4 renders a solution that is one more than the original number chosen, or in other words "x + 1."

THE SOLUTIONS

		Your Number	Another Number	Another Number	Algebraic Expression
TRICK #1	Pick a number	10	1	-5	x
	Multiply by 2	20	2	-10	2x
	Add 3	23	5	-7	2x + 3
	Subtract twice the original number	3	3	3	2x + 3 - 2x
	Common Result: always 3				3
TRICK #2	Pick a number	10	1	-5	x
	Add 4	14	5	-1	x + 4
	Multiply by 2	28	10	-2	2(x + 4)
	Subtract 7	21	3	-9	2(x + 4) - 7
	Subtract twice the original number	1	1	1	2(x + 4) - 7 - 2x
	Common Result: always 1				1
TRICK #3	Pick a number	10	1	-5	x
	Add 2	12	3	-3	x + 2
	Multiply by 3	36	9	-9	3(x + 2)
	Subtract 6	30	3	-15	3(x + 2) - 6
	Subtract twice the original number	10	1	-5	3(x + 2) - 6 - 2x
	Common Result: the same number that you started with				x
TRICK #4	Pick a number	10	1	-5	x
	Add 5	15	6	0	x + 5
	Multiply by 2	30	12	0	2(x + 5)
	Subtract 9	21	3	-9	2(x + 5) - 9
	Subtract the original number	11	2	-4	2(x + 5) - 9 - x
	Common Result: one more than the number you started with				x + 1

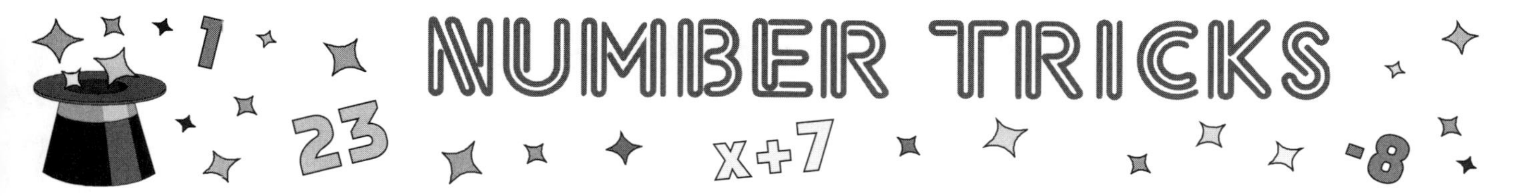

In the column titled "Your number," write the results of each step of the number trick listed on the left. Once other solutions are shown by your classmates, you may copy down two of the solutions in the two columns titled "Another Number." Analyze these three solutions, and make a conjecture to what you think the final result of the number trick will always be. Finally, in the fourth column, write the algebraic expression that represents the cumulative steps of the number trick. On the second to last line of this column should be the complete expression; then write the simplified version.

		Your Number	Another Number	Another Number	Algebraic Expression
TRICK #1	Pick a number	_____	_____	_____	__________
	Multiply by 2	_____	_____	_____	__________
	Add 3	_____	_____	_____	__________
	Subtract twice the original number	_____	_____	_____	__________
	Common Result: ____________________			**Simplified Expression**:	__________
TRICK #2	Pick a number	_____	_____	_____	__________
	Add 4	_____	_____	_____	__________
	Multiply by 2	_____	_____	_____	__________
	Subtract 7	_____	_____	_____	__________
	Subtract twice the original number	_____	_____	_____	__________
	Common Result: ____________________			**Simplified Expression**:	__________
TRICK #3	Pick a number	_____	_____	_____	__________
	Add 2	_____	_____	_____	__________
	Multiply by 3	_____	_____	_____	__________
	Subtract 6	_____	_____	_____	__________
	Subtract twice the original number	_____	_____	_____	__________
	Common Result: ____________________			**Simplified Expression**:	__________
TRICK #4	Pick a number	_____	_____	_____	__________
	Add 5	_____	_____	_____	__________
	Multiply by 2	_____	_____	_____	__________
	Subtract 9	_____	_____	_____	__________
	Subtract the original number	_____	_____	_____	__________
	Common Result: ____________________			**Simplified Expression**:	__________

ASSIGNMENT

Create a number trick similar to those shown above. The trick must be unique. Demonstrate the solution with three different numbers, and then show the algebraic expression that represents each step of the number trick. Finally, write the simplified expression that represents the result of each trick. The higher the degree of difficulty, the higher the grade.

PROJECT

I'm thinking of a number... what is it?

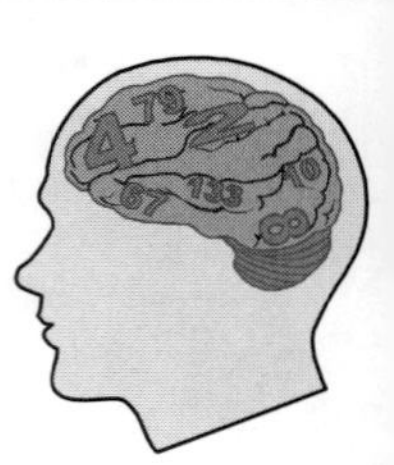

LESSON PLAN

Present students with the first "I'm Thinking of a Number" problem. If you are using the given handout, have students solve the problem by any means they wish. Instruct them to show their reasoning/calculations in the first column provided. Go through each of the problems in this fashion, soliciting students for their strategies. The students will almost unanimously tell you that they started at the end and worked backwards, doing the opposite of what the problem instructs them to do. Eureka! Reiterate that notion to them throughout the lesson. You may even want to place it on the board, "Work backwards, doing the inverse of the operations."

Concepts
Simplifying and solving algebraic equations

Time: One hours

Materials
Student handout (optional)

Preparation
None

Once all the problems are completed, start again with the first and lead the students through the various expressions that represent each step. Each problem should culminate with an equation. Emphasize that this very SHORT equation says the same thing as the LONG word problem. Then demonstrate the steps in solving each equation, emphasizing that the algebraic strategy is the same as their original arithmetic strategy. The only contrast is that the work be shown differently.

SPECIAL NOTES: (1) In Question #3, our strategy of "working backwards, doing the inverse" will avoid the need for the distributive property. It is left to the instructor to decide whether or not to show that method as well. (2) In the algebraic solution sets below, the intermediate steps are not shown (e.g. subtracting 8 from both sides). Again, it is left to the instructor whether or not to require these steps.

SOLUTIONS

	Arithmetic Solution	Algebraic Expressions	Algebraic Solutions
1. I'm thinking of a number, I then... add 8, to get 21.	$21 - 8 = 13$	x $x + 8$ $x + 8 = 21$	$x + 8 = 21$ $x = 13$
2. I'm thinking of a number, I then... multiply by 2, and add 3, to get 15.	$15 - 3 = 12$ $12 \div 2 = 6$	x $2x$ $2x + 3$ $2x + 3 = 15$	$2x + 3 = 15$ $2x = 12$ $x = 6$
3. I'm thinking of a number, I then... add 2, and multiply by 3, to get 18.	$18 \div 3 = 6$ $6 - 2 = 4$	x $x + 2$ $3(x + 2)$ $3(x + 2) = 18$	$3(x + 2) = 18$ $x + 2 = 6$ $x = 4$
4. I'm thinking of a number, I then... square the number, and add 1, to get 26.	$26 - 1 = 25$ $25 = 5$	x x^2 $x^2 + 1$ $x^2 + 1 = 26$	$x^2 + 1 = 26$ $x^2 = 25$ $x = \pm 5$

ASSIGNMENT

Have students create three "Thinking of a Number" problems similar to those shown in class. Each should be unique and involve at least two operations. The student should write the algebraic equation and show how to solve the equation to get the original number.

I'm thinking of a number...

what is it?

	Arithmetic Solution	Algebraic Expressions	Algebraic Solutions
1. I'm thinking of a number, I then..		______	
add 8,		______	
to get 21.		______	
2. I'm thinking of a number, I then...		______	
multiply by 2,		______	
and add 3,		______	
to get 15.		______	
3. I'm thinking of a number, I then...		______	
add 2,		______	
and multiply by 3,		______	
to get 18.		______	
4. I'm thinking of a number, I then...		______	
square the number,		______	
and add 1,		______	
to get 26.		______	

ASSIGNMENT

Create three "Thinking of a Number" problems similar to those shown above. Each should be unique and involve at least two operations. Be sure to write the algebraic equation and show how you solve the equation to get the original number that you were "thinking of."

Extra Credit:

Option 1 - Create a problem that involves the distributive property.
Option 2 - Create a problem that involves more than two operations.

Pig Pen Algebra

Adapted from a lesson in Carol McGehe's "Algebra for All" Training Workshop

LESSON PLAN

The following instructions assume that the students will be using manipulatives. If manipulatives are not available or desired, simply have the students draw the diagrams. Begin by explaining Farmer John's pig pen objective. He begins by enclosing one pig in a pen with 8 bales of hay. Ask the students to create the second pen for two pigs by moving and adding bales (blocks or tiles). Although they should all add two bales, how they move the existing bales may vary. Show the various methods, stressing that no matter how the pen is expanded, it will always require two more bales.

Now have students do the same thing for a pen of three pigs. Again, despite the strategy, two more bales will be needed. Have students create and record (draw) the pens that would hold 4 and 5 pigs. Emphasize again that two bales are needed for each additional pig. Once the students agree to the pattern of adding two bales for every one pig, have them complete the chart:

Concepts
Writing and solving equations

Time: 1-2 hours

Materials
Student handout and manipulatives (optional): any type of cubes or algebra tiles.

Preparation
None

P (pigs)	1	2	3	4	5	6	7	8	9	10	20	50
B (bales)	8	10	12	14	16	18	20	22	24	26	46	106

The key here is the discussion of strategies in finding the number of bales necessary for 20 & 50 pigs. Some students will just repeatedly added two, until they reached enough for twenty and fifty pigs. Several will start with the 26 bales at 10 pigs, and multiply two times another ten pigs, and then add to get 46. It is important to show that doubling the number of bales for ten pigs will not be the accurate amount of bales for twenty pigs.

Then comes the ultimate objective of the lesson. Have students write an equation for the number of bales B needed to pen P number of pigs. Some possible students responses are shown below:

$B = p + 2$ $B = 2p$ $B = 2p + 8$

Work with the students in understanding that although 2 bales for every pig is written as 2p, simply multiplying the number of pigs by two will not yield the correct answer. This is where most kids will offer to add 8 since that is the number of bales that we started with for ONE pig. What they should realize is that the equation needs to start at ZERO pigs. If they follow the pattern in the chart, they should see that zero pigs requires 6 bales (not zero bales, as common sense might suggest). Have the students move and remove the manipulatives, 2 per pig, until there are no pigs left. They will end up with a 3 x 2 matrix. Basically, Farmer John needs the three bales on the left and the three on the right (total of 6) plus two for every pig. So the class may now discuss the merits of the following equations and how they are equivalent.

$B = 2p + 6$ $B= 8 + 2(c -1)$

ASSIGNMENT

In regards to the tower activity. This may be used for practice, assessment or further instruction. The tower is expanded much like the pig pen. It starts with three blocks on the first story, with 4 blocks added to each consecutive story. The tricky part here is that following a pattern in the chart will yield a y-intercept of -1, which does not make practical sense in regards to the tower. However, the following equations do:

Horizontal perspective: $T = 3 + 4(s -1)$ Vertical perspective: $T = 3s + (s - 1)$

A discussion on how these both yield $T = 4s - 1$ (4 blocks for every story, minus the missing block) will be very worthwhile. Another valuable investigation is: Given the total number of blocks T, how many stories S can someone make. This yields the equation $(T + 1)/4 = S$. Students can then be asked to show how these two tower equations are equivalent.

Pig Pen Algebra

Farmer John is making a pig pen. He is short on materials so he is making the pen out of bales of hay. These bales are shaped as cubes. Farmer John likes to keep things simple, so whenever he gets another pig, he just extends the pen as shown below. Your job is to help Farmer John write a formula to tell him how many bales of hay he will need for a given number of pigs.

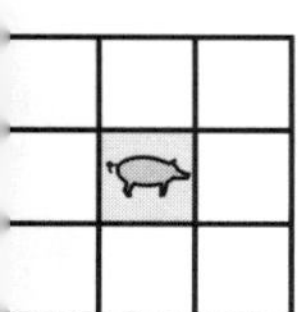

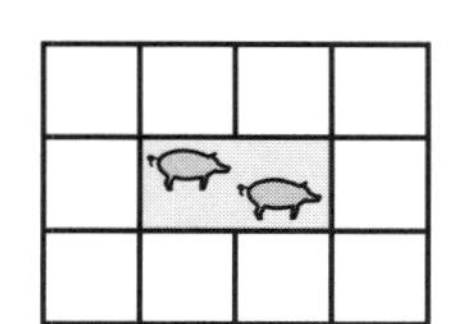

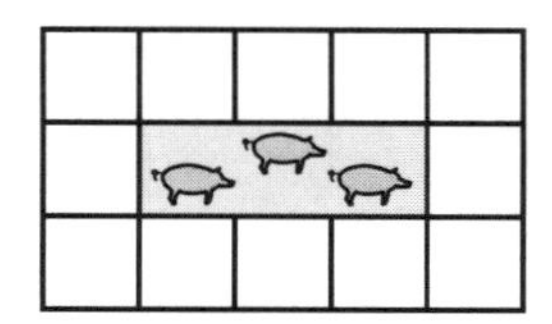

1. How would you move the existing bales to make room for another pig?

2. In the space above, draw pens that would hold 4 and 5 pigs respectively.

3. How many bales of hay must be added to an existing pen, to make room for the next pig?

4. Without anymore drawings, complete the chart below for P number of pigs and B number of bales.

P (pigs)	1	2	3	4	5	6	7	8	9	10	20	50
B (bales)	8	10	12									

5. How did you figure out your answers for 20 and 50 pigs?

6. According to the pattern in the chart, how many bales would you predict are needed for no pigs?

7. Write an equation that represents the number of bales B needed to pen P number of pigs.

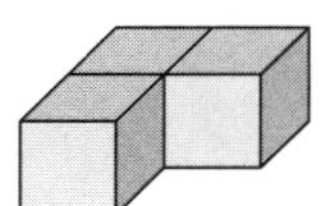

8. Show how to use your new equation to find how many bales are needed for 100 pigs.

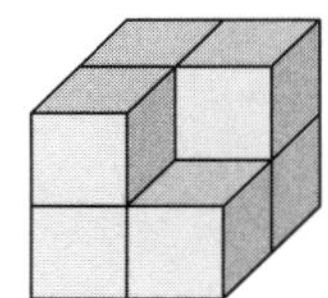

9. How many pigs could be penned by 96 bales of hay?

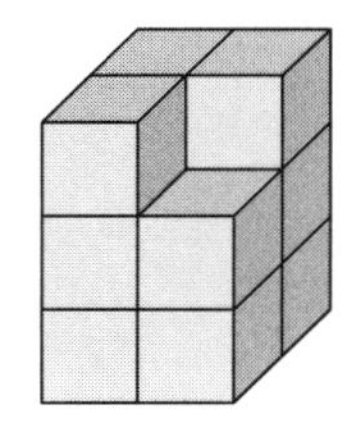

ASSIGNMENT

A tower is built up as shown on the right. Write an equation that represents the number of blocks T needed to build a tower S stories tall.

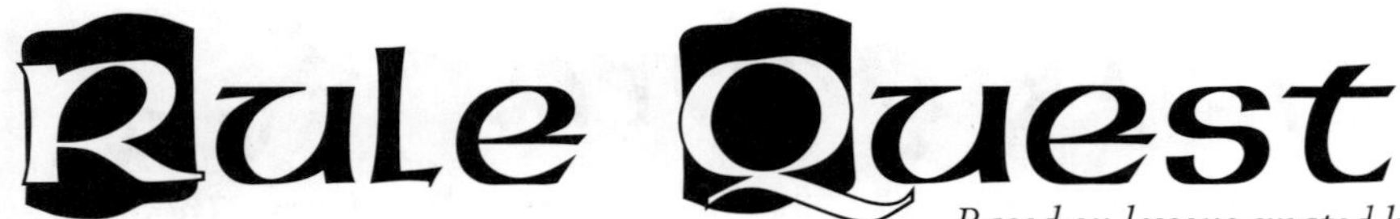

Based on lessons created by Geoff Giles, University of Stirling, Scotland for the DIME Project. Submitted by Michael Wheal, Southern Austrialia.

Concepts
Writing Equations, Pattern Recognition

Time: 1 hour per pattern

Materials
Student Handout

Preparation
None

LESSON PLAN

The point of emphasis of this lesson is to expose students to multiple representations of a mathematical pattern so they may better understand the algebraic models (namely variables and equations) that are used to represent them. Be sure students have a strong grasp of the concepts before they formalize the patterns with equations.

EDGE SQUARES

Introduce the students to the patterns of squares. Have them extend the pattern, recording the results in the table. While diagrams are not offered for edge lengths of 5 & 6, the students can figure the number of edge tiles by continuing the pattern in the chart. However, to get the results for edge lengths of 10 and 100, they will need some kind of strategy. Have them record this strategy in #3. Have the students then share their strategies.

Some of these strategies will be reflected in the diagrams of #4. These are geometric representations of the strategies — the first step in abstracting the pattern. Two diagrams are given; the students should generate two more. They may record strategies that they heard in class, or create their own. Once the geometric diagrams are recorded, the students are to describe these four strategies in words. These two steps (geometric drawing and verbal description) are intended to emphasize the mathematical pattern in the squares.

Once the verbal descriptions have been discussed, the students share the equations that match the diagrams. Having students write a variety of equations reflecting a variety of strategies for the same pattern helps them understand what the variables and equations represent. A particularly strong aspect of this lesson is the opportunity for the students to see that an expression inside parentheses represents a quantity — a principle that eludes most algebra students. The final question demonstrates that all the strategies lead to the same mathematical model ($n = 4s - 4$). Having students relate the coefficient and constant to the table of values is an excellent precursor to the future study of slope and y-intercept.

POSSIBLE SOLUTIONS

Diagram	Description	Equation
	a) Add the length of the side to itself, add two less than the length, and then add two less than the length again.	$n = s + s + (s - 2) + (s - 2)$
	b) Take one less than the length and add it four times, or multiply one less than the length of the side by four.	$n = (s - 1) + (s - 1) + (s - 1) + (s - 1)$ $= 4(s - 1)$
	c) Take the side length, add one less than the length twice, and then add two less than the length.	$n = s + (s - 1) + (s -1) + (s - 2)$
	d) Multiply the side length by four and subtract the four overlapping corners.	$n = 4s - 4$

EDGE TRIANGLES

Lead the students through the edge triangles in a similar fashion as was done for the edge squares. This time, however, exclude the geometric diagraming. Remember to have the class discuss each step publicly in order to share the various strategies that students devise. The lesson again culminates with all equations simplifying to the same formula: $n = 3w - 3$.

SOLUTIONS

Description	Equation
a) Take the side length, add one less than the length, and then add two less than the length.	$n = w + (w - 1) + (w - 2)$
b) Take one less than the length and add it three times, or multiply one less than the length of the side by three.	$n = (w - 1) + (w - 1) + (w - 1)$ $= 3(w - 1)$
c) Multiply the side length by three and subtract the three overlapping corners.	$n = 3w - 3$

MATCH STICKS

Here the pattern diverges significantly from the previous two. Still, the first steps are to have the students complete the table of values, which again requires them to develop a strategy for larger values in the domain, and to verbally describe their strategies in solving the pattern. The equations again simplify to the same formula ($n = 5w + 2$) which may be related back to the table of values. There is less discussion on this exercise, which allows the teacher to assess the understanding and skill level for each student.

This last pattern also offers two new twists on the lesson. Question #12 asks the students to describe the meaning of the value of w. This appears elementary, but, most students will state that it represents the number of "figure eights" of seven match sticks, when actually it represents the number of "figure threes" of five match sticks added to the original two match sticks on the left. The last question also offers the opportunity for students to begin exploration of solving for the independent variable given the dependent.

EXTENSIONS

The activity is rich in opportunities for further development. For example, it is a stepping off point for algebraic manipulation or for practical application. Whether or not it deals with a real world is immaterial, one of the characteristics of a good teacher is making every activity real. Here are a few areas for further exploration:

- Find a formula for the number of tiles in the two outside layers around a square array.
- Find a formula for the number of tiles around the outside of a rectangular array.
- Find a formula for the number of unit cubes in the outside layer of a cubic array.

And for the capable or courageous...

Create a series of questions that will generate formulas for the following scenario: A swimming pool has a rectangular top which is L metres by B metres. At one end it is 1.5 metres deep and it is 2 metres deep at the other. (e.g. How many tiles does it take to create a single row of tiles around the entire vertical edge of the pool?)

Rule Quest | edge squares

Mathematics is often referred to as the "science of patterning." The job of any good mathematician is to discover the patterns of the universe and communicate them in an easy, efficient manner. The following activity gives you the opportunity to find some of these patterns and write the algebraic rules that describe them.

1. How many edge squares are there around the arrangements below?

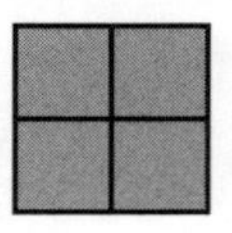
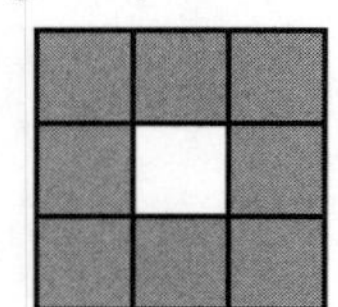
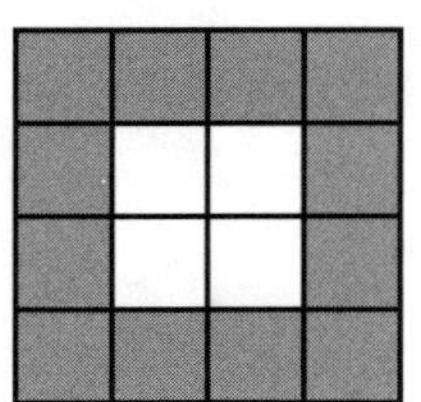

2. Complete this table. Use it to summarize and extend your results from question 1.

Length of side of square	2	3	4	5	6	10	100
Number of edge tiles							

3. Write down how you found the answer to the side of length 10.

4. Two of the diagrams below illustrate how you could show and calculate the number of edge tiles. Show two more ways in which the edge tiles could be illustrated

a) 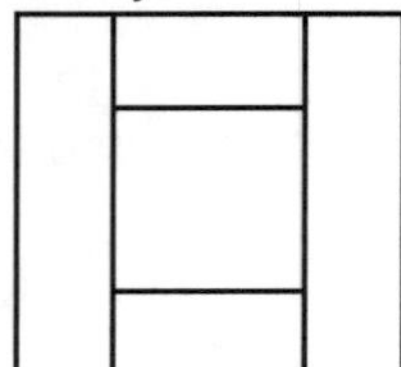
b) 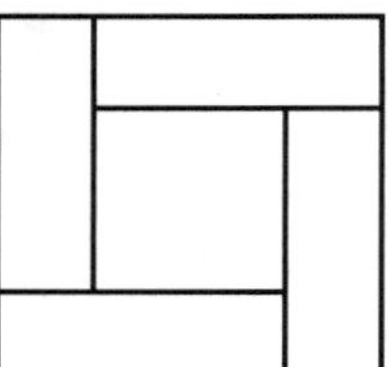
c) 
d) 

5. Write down in words the instructions for the four diagrams above. The first is given as an example.

	Verbal Description (#5)	Equation (#6)
a)	Add the length of the side to itself, and then add two less than the length, and then add two less than the length again.	$n = s + s + (s - 2) + (s - 2)$
b)		
c)		
d)		

6. Rewrite each of the instructions above using *n* and *s* to represent the total number of edge tiles and the number of tiles along one edge of the square.

7. What is the simplified formula for each answer in #6? How does this formula relate to the table in #2?

Rule Quest | edge triangles

Here is a similar pattern for you to study. The shapes are equilateral triangles of side length 2, 3, 4 and 5 respectively. The shaded triangles are edge tiles.

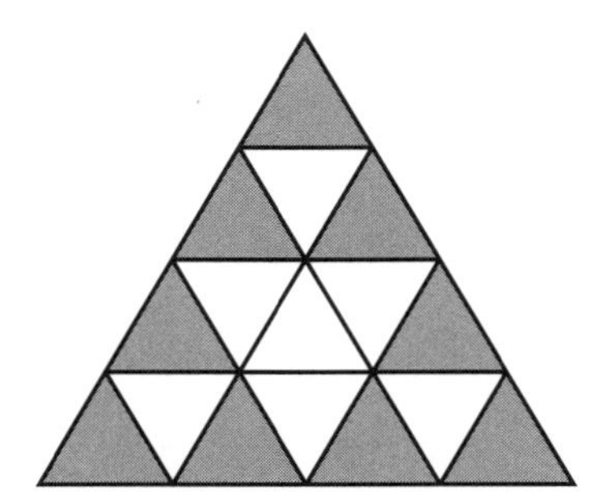
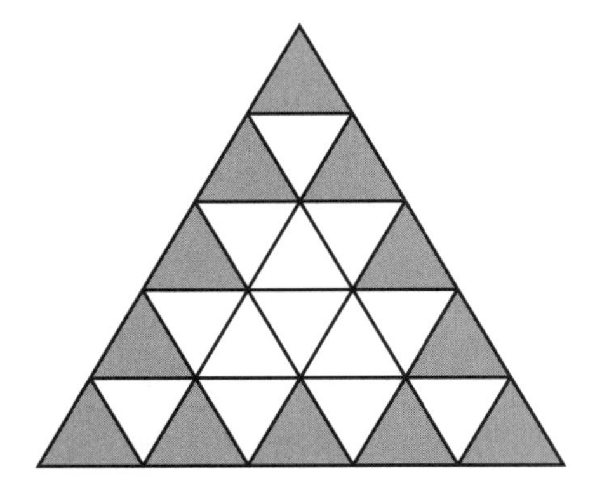

8. Complete this table and use it to show the numbers of edge tiles for the equilateral triangle arrangements.

Length of side of equilateral triangle	2	3	4	5	6	10	20	100
Number of edge tiles								
Total number of tiles								

9. Write down three different sets of instructions for finding the number of edge tiles around an equilateral triangle.

Verbal Description (#9) **Equation (#10)**

a)

b)

c)

10. Write the algebraic equations that summarize your steps in #9. What is the simplified formula that represents them all? How does this formula relate to your table above?

11. What is the formula for the total number of tiles in each triangle?

Rule Quest | match sticks

The next investigation is concerned with the following match stick diagrams.

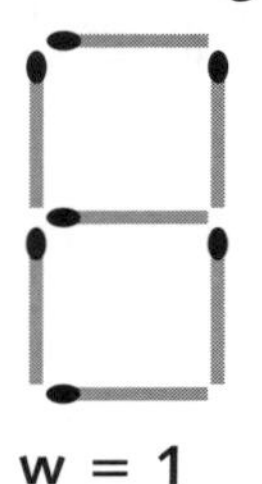
w = 1

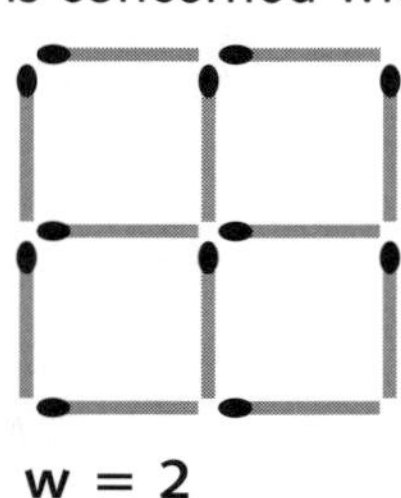
w = 2

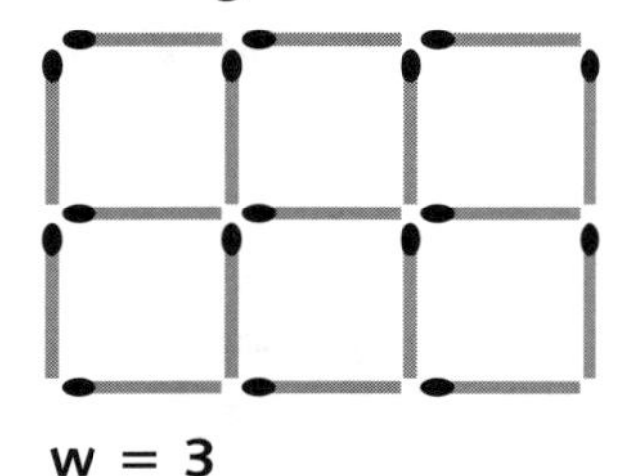
w = 3

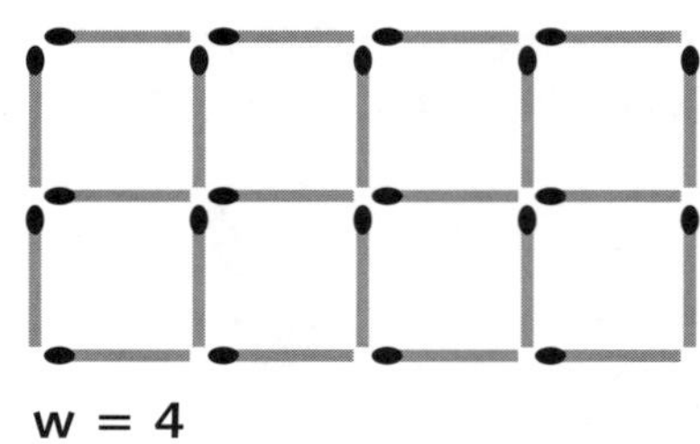
w = 4

11. Complete the table below.

Value of w	2	3	4	5	6	10		100
Number of matches							82	

12. Write down a sentence that explains the meaning of w.

13. Write down in words a set of instructions that would enable you to calculate the number of matches used when you know the value of w.

14. Write down a formula that tells you how to calculate the number of matches needed for a given value of w.

15. Check that your formula is correct by using it to calculate the number of matches used when w =
a) 1 b) 4 c) 10

16. Use your formula to calculate w for a diagram containing 172 matches.

PROJECT

Submitted by teachers of Sri Atmananda Memorial School, KPM Approach to Children, Atma Vidya Educational Foundation, India

OBJECTIVE

The aim of this activity is to let the students experience that equations and lines (graphs) are just different ways of expressing mathematical relationships.

Concepts
Linear relationships, dependent and independent variables, linear equations and graphs, slope, x- and y-intercepts, slope-intercept form of a linear equation

Time: 2 - 3 hours

Materials
Student handout, graph paper

Preparation
Graph of the pool example.

LESSON PLAN

Part One

Discuss a situation in which the quantities have a linear relationship. For example: think about a nearly empty swimming pool that is being filled with water. The pool begins with a water depth of 10 cm; the water rises at the rate of 3 cm per minute until it is full.

Lead the discussion in such a way that the students discover which quantities are related and how they are related — which quantity depends on the other one (in order to bring up dependent and independent variables). Often it is helpful to draw figures on the board to help the students visualize the situation. Once they have expressed the relationship in words, help them to use mathematical symbols to express the relationship. For example:

height = 3 • number of minutes or $H = 3M$ or $y = 3x$

Then let the students take several values for M (or x) and find corresponding values for H (or y). Taking these values together as ordered pairs (number of minutes, height) we get points. Have each student record these points in the table. Then plot the points on graph paper and let them find out what figure they get when all the points are connected. So we started with a relationship, turned it into an equation, then into a table of values and finally into a line! This is the crux of the lesson; students will be repeating this progression numerous times on the student handout.

At this point have the students think of several other situations in which two quantities are related. Again, have them express each relationship in words, symbols, a table of values, and a graph. Then the students should analyze the graph to see what shape is created. It may be useful if the class is divided into smaller groups to think of situations and their corresponding relationships. This gives more students a chance to present their ideas, and the teacher ends up with more situations in a shorter time. Bringing the class back together, let the students share all their ideas and split the figures they generated into two groups — those that are lines and those that are not. Then have a discussion in which the students describe what is common about all the equations whose figures turned out to be lines.

Now they know what a linear relationship is — a physical situation, a mathematical equation, a table of values, and a visual graph all at the same time!

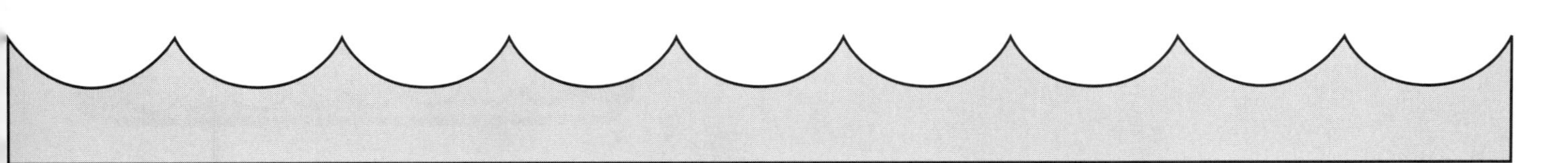

LESSON PLAN (continued)

Part Two

Having found the relationship, equation, table and line for each situation, it is time to compare all the linear graphs. For this part of the lesson, it is very helpful to have all the graphs drawn on large chart papers and hung in front of the class. The aim of this discussion is to use the graphs to bring out the meanings of terms like slope (gradient), x-intercept, y-intercept and the point-slope form of the equation of a line. The discussion can be precipitated with questions such as these.

In order to bring out the meaning of the term intercept:

1. What was the original height of the water in the given example? Or, at what point did the line cross the y-axis? What is the significance of that point in the given situation?

2. Did any of the lines cross through, or intercept, the x-axis? If so, what did this point represent in that particular situation? The answers to these questions can lead to a discussion on the meaning of the term 'x-intercept'.

In order to lead to a discussion on the meaning of slope:

3. In the given example, what was the height of the water at the first minute, the second minute, the third minute, and so forth? What was the difference in heights between each pair of minutes? So, what was the change in height of the water level per minute?

For the topic of slope, make sure to focus your discussion around two central issues:

1. For a given change in x, how much does y change? In other words, what is the change in y for a unit change in x?

2. Does the value of y increase or decrease for a given change in x? In other words, is the slope positive or negative?

After filling in the tables on the student handouts, the students can compare the constants in the equations to the information from the graph and form generalizations about the meaning of each constant.

NOTE: There are two reasons for having the students come up with their own situations. One reason is that the students have to make decisions, be creative and think about the concepts more openly. The other is that when the situation is their own, they identify with it; they possess it. This makes them more likely to be interested in solving it. They aren't trying to find out the solution to the teacher's problem; it is their own. They are solving something that belongs to them. So the knowledge they gain belongs to them. It isn't filtered through the teacher.

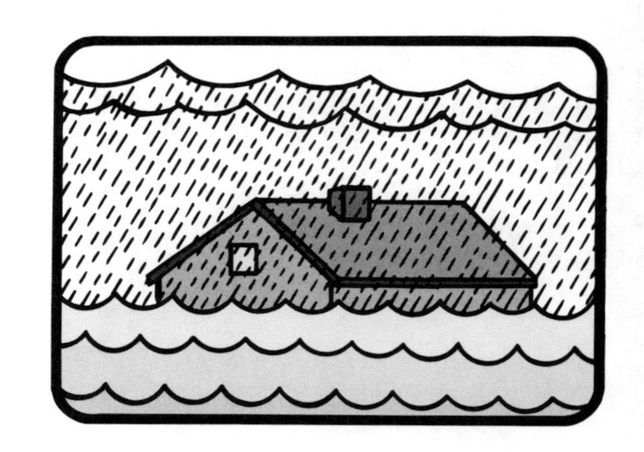

Part One

1. Using the example of the pool, describe the relationship of the rising water to time.

Relationship (in words): ____________________

____

____

Relationship (using symbols): ____________________

Relationship (table of values):

Relationship (by graphing):

2. What were the situations and relationships that you thought of?

Situation 1

Relationship (in words): ____________________

____

____

Relationship (using symbols): ____________________

Relationship (table of values):

Relationship (by graphing):

Situation 2

Relationship (in words): ____________________

____

____

Relationship (using symbols): ____________________

Relationship (table of values):

Relationship (by graphing):

3. Fill in the following table using all the equations that came up during the discussion:

Linear Equations			
Non-Linear Equations			

4. What is a linear equation according to you? ____________________

____

PART TWO

1. Which of the situations/relationships that you thought of formed linear equations? ____________________
____
____

2. What is the significance of the y-intercept (the point at which the line crosses the y-axis) in each situation?
____
____

3. What is the significance of the x-intercept (the point at which the line crosses the x-axis) in each situation?
____
____

4. What is the slope in each situation? (How much does the dependent variable change for each unit change in the independent variable?) ____________________
____

5. Fill in the following table with the linear equations offered by you and the class :

Equation				
Slope				
y-intercept				

6. Explain how you could find the y-intercept by:
   a) looking at a graph ____________________
   ____
   b) looking at an equation ____________________
   ____

7. Explain how you could find the slope by:
   a) looking at a graph ____________________
   ____
   b) looking at an equation ____________________
   ____

8. On a separate sheet of paper, create two examples of linear relationships: one in which the slope is positive and another in which it is negative. Draw rough figures of those situations.

EXTRA CHALLENGE

Have the classmate to your right give you a number, m, and the classmate to your left give you another number, b. Using m as the slope and b as the y-intercept, sketch a graph and think of a situation which it could represent.

The Student-Generated Word Problem

Submitted by Jacqui Ochoa, Kurt Walker, and Randy Davis, Trabuco Hills H.S., Mission Viejo, CA.

Concepts
Use and meaning of variables, writing and solving equations.

Time: 1 hour

Materials
None

Preparation
Student should have prior exposure to assigning variables and writing and solving equations.

One of the reasons students have a tough time learning to solve equations is that many do not really know what the algebraic equations represent. In particular, they do not understand the concept of a variable. We discovered this through a new assignment called the "Student-Generated Word Problem," The student is asked to: **write a word problem that can be solved by using an equation, write and solve the equation relating the solution to the original word problem.**

Here are some examples of the traditional problems the students had previously experienced:

Question #1: You have three boxes of pencils and you lose 5 pencils. You then buy another box. If you now have a total of 37 pencils, how many are in each box?
Equation #1: $3x - 5 + x = 37$

Question #2: The length of a rectangle is 2 more than 3 times the width. If the perimeter is 36, what are the dimensions of the rectangle?
Equation #2: $2w + 2(3w + 2) = 36$

The students were familiar with a variety of word problems for which they would be asked to write and solve the equations. Our fear was that they would simply copy one and submit it. Here is what we received instead:

Student #1: I have 5 Great Danes and 4 Poodles. I get another Great Dane. How many dogs do I have?
Equation #1: $5g + 4p + 1g$
Answer #1: 10

Student #2: I have 4 shirts. I buy three more. I sell ten. I get left with 4 shirts. How did I get 4?
Equation #2: $4s + 3s - 10 = 4$
Answer #2: $s = 2$

In the first example, the student created a logical question that did not need algebra to solve. The second student created an illogical question, then correctly solved an unrelated equation. In both cases, there was no cognitive connection among the word problem, the equation, and the solution.

This assignment told us volumes about how little our students really understood algebra, even if they could do algebra.

Consequently, we offered them equations for which they were required to write a problem. Eventually our students developed a stronger understanding as demonstrated below.
Equation: $5x - 3 = 3x + 10$
Answer: On Monday, Joe bought 5 boxes of apples and ate three on the way home. On Tuesday, he bought 3 boxes and a friend gave him 10 more. He noticed that at the end of each day he had the same amount of apples. How many apples were in each box?

Through these assignments, the students' understanding of the word problems improved, as did their ability to solve algebraic equations. The Student-Generated Word Problem was not only a powerful assessment tool, it proved to be an effective instructional instrument as well.

Adapted from CPM Educational Program and the MCTP Professional Development package, written by Australian mathematics teachers

PREPARATION

1. Find a good "event site" on your campus. This place should allow students to view their classmates from above, e.g. a hillside next to the blacktop or the bleachers in the gym.
2. Create a portable coordinate plane. For the axes, use two pieces of string, each 60 feet long. You will also need two sets of 21 index cards; each set of cards should be labeled with large numbers ranging from -10 to 10. On the x-axis, tape each card, approximately 3 feet apart, to the string; then do the same for the y-axis.
3. Create sets of input cards from the colored index cards. Designated by different colors, create the following sets below. Note, only the numbers should appear on the card, not the equations.

Concepts
Graphing, writing and solving linear equations

Time: one hour

Materials
Student handout, portable coordinate plane (see below), sets of colored index cards

COLOR	NUMBERS	EQUATION (not shown on card)
White	-6, -5, -4, -3, -2, -1, 0, 1, 3, 4, 5, 6	$y = x$
Blue	-6, -4, -4, -3, -2, -1, 0, 1, 3, 4, 5, 6	$y = x + 3$
Green	-4, -3, -2, -1, 0, 1, 3, 4	$y = 2x$
Yellow	-4, -3, -2, -1, 0, 1, 3, 4	$y = -2x + 1$
Pink	-3, -2, -1, 0, 1, 3	$y = x^2$
Orange	0, 1, 4, 9	$y = \sqrt{x}$

LESSON PLAN

While in the classroom, distribute the handout and at least one input card to each student, making sure that no student has two cards of the same color. For each input card, the student is to write the number shown in the appropriate IN box on the handout. For instance, if a student has a white input card with number 3 on it, the student should then write 3 in the IN box labeled "white." Then, the student should follow the description given on the handout for that colored set, and write the result in the OUT box. For example, the student with the white input card labeled 3 will read "the output is the same as the input," and write 3 again in the OUT box. The students should complete this for each card they receive.

Proceed to the event site and lay out the coordinate plane (made from string as depicted above) on the ground facing the group of students. Once you are ready, call the first set of students, e.g. everyone with a white card, to stand along the graph. Instruct them to stand along the x-axis on the number corresponding to the value on their card, with their backs to the class. In other words, they will all be facing in the positive-y direction. Our sample student with the number 3 on the card will stand then at 3 on the x-axis. Tell them that when you say "go" they are to walk either straight forward or backward until they are next to their output value on the y-axis. Again, our sample student should walk forward until he is "even with" or next to the number 3 on the y-axis.

"Ready? Go!" All the students should walk, and miraculously (at least in the eyes of your students) they form a straight line. Have all the other students draw what they see on the graph provided on the handout. When everyone finishes, have the white group step aside. Repeat this activity with each of the colored groups.

The objective of the Algebra Walk is to have students understand that the values along the x-axis and y-axis serve as the input and output, respectively. The objective of the classroom portion of the lesson is that students understand the relationship between the equation and the graph and know how to write and solve an equation.

WHITE
The output is the same as the input.

IN (x) OUT (y)

y = ______________________

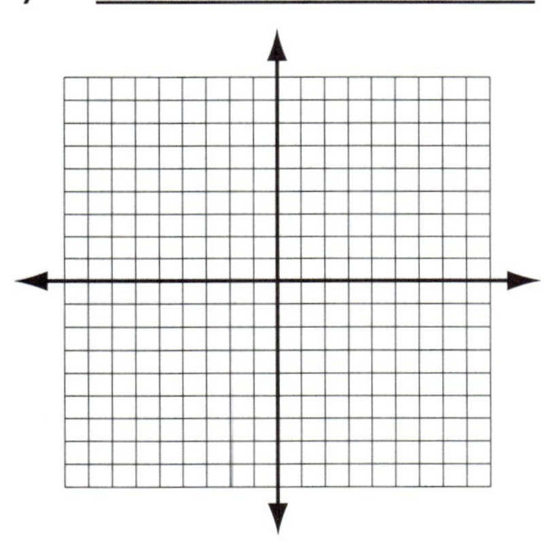

BLUE
The output is equal to the input plus 3.

IN (x) OUT (y)

y = ______________________

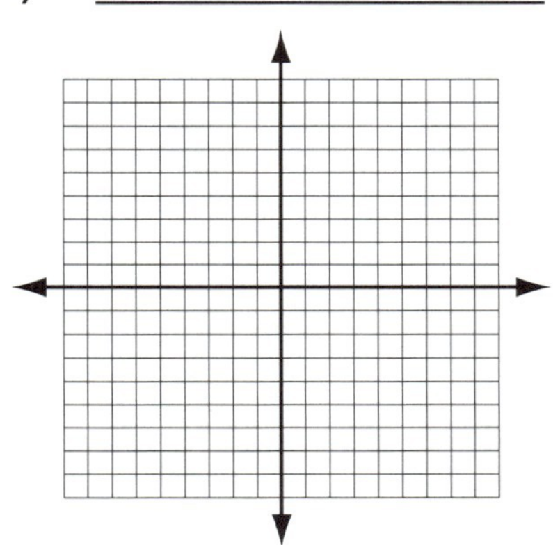

PURPLE
The output is twice the input.

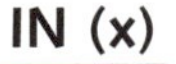

IN (x) OUT (y)

y = ______________________

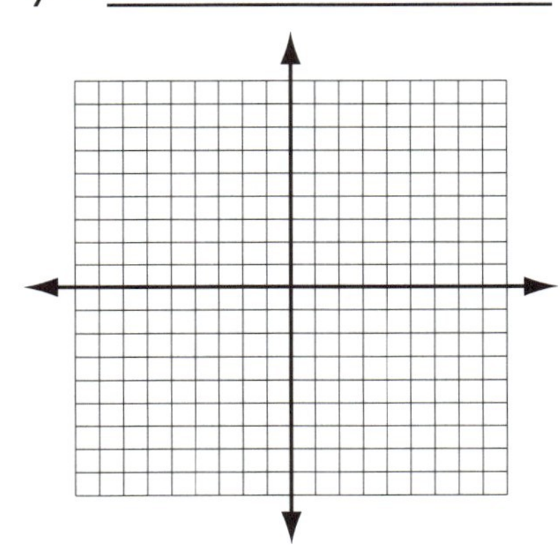

YELLOW
The output is one more than negative two times the input.

IN (x) OUT (y)

y = ______________________

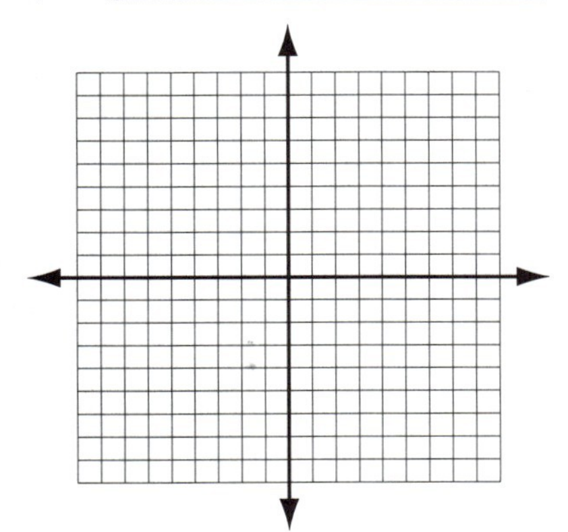

PINK
The output is equal to the input times itself.

IN (x) OUT (y)

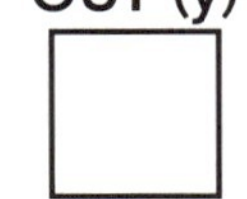

y = ______________________

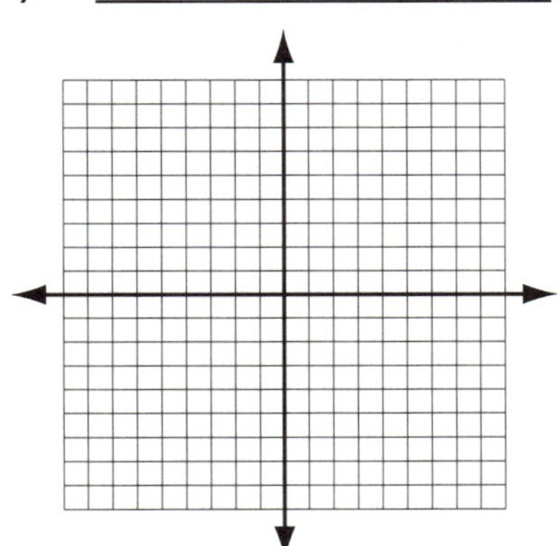

ORANGE
The output is equal to the square root of the input.

IN (x) OUT (y)

y = ______________________

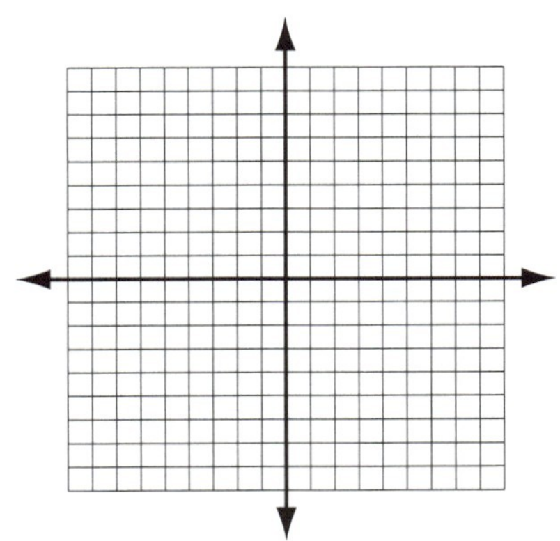

STUDENT HANDOUT

1. During the "Algebra Walk," why was it (would it be) easy to spot someone that was out of place?

2. For each of the graphs from the "Algebra Walk," predict the output (y-value) for an input (x-value) of 2. Then for each equation, find y for an x-value of 2. (show your solutions below) What do you notice?

   WHITE ____________ BLUE ____________ GREEN ____________

   YELLOW ____________ PINK ____________ ORANGE ____________

3. a) For the Orange set, complete the graph for the following inputs.

In (x)	0	1	2	3	4	5	6	7	8	9
Out (y)										

   b) What do you notice?

   c) What would happen if you used negative numbers for inputs? What does your graph tell you about this?

4. For each of the following equations, copy and complete the chart below, graph the equation on graph paper, and express in words the rule that the equation represents.

In (x)	0	1	2	3	4	5	6	7	8	9
Out (y)										

   a) $y = 2x + 1$ b) $y = -2x$ c) $y = x - 4$ d) $y = -x + 4$ e) $y = x^2 - 1$

5. Compare the graphs in #4 with the ones from the Algebra Walk. How are they similar? How are they different?

6. How can the graph for the rule $y = 2x + 1$ be used to predict the result for an input (x-value) of 7? How can the graph be used to predict the result for an input (x-value) of 3.5?

7. If you wanted an output of 7 for the rule $y = -x + 4$, what would you need as an input?

8) For each of the equations in #4, as well as the graphs from "Algebra Walk," where does the graph cross the y-axis? Describe any patterns that you notice.

PROJECT

Submitted by Dan Brutlag
University of California, Office of the President

PART ONE: The Carpenter's Staircase

The purpose of this lesson is to help students understand the concept of slope by offering a physical model for slope. It offers an alternative to the abstract models that are typically used in math classes, such as time vs. distance. This physical model demonstrates that slope is a ratio of vertical change to horizontal change ("rise over run"); therefore, larger values represent larger slopes. The intention here is that students will go beyond simply counting "up and over" and actually understand slope/rate in its various contexts.

Concepts
Slope

Time: 2-3 days

Materials
Rulers/tape measures, graph paper, student handout (optional)

Preparation
Determine the venue of the activity portion of the lesson by finding two staircases on campus that have different slopes. You will also need to find some type of ramp to use for the project, such as one that offers wheelchair/handicap access or one that is used for receiving deliveries.

1. Prompt the students to draw three staircases (on graph paper) that have a slope of $^2/_5$. Up 2 and over 5, will suffice for one example, then students will most likely explore $^4/_{10}$, and $^1/_{2.5}$. However, many erroneous attempts such as $^5/_2$ are often posed as possible solutions by the students. Share them all publicly, both the correct and incorrect, as a means of emphasizing the notion of equivalent ratios and how it applies to slope.

2. Now ask the students to draw a staircase with a slope of $^2/_5$, but with a tread of 12. Have the students draw a horizontal line 12 units long on the graph paper and challenge them to find the corresponding rise that will produce a slope of $^2/_5$. The students will find that the difficulty here is that the rise is not a whole number. Several students may utilize a proportion, while other students will apply an estimation strategy. The correct answer is 4.8.

3. Have the students articulate, in writing, the use of a proportion to find either the rise or the tread (run) of a staircase.

4. By completing the chart, students will have adequate practice at applying the method formalized in #3.

5. Here the students are exposed to the idea of negative slope and to the idea of rate presented in decimal form. Show the sketch of each of these graphs on the classroom board, as a point of discussion for #6.

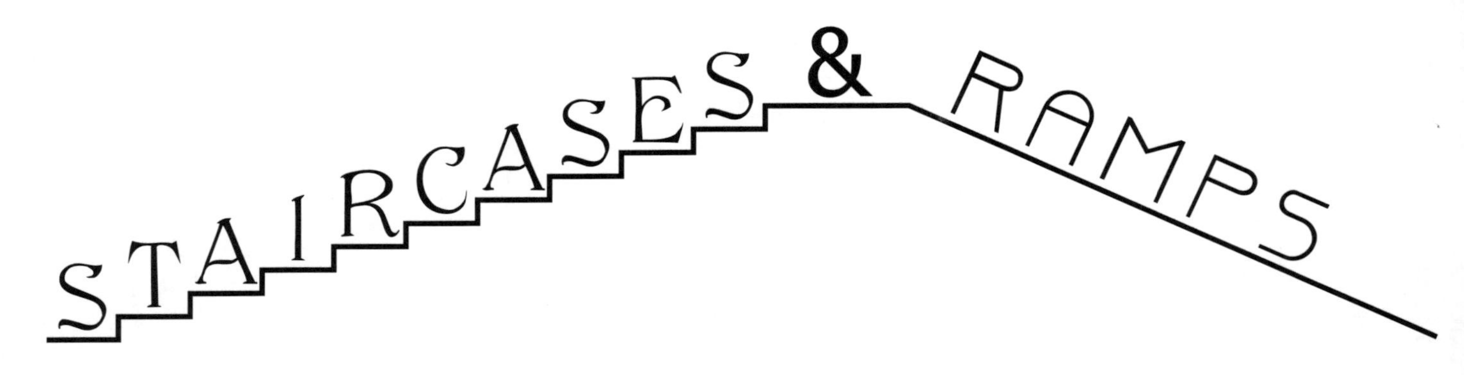

PART ONE (continued)

6. Defining slope as "rise over run" rather than "run over rise" is easier to understand, since the larger slopes (absolute value) will correspond to a steeper line, and smaller slopes will correspond to a flatter line. Prompt the students to verify this concept with the lines that they just drew in #5.

7. This activity portion of the lesson offers the students an opportunity to physically investigate slope by measuring 2 staircases. The students are able to measure the rise and run (tread) of a single step. From this information, they project the vertical height and horizontal length of the overall staircase. The students will naturally do this by multiplying the rise and the run of one step by the number of steps. From these new values the students are to calculate the slope of the entire staircase, and compare it to the individual step. They are the same! This is the key point of the activity, one step, two steps, or 100 steps of the same staircase will all have the same slope.

PART TWO: The Wheelchair Ramp

We now move from counting up and over to calculating/reading linear slope with the wheelchair ramp. This lesson emphasizes the relationship of the vertical to horizontal change of a line.

1. Once the students have measured the height and length of the ramp that has been designated by the instructor, they are to use these measurements to calculate the slope of the ramp. The simplified fraction representing the height over the length should be used for the slope. This fraction converted to decimal form will be the percent grade.

2. Here, students need to convert percent to decimal, then to a simplified fraction. This fraction is then to be drawn as a slope on the 5 x 5 grid. Then, assuming the ramp were to have that given slope with its current height, the students are to calculate the new length. For example, assume the ramp had a height of 2 feet, and a 20% grade were considered to be a slope of $^1/_5$. The student should draw a LINE (not another staircase) with a slope of $^1/_5$. Then the length of the ramp, with a height of 2 feet, would be 10 feet.

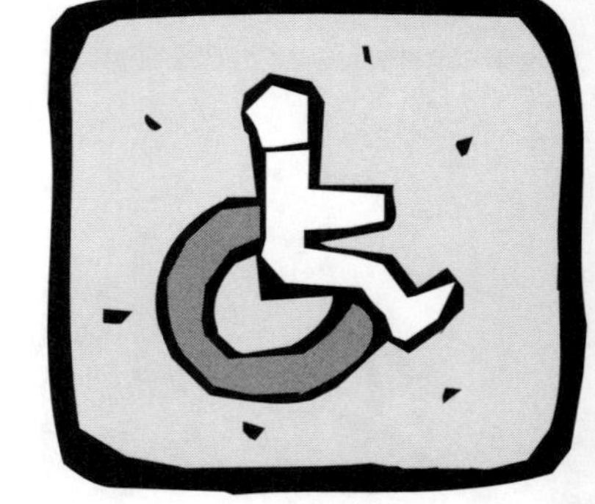

3. This problem offers students the opportunity to combine the various representations of slope. For example, assume that the chosen staircase is 60 inches high. For a slope of $^5/_{12}$, the horizontal length of a ramp with a height of 60 inches would be 144 inches. For a grade of 8%, the new horizontal length of the ramp would be 750 inches.

STAIRCASES & RAMPS

The Carpenter's Staircase

1. A carpenter is building a staircase. The slope of the staircase must be $2/5$. On graph paper, draw three examples of a staircase that has a slope of $2/5$.

2. Each step of the staircase is to have a 12" tread. Accurately draw this on the graph paper (scale: 1 unit = 1 inch). What is the vertical rise of each step of a 12" tread, for a slope of $2/5$?

3. Choose a method to quickly find the rise of a 12-inch tread for any slope ratio. Explain how the method works.

4. Use your method to find the rise that goes with a 12-inch tread, and a 10-inch tread for each slope ratio in the chart on the right.

Slope Ratio	Rise of a 12-inch tread	Rise of a 10-inch tread
$2/5$		
$5/8$		
$3/4$		
$5/6$		
$2/2$		
$6/5$		
$4/3$		
$8/5$		
$5/2$		

5. On a separate sheet of graph paper, draw two different staircase examples of each slope:
a) $5/2$ b) 1 c) $-3/4$ d) -2 e) 0.5

6. Why do you think slope is defined as "rise over run," instead of "run over rise?"

7. Find two staircases on your campus. Measure the attributes listed below and answer the questions.

		Staircase #1	Staircase #2
a) What are the measurements of the tread and the rise of each step of the staircase?	Tread:	________	________
	Rise:	________	________
b) What is the slope of one step?	Slope of a Step:	________	________
c) What are the overall height and length of the staircase?	Height:	________	________
	Length:	________	________
d) What is the slope of the entire staircase? How does this compare to your answer in part (b)?	Entire Slope:	________	________
e) What is the percent grade of your staircase?	Grade:	________%	________%

The Wheelchair Ramp

Wheelchair ramps are traditionally built at an 8% grade. Investigate the grade of a ramp at your school.

1. Ramp: ______________________

   a) What are the overall height and length of the ramp? Height: _______ Length: _______

   b) What is the slope of the ramp? _______ c) What is the percent grade of the ramp? _______

2. For each percent grade below, find its corresponding slope and create a scale drawing to represent a ramp with that particular slope. Assume the ramp has the same height as the ramp that you measured the other day, and determine the new length for the given percent grade. Height of ramp:_______

   a) 20%, slope = ______
   New length = ______

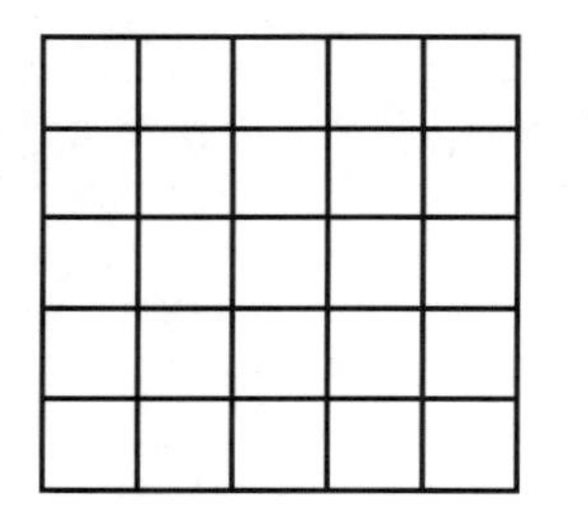

   b) 50%, slope = ______
   New length = ______

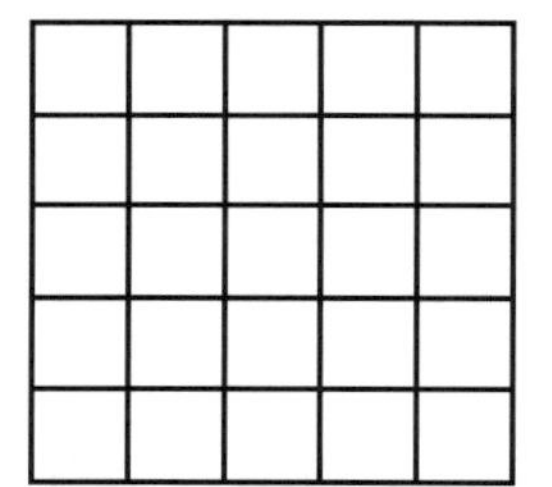

   c) 75%, slope = ______
   New length = ______

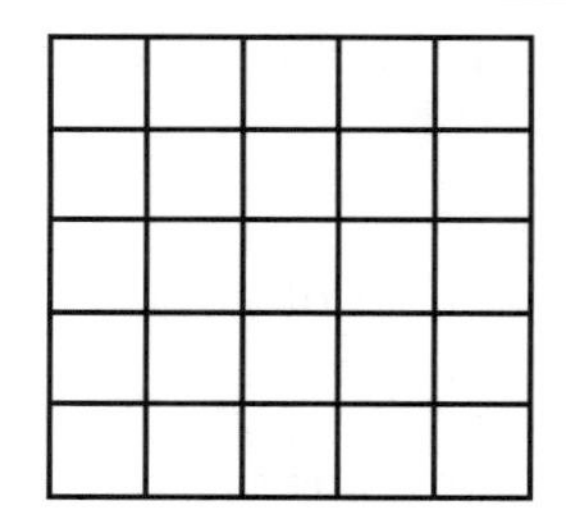

   d) 100%, slope = ______
   New length = ______

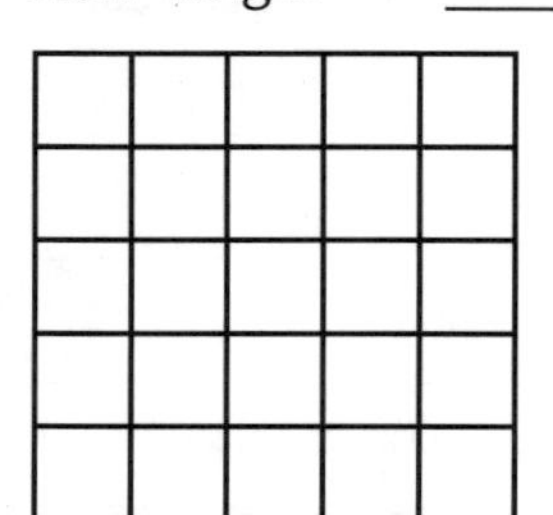

   e) 150%, slope = ______
   New length = ______

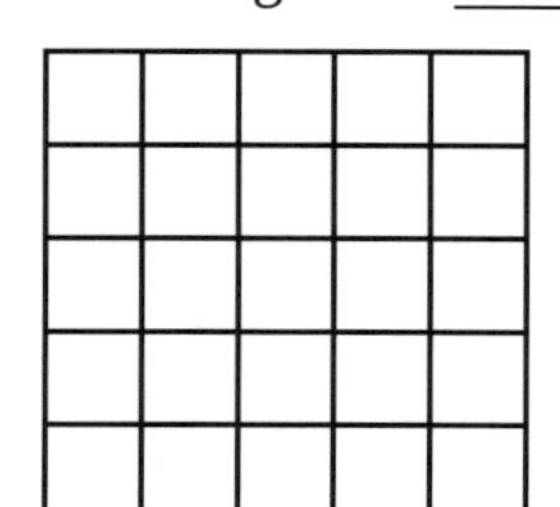

   f) 200%, slope = ______
   New length = ______

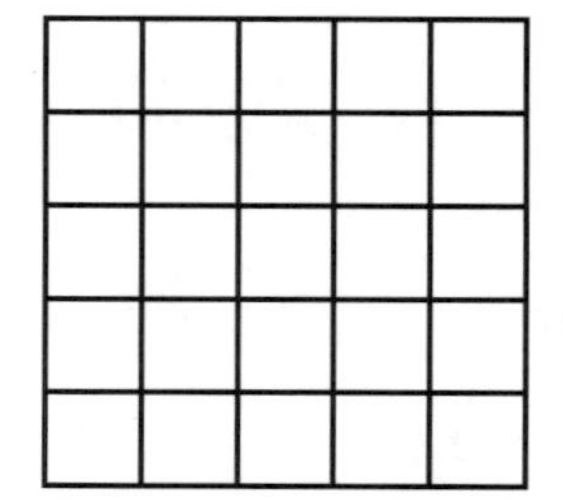

3. Choose the height of one of the staircases that you measured. Staircase #_____ Height: _________
   What would be the horizontal length of a wheelchair ramp that has the same height as the staircase with a...

   a) ...slope of $5/12$?

   b) ...percent grade of 8%?

PROJECT

TUMBLING CARS

Submitted by Dan Brutlag
University of California, Office of the President

he objective of this lesson is to offer a physical model of slope. This relates losely with the graphic model of slope ("rise over run"). The students will nvestigate various values for slope and its relationship to the steepness of a vooden board (line). This lesson also relates the representation of slope as a raction with its representation as a decimal (percent). This lesson is best sed as a precursor to instruction on linear equations.

ssign students to each one of the stations, consisting of a vertical ruler and board. It is ideal to have one station per group, but for large classes, there an be one station for every other group. While part of the class works with he cars, the others can work on the steep grade problem (in part three) nd then the groups can switch halfway through the class period.

Concepts
Slope

Time: 1 day

Materials + 4-5 cars
Wooden boards (3′ x 1.5′) and one toy car per station. The cars must have free-spinning wheels. Consider buying battery-operated cars so that they can also be used in future lessons on linear equations.

Preparation
Create vertical rulers on the classroom walls. Place 30-inch strips of masking tape on the wall vertically starting from the floor. Number these strips with hash marks at every inch. Starting at zero along the floor, number every other hash mark (2, 4, 6, etc.). Then lean a wooden board against each ruler. You will need one ruler for each group, or in large classes, for every other group.

ART ONE: Car Rolls Over
he students place the car sideways on the board, and test various slopes in rder to determine which slopes make the car topple. The students will robably experiment with many slopes in order to find the five that they are equired to record. Generally, they consist of two that do not make the car umble, two that do make it tumble, and one that puts the car on the verge f tumbling. One example should have a slope of 1 (rise = run), and one hould have a slope that is greater than 1 (rise > run).

he vertical ruler displays the rise of the board, the students will need to neasure the run (horizontal distance from the wall to the board along the loor). In the chart, the run is intentionally listed before the rise, so the tudents must think of dividing the rise by the run (so that larger numbers epresent steeper slopes). The slope should be written as a simplified raction. This fraction can then be converted to decimal form to get the ercent grade. In other words, a slope of $^1/_2$ is a 50% grade.

ART TWO: Car Races Downhill
he next phase of the lesson is just another method of exploring slope. The teacher is not really interested in the ptimal rolling distance, although the students will be. The students should begin with a relatively flat slope, and oll the car down the board. Draw a line on the board so that the students start the car from the same place each rial. The students no longer record both rise and run, only the slope and percent grade. As the students test teeper slopes, there will be a point at which the car crashes rather than rolls. This point will represent the teepest slope that the students will need to test.

ART THREE: Steep Grade
his is an excellent problem to help students visualize slope. The warning sign implies that the a 7% grade is as teep as the hill shown. Have the students actually measure, with a ruler, the height and base of the triangle in he sign, and determine its slope and grade (about $^1/_2$, 50%). It's definitely not 7%. To get an idea of what a 7% rade looks like, have the students draw it on the grid. Let them wrestle with converting from percent to fraction $^7/_{100}$), and with fitting this fraction on the 10 by 10 grid. With a run of only 10, the rise should be 0.7. In other vords, a 7% grade is not all that steep from a side view, but can be dangerous to a fully loaded truck.

STUDENT HANDOUT

Tumbling Cars

You are designing the embankment of a racetrack. You need to make the turns as steep as possible without making the cars tumble. You also need to know how the slope of the embankment affects the speed of the car.

PART ONE: Car Rolls Over

First determine at which slope the car tumbles, by setting the car sideways along the board. Test five slopes: two that do not make the car tumble, two that do make the car tumble, and the slope that appears to put the car on the verge of tumbling. Also, be sure to test at least one example for which the rise and run are equal, and one for which the rise is greater than the run.

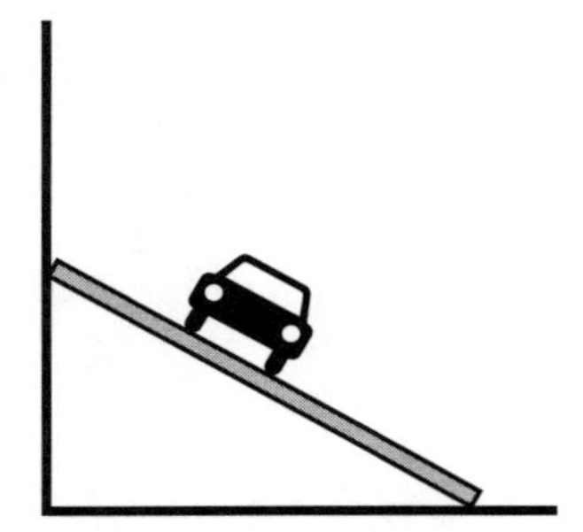

Trial	Run	Rise	Slope	% Grade	Roll? Y/N
1					
2					
3					
4					
5					

PART TWO: Car Races Downhill

Now determine the rolling distance of various slopes. Test five slopes. Is there an optimal slope for maximizing the rolling distance?

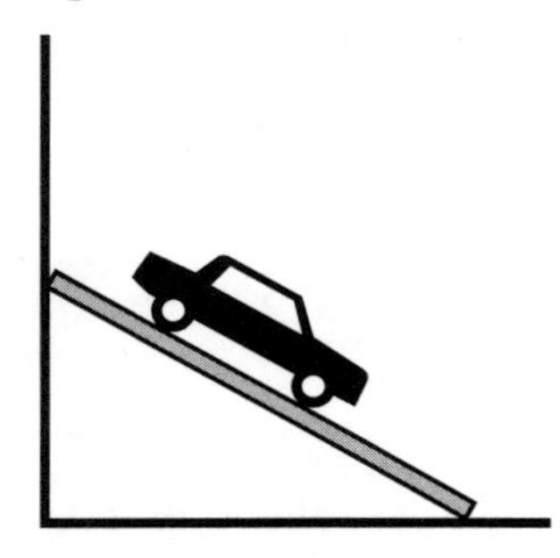

Trial	Slope	% Grade	Distance
1			
2			
3			
4			
5			

PART THREE: Steep Grade

To better understand the affect that the grade has on the speed of a vehicle, consider the sign on the right.

a. Measure the diagram to determine the slope and grade of the hill in the road sign. ____ = ___%

b. Make a scale drawing of a hill with a true 7% grade. .07 = 7/100

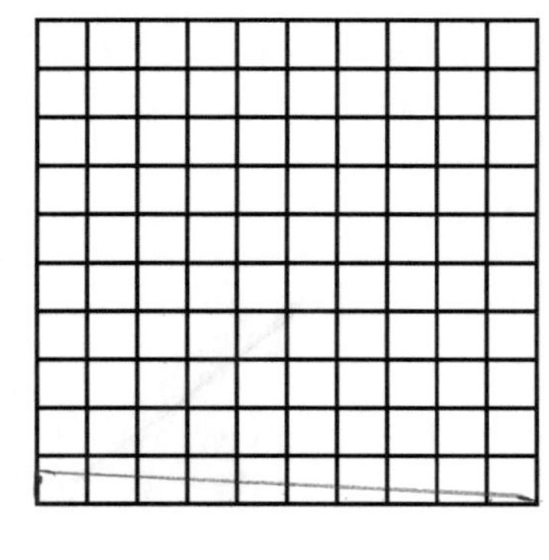

PROJECT

MONSTER CARS

OBJECTIVE

The purpose of this lesson is have students understand how rate (slope) is calculated from two data points. This is done by collecting data (time, distance) on battery operated cars. The students will also relate the y-intercept to the starting point of the car. This activity also offers a context for the students to understand the variables and various solutions of the equation that they will be learning to write.

ESTABLISHING THE RACE COURSE

The cars need a smooth flat surface (classroom carpet, or hallway floor) that is about twenty feet long and about ten feet wide. Choose a place where the students can take measurements, record data, and complete calculations. The classroom works best if there is enough room. On the floor, place seven parallel lines of masking tape every three feet. Write the assigned distance on each line, ranging from -6 feet to 12 feet. (See diagram on student handout.)

Assign students to pairs or groups. When the students are timing the cars, make sure that the person starting the car says go, and then the timekeeper starts the watch. If the timekeeper says go, there will be a lag before the car is actually released.

Concepts
Rate; y-intercept; writing, solving and graphing equations of lines in slope-intercept form

Time: 3-4 hours

Materials
Battery operated cars (two types/speeds for each group of students), stop watches, student handout, graph paper.

Preparation
Find a venue for the lesson, and establish the "race course" by placing masking tape at the designated intervals.

PART ONE: Finding the Rate of the Car

The students start the car at the starting line (distance of zero), and time the car to the first mark (3 feet). They should record this data as an ordered pair. For example, if the car took 1.5 seconds to get to the first mark, this data would be represented as (1.5, 3). The group should time the car again, this time to the second mark and record the data, again as an ordered pair. The first mark is then used to calculate the rate of the car. In our example above, the car traveled 3 feet in 1.5 seconds, or at a rate of 2 feet per second. The rate is then calculated using the second data point. The two calculated rates should be nearly equal.

Writing the equation to relate the car's distance to time follows the well-known formula: $d= rt$. For our hypothetical car, the students should write $d = 2t$. This equation is then used to answer the next two questions. In order to predict the distance that the car will travel in 10 seconds, we substitute 10 in for t and find that the car should travel 20 feet. In order to predict the time it will take the car to reach the third mark, the students should substitute 9 in for d, and get a time of 4.5 seconds. Once the students have calculated their predictions algebraically (showing all their work, of course), then they are to test their predictions by timing the car for 10 seconds, and timing how long it takes the car to reach the third mark.

For the graph, the students should plot the four ordered pairs that they produced for the lesson. The points will nearly form a straight line, although error in the data collection may produce some slight variance. The students will get the point though: the equation is linear because the car's rate is constant. The class discussion should center around the following concepts: 1) the slope of the line, the coefficient of x and the rate of the car are all the same thing, 2) the y-intercept is zero because the car started at the starting line, 3) the domain and range of the data determine the scales for the axes.

MONSTER CARS

LESSON PLAN (continued)

PART TWO: Finding the Starting Point of the Car

This next phase focuses on calculating the slope from two data points (rather than one) and on calculating the y-intercept. The students are now to time a different car, beginning at a location other than the starting line. Mark the starting point of each car with a piece of masking tape, and mark the tape with a number designating the group. Note that now the distance will be recorded in inches, and the time will be recorded for the first and the third mark. When the students attempt to find the rate, though, many of them will make the mistake of using the two data points separately, as was done with the previous car, rather than finding the difference between the two points. This is exactly the mistake that we hope they make! This common error offers the prime teachable moment: Did the car really travel the entire distance of the mark in the recorded time?

For example, let us assume that a car starts an unknown distance in front of the starting line, and reaches the first mark in one and a half seconds, and reaches the third mark in ten and a half seconds. This would yield the following data points: (1.5, 36) and (10.5, 108). We can show the students that, no, the car did not travel all 36 inches in 1.5 seconds, nor did it travel all 108 inches in 10.5 seconds. Where on the course, then, can we find a corresponding distance and time? Between the two marks! Yes, and we can find that distance and time by subtracting the coordinates of the data points. The car actually traveled 72 inches (108 - 36) in 9 seconds (10.5 - 1.5). By dividing as we have before we get a rate of 8 inches per second. The students are then prompted to formalize this process with an equation: $m = (y_2 - y_1)/(x_2 - x_1)$

Once the students calculate this rate, they will then be asked to find the car's starting point. Allow the students to wrestle with this one for awhile, until they understand that they can figure the distance forward to the first mark, then subtract to get the distance backward to the starting line. In our example, the car traveled at 8 in/sec for 1.5 seconds, which means, it traveled 12 inches. Therefore, it started 24 inches in front of the starting line. The students will follow the reasoning rather easily, but they must be pressed to communicate it algebraically and to write the equation as follows.

$$M = \frac{108 - 36}{10.5 - 1.5} = \frac{72}{9} = 8$$

$$36 = 8(1.5) + b$$
$$36 = 12 + b$$
$$24 = b$$

$$d = 8t + 24$$

Once the students calculate the starting point, they should use a ruler to measure the actual starting distance. They will be impressed by their own accuracy. In the subsequent questions, making the predictions, taking the actual measurements and drawing the graphs are all handled similarly to those with the first car. The obvious point of emphasis here is the relation of the y-intercept to the starting distance of the car.

PART THREE: Starting Behind the Starting Line

If possible, have the students use a different car (different rate) for the final phase of the lesson. This phase runs identically to the previous one, with the simple exception that the car begins behind the starting line yielding a negative y-intercept. Again, push them to show the calculations algebraically.

MONSTER CARS

You have a toy car. Determine its speed, and use that data to determine other valuable information of the car. Gather your data using the established race course. Each mark is 3 feet apart as shown in the diagram below.

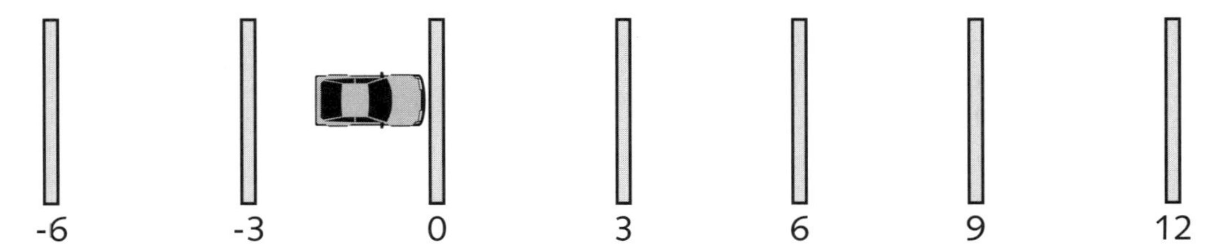

PART ONE: Finding the Rate of the Car

1. Place Car #1 on the starting line and time how long it takes to reach the first mark. Repeat this process for the second mark. Write the data as an ordered pair (t, d).

   1st Mark (,) **2nd Mark** (,)

2. Calculate the rate of the car twice, once with your first data point, then again with the other. If the two rates differ dramatically, feel free to time your car again.

   Rate: ________ ft/sec (1st mark) **Rate:** ________ ft/sec (2nd mark)

3. Write an equation to represent the relationship of the car's distance, d, to time, t. ____________________

4. Use your equation to predict how far the car will go in 10 sec. Then test your result.

   Prediction: (,) **Actual Distance:** ____________

5. Use your equation to predict how long it would take to get to the third mark. Test your result.

   Prediction: (,) **Actual Time:** ____________

6. Graph your data points from numbers 1, 4, & 5. Draw a line through these data points. Show the slope of the line. How does this relate to your answer in number 2? What does the y-intercept of the graph represent?

MONSTER CARS

PART TWO: Finding the Starting Point of the Car.

7. Place Car #2 somewhere between the starting line and the first mark. Time how long it takes the car to get to the first mark. Then place the car at the same starting point and time how long it takes to reach the third mark. Write the data as an ordered pair (t, d).

   1st Mark (,) **3rd Mark** (,)

8. Calculate the rate of the car. Then start the car at the starting line and test your prediction.

   Prediction **Rate**: ________ in/sec

   Actual **Rate**: ________ in/sec

9. Describe how to calculate rate when the starting point is not the starting line, given two data points.

10. Write a formula to represent your explanation above. ________________

11. Calculate the starting point of the car, then measure the actual starting point.

    Calculated: ________ **Measured:** ________

12. Write an equation to represent the relationship of the car's distance, d, to time, t. ________

13. Use your equation to predict how far the car will go in 15 sec.

    Prediction: (,) **Actual Distance:** ________

14. Use your equation to predict how long it would take to get to the second mark. Test your result.

    Prediction: (,) **Actual Time:** ________

15. Graph your data points from numbers 7, 11, 13, & 14. Draw a line through these data points. Show the slope of the line. How does this relate to your answer in number 8? What does the y-intercept of the graph represent? How does it relate to your answer in number 9?

MONSTER CARS

PART THREE: Starting Behind the Starting Line

16. Place Car #2 somewhere BEHIND the starting line, but not at any of the established marks. Determine two data points (t, d).

1st Point (,) **2nd Point** (,)

17. Calculate the rate of the car.

Rate: __________ in/sec

18. Calculate the starting point of the car, then measure the actual starting point.

Calculated: __________ **Measured:** __________

19. Write an equation to represent the relationship of the car's distance, d, to time, t. ____________________

20. Use your equation to predict how far the car will go in 15 sec.

Prediction: (,) **Actual Distance:** __________

21. Use your equation to predict how long it would take to cross the starting line. Then test your result. Where is this point on the graph?

Prediction: (,) **Actual Time:** __________

22. Graph your data points from numbers 16, 20, & 21. Draw a line through these data points. Show the slope of the line. How does this relate to your answer in number 17? What does the y-intercept of the graph represent? How does it relate to your answer in number 18?

Y=MX+B

PROJECT

The Jogging Hare

THE STORY

The Hare is training for his long awaited rematch with The Tortoise. He begins his morning jog an unknown distance in front of his burrow. He runs at a constant rate (assume instantaneous acceleration), in a straight line away from his burrow. After 35 seconds he passes a boulder that he knows is 220 feet in front of his home. After two and half minutes (total time), he passes a tree that he knows is 680 feet from his home.

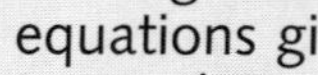

Concepts
Writing and graphing linear equations given two points. Interpolation & extrapolation. Slope and y-intercept.

Time: 2- 3 hours

Materials
Graph paper, straight edge, student handout

Preparation
Give to students before any formal instruction of algorithms of writing equations from two points.

THE TASK

With this information about the Jogging Hare, answer the following:

1. How fast is the Hare jogging in feet per second?
2. How far in front of his burrow did he start?
3. How far will the Hare be in 3 minutes?
4. When will the Hare reach the stream that is a 1000 feet away? Once you have answered the above questions, complete the following:
5. Write the numerical information given as two data points (ordered pairs). Set the time in seconds as your domain, and distance in feet as your range. Also write your answers to #2-4 above as ordered pairs.
6. Write an equation for the scenario and show that your equation supports your answers to #1-4 above.
7. Graph the scenario above by plotting your five data points. Then graph your equation to show that your data concurs with your equation.

LESSON PLAN

1. Give students the scenario written above and allow them to wrestle with the first two questions regarding the Hare's rate and starting point. Once they have been given ample time, collectively discuss the various strategies and answers. Help the students connect their intuitive strategy of finding the rate with the standard formula of slope. Be sure to close this phase of the lesson with a class consensus on the two answers. (Note: this phase alone may take an entire hour.)
2. Using the rate and starting point found in the first phase of the lesson, have the students respond to questions 3 and 4. Again let them solve it by any means they can and then share responses publicly. Once this is done, demonstrate how to solve an equation for one variable when given the other.
3. Then have students graph the two data points and draw the line through it. It is best to let them determine the scale of each axis.
4. Have students display their answers to the questions as data points on the graph. They should also be able to identify the y-intercept as the starting point for the Hare and graphically demonstrate the slope.
5. Be sure to revisit this problem often. In other words, give the same story with different data as homework, or quiz and test questions.

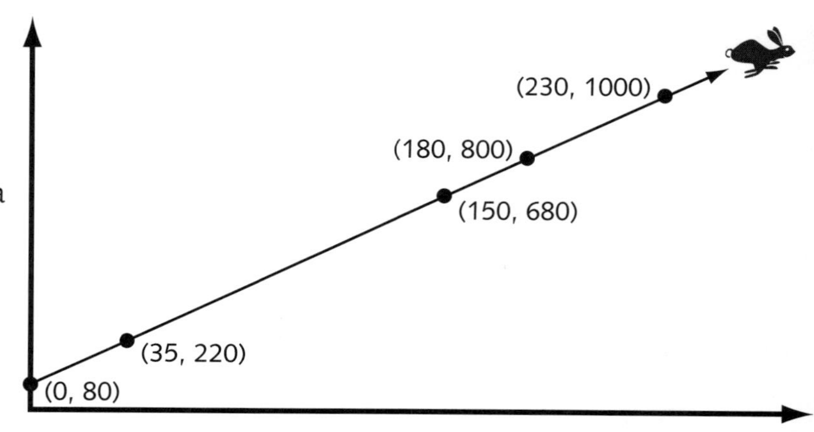

SOLUTIONS

1. 4 ft/sec
2. 80 feet
3. 800 feet
4. 230 sec
5. (35, 220), (150, 680), (0, 80), (180, 800) & (230, 1000)
6. $d = 4t + 80$
7. See graph

The Jogging Hare

The Hare is training for his long awaited rematch with The Tortoise. He begins his morning jog an unknown distance in front of his burrow. He runs at a constant rate (assume instantaneous acceleration), in a straight line away from his burrow. After 35 seconds he passes a boulder that he knows is 220 feet in front of his home. After two and a half minutes (total time), he passes a tree that he knows is 680 feet from his home.

With the information that you now have regarding the Jogging Hare, answer the following four questions.

1. How fast is the Hare jogging in feet per second?

2. How far in front of his burrow did he start?

3. How far will the Hare be in 3 minutes?

4. When will the Hare reach the stream that is a 1000 feet away?

Once you have answered the above questions, complete the following.

5. Write the numerical information given as two data points (ordered pairs). Set the time in seconds as your domain, and distance in feet as your range. Also write your answers to #2-4 above as ordered pairs.

(,   )   (,   )

(,   )   (,   )   (,   )

6. Write an equation for the scenario and show that your equation supports your answers to #1-4 above.

7. Graph the scenario above by plotting your five data points. Then graph your equation to show that your data concurs with your equation.

PROJECT

Adapted from CPM Educational Program's project "Bee Bopper Shoe Store"

LESSON PLAN

The extent to which a teacher uses this lesson depends on the level of the course. A pre-algebra student may only be capable of plotting points, recognizing the correlation, and estimating a line of best fit. Beginning or advanced algebra students will be able to find the slope, y-intercept, and write the equations. All students should be able to solve the equations for one variable given a value for the other. Therefore, for a pre-algebra class, the teacher may give the students the equations, discuss the values of slope and y-intercept, and then have the students plug in their numbers.

Begin the class by having students come to the board, one boy and one girl at a time, and write their height (in inches) and their shoe size in the charts provided. While the others are waiting their turn, they should be copying the information written in the charts. If you have a large class, you may also want to have some additional warm-up problems on the board.

Next, have the students plot the points. They must first establish the scale for each axis — the height is the domain, and the students will instantly recognize that if they simply start at zero and count up by ones to seventy, they will run out of room on their papers. Discuss with them the fact that they really don't need any numbers less than 58; so they can establish a broken scale, but they must show this (—√—) on the axis. The shoe size is the range, so we label that axis from 0 to 13, in increments of one-half.

> ***Concepts***
> Plotting points, lines of best fit, writing equations
>
> ***Time:*** 1-2 hours
>
> ***Materials***
> Student handout (optional)
>
> ***Preparation***
> Students will need a means by which to measure their height. Establish a vertical "height line" on the board, so that students simply stand with their backs to it and have a classmate read the height. Also, have the boys and girls charts as well as the one-quadrant coordinate plane on the board when class begins.

Then have the students plot the boys data using a certain color or symbol (+). Then have them plot the girls data using a different color or symbol (*). Once the data is plotted, prompt the students to articulate the relationship between the height of a person and their shoe size. Is there a correlation?

Once the students agree that taller people generally have larger shoe sizes, have them draw an approximate line of best fit for each gender. Students should then find the rate of change of shoe size to height (slope). Help them articulate their answers. For instance, if a student calculates the change in y of 3 shoe sizes and a change in x of 6 inches, they should say that there is change of 3 sizes for every 6 inches, or a half shoe size per inch.

Have the students calculate the y-intercept of the line. They should understand that this would be the shoe size for a theoretical person of zero inches tall. The answer will not make sense according to where their projected line intersects the y-axis. In other words, their line looks like it has a y-intercept of 1, but their calculated intercept may come out to be -24. This contradiction is due to the broken scale along the x-axis. Have students mentally stretch this axis until it is continuous, and they will more easily see how the line "gets down that far."

Once the students have determined their equations, have them choose values for x and y (height and shoe size) that do not appear on the charts. The students can then plug the values into their equations and check that the solution appears on the line of best fit. For instance, assume that no girl in the class is exactly 68 inches tall. Have the students use their equation of the line of best fit for the girls data to find the appropriate shoe size for a 68 inch tall girl. This solution should lie on the line. Students should repeat this process for a missing shoe size, as well as. This allows them to practice solving for y given x; and also the more difficult task of solving for x, given y.

CoolShoes.com

You own and operate CoolShoes.com, an online shoe store. Many people want to order shoes for friends and relatives, but do not know their shoe size. Since it is easier to estimate a person's height than shoe size, you want the customer to be able to enter a person's height and calculate the appropriate shoe size (approximate). You must have either a graph or equation in order to do this. So, your task here is to create both, using sample data from your class.

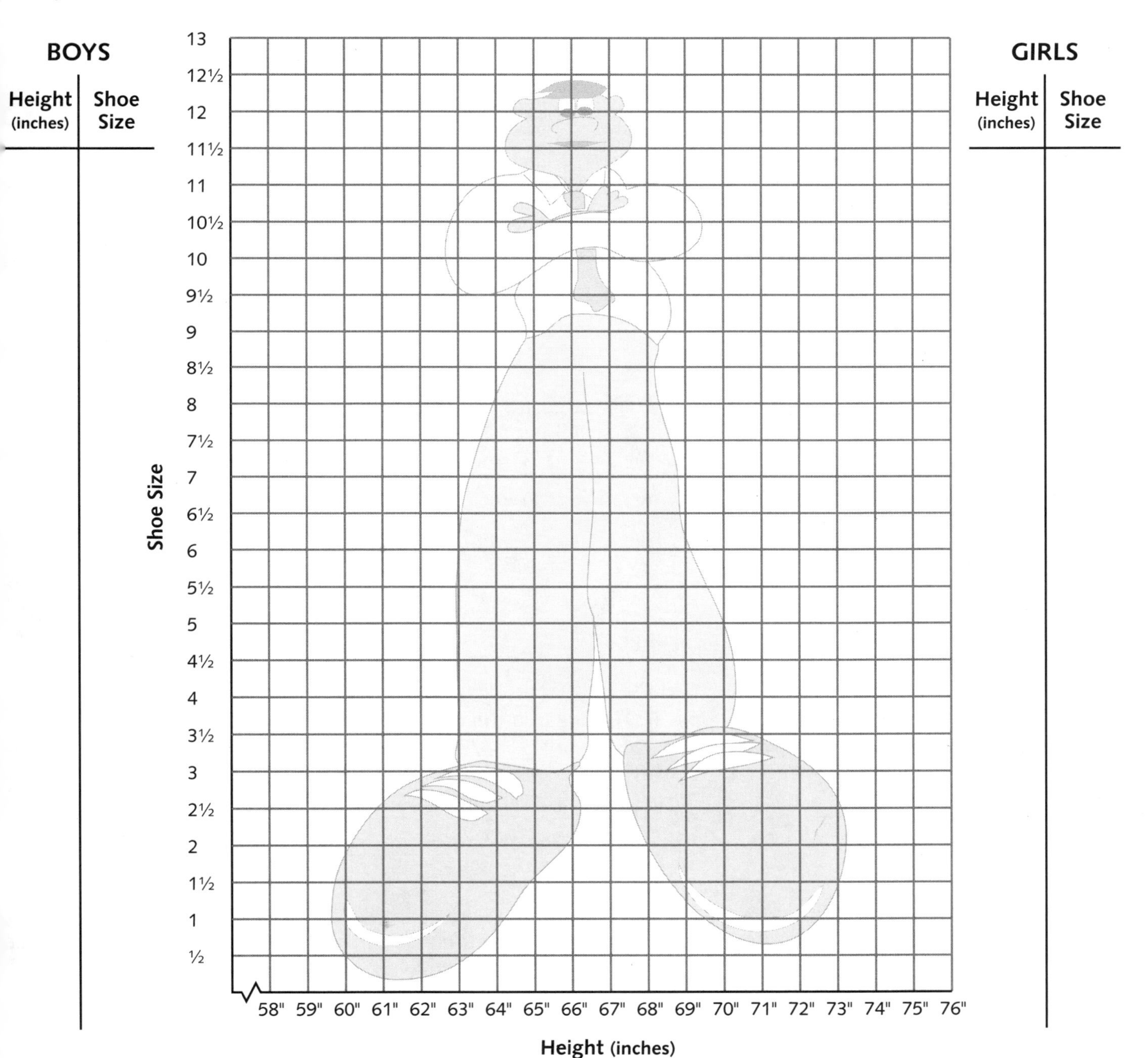

STUDENT HANDOUT

CoolShoes.com

1. Fill in the charts with data from your class. Record each person's height and corresponding shoe size.
2. Plot the data points from the charts. Use one color or symbol (+) for boys and a different one for girls (*).
3. Do you notice any relationship between people's height and their shoe size? What kind of correlation is it?
4. Draw an approximate line of best fit for each set of data (one for the boys, one for the girls).
5. For each line, calculate the rate of change (slope).

   BOYS: There exists a change of _____ sizes for every ________ inches of height, or _____ sizes per every one inch.

   GIRLS: There exists a change of _____ sizes for every ________ inches of height, or _____ sizes per every one inch.

6. a) Calculate the y-intercept of each line. **BOYS:** ________ **GIRLS:** ________

   b) What do these intercepts imply? Do they match your graph?

7. Write the equations of each line.

   BOYS: ________________ **GIRLS:** ________________

8. For each set of data, find a height that does NOT appear in the chart. For instance, if no girl in the class is exactly 68 inches tall, then choose 68 inches for the girls. Use your equation and your chosen value for height to find the corresponding shoe size at that height. Do your solutions match the graphs?

   BOYS: Height = ________ **GIRLS:** Height = ________

   Shoe Size = ________ Shoe Size = ________

9. For each set of data, find a shoe size that does NOT appear in the chart. For instance, if no boy in the class has a shoe size of 13.5, then choose 13.5 for the boys. Use your equation and your chosen value for shoe size to find the corresponding height. Do your solutions match the graphs?

   BOYS: Height = ________ **GIRLS:** Height = ________

   Shoe Size = ________ Shoe Size = ________

PROJECT

LAND CRUISER

Announce that there is going to be a contest to determine which group of students can build the best wind-powered Lego™ Land Cruiser. The contest will determine which of the cruisers is the FASTEST, and which goes the FARTHEST. Using only the materials supplied—the Legos, 2 straws, a piece of aluminum foil, and adhesive tape—the students will build a vehicle with a sail. An electric fan will be used to simulate the wind. Explain or show the venue to the students, as well as the procedure for building, modifying, and running their Land Cruiser, as described below.

PART ONE: Design & Build

Students may reduce the size of the straws and the foil, but they should not be given any more. Students should sketch their Cruiser and describe the reason that they chose their particular design.

PART TWO: Test & Modify

Allow each Cruiser three trial runs with the fan. After each run the students may make modifications. They should describe these modifications, and the reasoning for them. Once the trials are complete, display the Cruisers for the class, and have each group predict which will go the farthest and fastest.

PARTS THREE & FOUR: Compete & Collect Data

Each group is allowed one attempt in the final competition. Have a student hold the Cruiser on the starting line before turning the fan on. Then have the student release the Cruiser once the wind reaches maximum velocity. Begin timing as soon as the Cruiser begins to move (not all cruisers begin once they are released). Stop timing as soon as the Cruiser comes to rest. Have the students measure the distance from the starting line to the Cruiser. Groups record the results of both time and distance for all Cruisers.

PART FIVE: Calculate Speed

Example: A car traveled 15 feet in 10 seconds

Feet per Second

$$\frac{15 \text{ feet}}{10 \text{ seconds}} = 1.5 \text{ ft/sec}$$

Miles per Hour

$$\frac{15 \text{ feet}}{10 \text{ seconds}} \bullet \frac{1 \text{ mile}}{5280 \text{ feet}} \bullet \frac{60 \text{ sec}}{1 \text{ min}} \bullet \frac{60 \text{ min}}{1 \text{ hour}} \approx 1.02 \text{ mph}$$

PART SIX: Analyze

Graph the results to test their conjectures. On the samples given below, the students have made a conjecture that the Cruisers with relatively long bodies go farther (this is also true of wide bodies, and large sails), and that the small cruisers go faster. The graphs below support both conjectures.

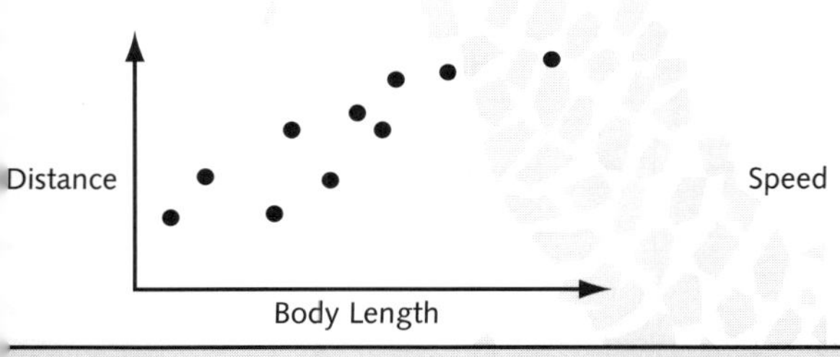

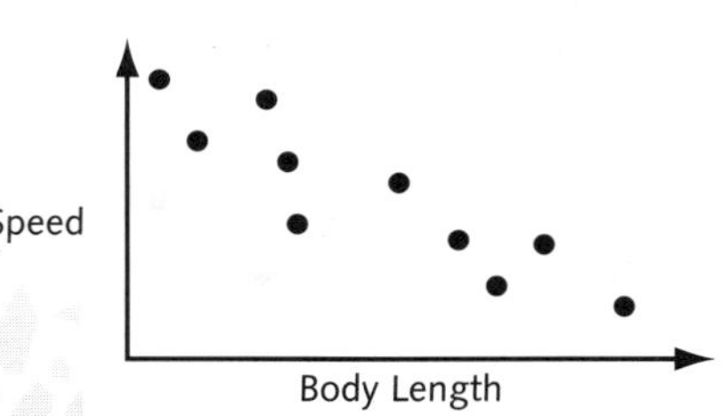

Concepts

Distance-rate-time relationships; graphing, making conjectures.

Time: 2 hours

Materials

Student handout, 10 bags of Legos, 20 straws, 10 one-foot square pieces of aluminum foil, an electric fan, masking tape, adhesive tape, a tape measure, stop watch, and class set of scissors. Lego™ manufactures several small, inexpensive car kits which can be purchased at any toy store.

Preparation

The Materials: Open each box of legos and place its contents in a plastic baggie. Make an inventory list for each baggie. Cut the aluminum foil sails into 1-foot squares. The bag of Legos will be reusable from class to class, but the foil and straws will not.

The Venue: Find a location with a slick floor (linoleum hallway or gymnasium). The run needs to be at least sixty feet long. Place the electric fan at one end of the floor, low to the ground. Mark the starting line with a strip of masking tape about twelve inches in front of the fan. It is strongly recommended that you build a land cruiser and test it yourself, so that you know the peculiarities of your venue and fan before you conduct this activity with students.

LAND CRUISER

Your group's task is to design, build, test and demonstrate a wind-powered Lego™ Land Cruiser. Each Cruiser will have a foil sail and be placed in front of an electric fan. The groups will compete to determine which Cruiser travels the longest distance and which is the fastest. You will also be required to collect data, calculate speeds and make conjectures.

PART ONE: Design & Build

Design and build a land cruiser out of the following given materials ONLY: one bag of Legos, two straws, and a small sheet of aluminum foil. Give a sketch and a brief description of your design. Be sure to discuss the reasoning behind your design.

PART TWO: Test & Modify

Perform three trial runs of your cruiser and make any modifications to your design based on your observations. Discuss the nature and purpose of any modifications you make.

PARTS THREE & FOUR: Compete

Place and release your Cruiser at the starting line in front of the fan. You will be allowed only ONE run, which will be measured for distance and time traveled. Be sure to record how far your Cruiser traveled, as well as how long it took to come to a complete stop. Did your Cruiser perform as expected? Why or why not?

PARTS THREE & FOUR: Collect Data

Record the results of the contest in the chart below.

CRUISER	1	2	3	4	5	6	7	8	9	10
Distance (ft)										
Time (sec)										

LAND CRUISER

PART FIVE: Calculate Speed

Calculate the speed of each cruiser in both feet per second (ft/sec) and miles per hour (mph). Record your results in the chart below. Show your calculations in the space provided.

CRUISER	1	2	3	4	5	6	7	8	9	10
Speed (ft/sec)										
Speed (mph)										

PART SIX: Analyze

Examine the winning cruisers and make conjectures about how the design of the cruisers helped them win. Then, verify or disprove your conjecture by completing the graphs below. On the graph, be sure to label the domain with the factor which you are considering (such as weight, number of bricks, size of sail, etc.)

We thought that Land Cruiser # ____ would travel the farthest, and # ____ would go the fastest, because:

Conjecture regarding DISTANCE: The cruiser that went the farthest was

____.

Conjecture regarding RATE: The cruiser that went the fastest was

____.

______________________ ______________________

Understanding Mixture Problems

LESSON PLAN & SOLUTIONS

Hook students with the Lab Tech question. As always, we want students to understand the problem before they represent it algebraically. Students will learn how to construct solutions, before they learn how to deconstruct them.

> ***Concepts***
> Mixture problems; fractions and percentages; geometric and algebraic representations
>
> ***Time:*** 1-2 hours
>
> ***Materials***
> Multi-link cubes or colored blocks
>
> ***Preparation***
> Each pair of students needs 20 cubes of one color, and 20 of another color.

1. Have students use the multi-link cubes to represent 2L of 50% acid solution. Define one multi-link cube as one liter of solution. One color represents water, the other color represents acid.The students then represent the 5L of 20%. Record both of these on the student handout. Then combine the two sticks to form one long stick. Common colors should all be adjacent to one another, so the percentage of acid is easily discerned. The combined mixture will have 2 of 7 liters being acid, thus $2/7 = .285 = 28.5\%$ acid. Record this third stick diagram, then have the students repeat this process for problem #2 on their own.

2. Ask the students why the combined percentage is not the average of the two percentages. The students see that the greater initial amount has a greater influence on the outcome (weighted averages). Then lead the students through the algebraic representation of their sticks. Show the students that the first step in solving the equations represents the number of acid cubes in each stick. The second step shows the number of acid cubes in the combined stick. The third step shows the final percentage.

   #1: $.5(2) + .2(5) = x(7)$; $1 + 1 = 7x$; $2 = 7x$; $x = 2/7 =$ **28.5%**
   #2: $.6(10) + .25(4) = x(14)$; $6 + 1 = 14x$; $7 = 14x$; $x = 0.5 =$ **50%**

3. Now that the students understand the dynamics of mixing solutions, it is time to deconstruct one solution into the initial two solutions (Separate one large stick into two smaller sticks). Have students first use the multi-link cubes to represent the 10L of 40% solution. Then challenge them to separate this stick into two smaller sticks, one of which is 50% acid, another that is 25%. Don't be surprised how easily they accomplish this! Have them repeat this process for problem #4. They should record each stick diagram.

4. The ultimate task now, of course, is to get the students to represent these problems algebraically. The big question in splitting the sticks is "How many cubes (liters) are in each stick?" So, let x = number of liters in the first solution. Once they can solve the problem algebraically, go back and tackle the original question regarding the Lab Tech.

   #3: $.5(x) + .25(10 - x) = .4(10)$; $.5x + 2.5 - .25x = 4$; $.25x = 1.5$; $x = 6$; **6L of 50% & 4L of 25%**
   #4: $.4(x) + .1(15 - x) = .2(15)$; $.4x + 1.5 - .1x = 3$; $.3x = 1.5$; $x = 5$; **5L of 40% & 10L of 10%**

DIAGRAMS (multi-link cube models)

#1 #2 #3 #4

A lab technician has a solution that is 60% acid, and another that is 40% acid. How much of each solution should be mixed to produce 100L of 50% acid solution?

Representation with multi-link cubes: 1 cube equals 1 liter of solution or 1 liter of acid (depending on color).

1. Start with 2L of 50% solution and 5L of 20% solution. What percent of the 7L mixture will be acid?

   a) Use multi-link cubes to demonstrate the 2L of 50%.
   b) Use multi-link cubes to demonstrate the 5L of 20%.
   c) Combined the two to show the percentage of 7L solution.

2. Try again with 10L of 60% solution and 4L of 25% solution. What percent of the 14L mixture will be acid?

3. Now, some quantity of 50% acid solution and another quantity of 25% acid solution must be mixed to form 10L of solution that is 40% acid. How much of each must be mixed?

   a) Use multi-link cubes to represent the 10L of 40%.
   b) Split the 10L into two portions, one that is 50% and another that is 25% solution.

4. Finally, some quantity of 40% acid solution and another quantity of 10% acid solution must be mixed to form 15L of solution that 20% acid. How much of each must be mixed?

THE TORTOISE AND THE HARE

Inspired by questions found in the UCMP Algebra (Chicago), second edition

In this problem, students extract data from a story in order to write, manipulate, and graph systems of equations. It offers students a context to understand the relationships among data, equations, graphs and solutions.

Concepts
Writing, graphing and solving systems of equations. Rate and unit conversion.

Time: 2-3 hours

Materials
Graph paper and straightedge.

Preparation
Students should be able to write, manipulate and graph linear equations. They should also have had an introduction to solving systems.

THE STORY

You know the fable — the tortoise and the hare have a race. In the algebra version, the hare gives the tortoise a 1,000 foot lead. The tortoise runs at a rate of 9 inches per second, while the hare runs at 6 feet per second. There is also a rat in the story. The rat starts 1,200 feet ahead of the hare, but runs back towards the starting line at 2 ft/sec (-2 ft/sec to be precise).

LESSON PLAN

1. Be sure the students correctly write the equations before they move on. Otherwise, their answers for the remainder of the lesson will be incorrect.

2. Allow the students to work freely on the rest of the assignment. Stop and address the class as a whole only when you notice a common problem. For instance, they should know how to solve for x given y, but they may not know that in order to find the time that the rat crosses the finish line, they simply assign zero for the distance and solve the time. It will take students at least two full hours to respond correctly to all components of this problem.

3. Once the students find each of the answers, they should graph them as ordered pairs. If they include the coordinates for each starting point, there should be twelve data points. From here they should see that the points hint at the lines that represent each of the five equations. Also, be sure that they set their domain and range to the proper limits and at a proper scale. Encourage them to fill most of the graph paper.

4. Once the graph is complete, assign the story writing. Stress that the events of the story should be in chronological order. In order to do this, students just need to read the time value for each answer. However, since many won't understand this concept, this is an excellent teachable moment. After the students have submitted their completed stories, discuss the chronology of the story in accordance to the graph. This is the crux of the lesson. Many students do not see the graph as an abstract representation of the relationship between time and distance. They see it as an aerial view of the race. In other words, the critters are running in an open field and the intersections represent when they will collide with each other. You can place a transparency of the graph on the overhead and cover it with a sheet of paper. Incrementally, slide the paper to the right, allowing the left edge to reveal the graph *moment* by *moment*. This trains the students to read the graph from left to right and also establishes the sense of a relationship between the time of the race and the distance of the runners.

SOLUTIONS

1. 72 sec at 1054.5 ft
2. 150 sec at 900 ft
3. 190.5 sec at 1142.9 ft
4. T=1045 ft, H=360, R=1080
5. 600 sec
6. The Hare wins in 220 sec.
   The Hare wins by 155 ft and is 206 sec in front of the Tortoise.

EQUATIONS:
Tortoise: d=0.75t + 1000
Hare: d=6t
Rat: d=-2t + 1200
Finish Line: d=1320
One Minute Mark: t = 60

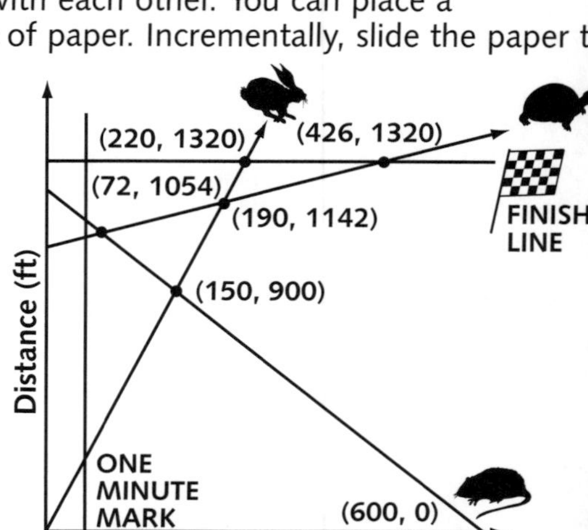

THE TORTOISE AND THE HARE

THE SCENARIO

The Tortoise and the Hare finally have their long awaited rematch. The Tortoise gets a 1,000 foot lead and runs at 9 inches per second. The Hare begins at the starting line and runs at a rate of 6 feet per second. There is also a rat in this race. The Rat starts 1,200 feet ahead of the Hare and runs back towards the starting line at a rate of 2 feet per second.

THE ASSIGNMENT

Write a story about the race. The story should contain the following events and information in chronological order:

1. When will the Tortoise and Hare pass each other and how far will they be from the starting line?
2. When will the Tortoise and Rat pass each other and how far will they be from the starting line?
3. When will the Rat and Hare pass each other and how far will they be from the starting line?
4. After one minute into the race, how far will each runner be?
5. When will the rat cross the starting line?
6. If the race is a quarter-mile long, who will win and what will be the margin of victory (both time and distance)?

Accompanying your story will be the following:

I. Equations for each of the runners, relating time t to distance from the starting line d.
II. A graph of all three equations on the same coordinate plane, with a domain of $0 \leq t \leq 650$ seconds, and a range of $0 \leq d \leq 1500$ feet. Be sure the graph shows all significant data points.
III. An equation and graph for both the one-minute mark and the finish line.

THE CALCULATIONS

Attach your story and graph. Make sure the story is in chronological order.

Write equations for each of the runners, relating time t to distance from the starting line d. Also, include an equation for both the one minute mark and the finish line.

Tortoise: ____________________

Hare: ____________________

Rat: ____________________

One Minute Mark: ____________________

Finish Line: ____________________

1. When will the Tortoise and Hare pass each other and how far will they be from the starting line?

Time: ______________ seconds

Distance From Start: ________________ feet

2. When will the Tortoise and Rat pass each other and how far will they be from the starting line?

Time: ______________ seconds

Distance From Start: _________________ feet

3. When will the Rat and Hare pass each other and how far will they be from the starting line?

Time: ______________ seconds

Distance From Start: _________________ feet

4. After one minute into the race, how far will each runner be?

Tortoise: _________________ feet

Hare: _________________ feet

Rat: _________________ feet

5. When will the rat cross the starting line?

Time: ______________ seconds

6. If the race is a quarter-mile long, who will win, and what will be the margin of victory (both time and distance)?

Winner: ______________

Margin of Victory: ______________ seconds

______________ feet

Rescue Mission

Submitted by Mike Cornelius, Shaun Mc Bride, and Cathy Ralston
Temecula Valley High School, Temecula, California

Each student is one of two pilots whose planes have been shot down behind enemy lines. The pilots have safely parachuted down in two different locations. Using the information provided on the handouts, the pilots have to determine a slope according to the compass and then use the landing point to write a linear code for their directions to the rescue point. Once they are given the other pilot's equation, the students solve the system of equations to determine the point where the rescue helicopter will pick them up.

Concepts
Slope, writing linear equations, solving systems of equations

Time: 1 hour

Materials
Student handouts for Pilots A and B, which include scenario, landing point and compass; topographical map (latitude-longitude grid); index cards (one for every two students) half of which have "Pilot A" written on the back, the half with "Pilot B."

Preparation
Students need mastery of writing linear equations and an understanding of solving systems of equations using either substitution or linear combinations.

LESSON PLAN

1. Split the class into two groups (A & B) on opposing sides of the room. Introduce the lesson by reading the scenario provided on the student handout. Keep the orders for pilots A & B (the student handout) from the class until after the brainstorming session.
2. Brainstorm regarding ways to resolve the situation. Have the students write down and share their ideas. Write ideas on the board for all to see and discuss which ideas will be the most appropriate.
3. Disseminate the pilot orders (student handout) to the corresponding groups. Group A will write the linear code for Pilot A; group B will write the linear code for Pilot B. Students are permitted to work in pairs in order to write the linear equations in slope-intercept. The pilot orders include a directional compass, by which they can derive the slope (students might need assistance using the compass). Encourage them to count the change in latitude over the change in longitude. Many students find difficulty in dealing with the fractional slopes. Discussion on the difference between a slope of 2 versus $1/2$ can be very helpful. With this slope, the students should be able use the landing point to write the linear equation.
4. Have a student who is finished collect one linear code from each pair of students in group A and redistribute them to group B and vice-versa. Students then find the pick-up point by solving the system formed by the equations of both pilot A and B.
5. Have the students graph both equations. For their own equations, encourage the students to use the landing point and slope to graph the line. For the other line, they will not be able to determine the starting point of the other pilot, so they should plot points by using the equation.
6. Review the lesson by discussing any new ideas that were not offered during the brainstorming session. Emphasize the concepts involved in finding the slope using the compass, and the fact that the x- and y-axes are not part of the map.

SOLUTIONS

Linear Code for Pilot A

$y = -\frac{1}{2}x + b$
$82 = -\frac{1}{2}(48) + b$
$82 = -24 + b$
$106 = b$

$\mathbf{y = -\frac{1}{2}x + 106}$

Linear Code for Pilot B

$y = x + b$
$36 = 56 + b$
$-20 = b$

$\mathbf{y = x - 20}$

Pick Up Point

$-\frac{1}{2}x + 106 = x - 20$
$-\frac{3}{2}x = -126$
$x = 84$

$y = 84 - 20$
$y = 64$

84° Longitude, 64° Latitude

Rescue Mission

Pilot A

You are a fighter pilot and your plane has been shot down behind enemy lines. As you parachute down, you see another plane in your squadron being shot down. You land safely and begin to check your equipment. You have a map with a compass; your radio is working, but if you transmit messages, the enemy will be able to pick up on your location. Therefore, your messages must be short and in code. You also realize that the other pilot will need to be rescued and that it would be safer if the rescue helicopter only had to make one pickup. In a case where two people need to be rescued, your orders are to radio headquarters asking for the direction you should travel to meet the other person. Before proceeding on this course, you must give headquarters your path in the form of a LINEAR code. Headquarters will radio back to you the other pilot's linear code. Once you have both codes, you will determine the POINT where the two paths will meet. This is the place where the rescue helicopter will land. If your linear code is wrong, you and the other pilot may miss the pickup point.

PART ONE

When you land, you are at 48 degrees longitude and 82 degrees latitude. Headquarters determines that you will need to walk in an East Southeast (ESE) direction. Use your compass and your algebra skills to help you write a LINEAR code for the path you will be walking.

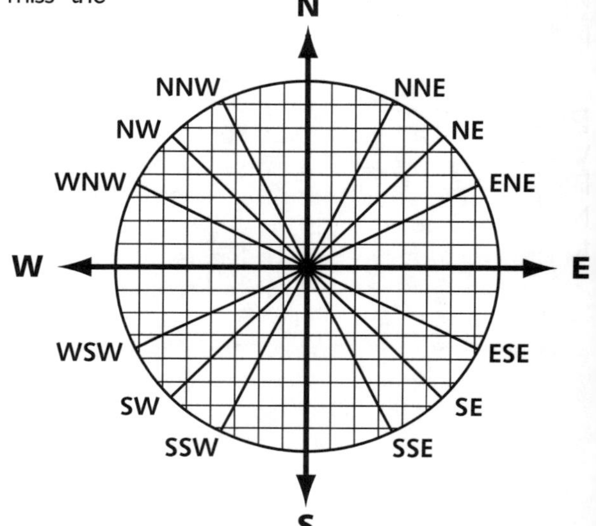

Linear Code for you, Pilot A: ______________________________

PART TWO

a) When your linear code is finished, write it on the index card provided and exchange it with Pilot B.

Linear Code of Pilot B: ______________________________

b) Now determine the location of the pick-up point.

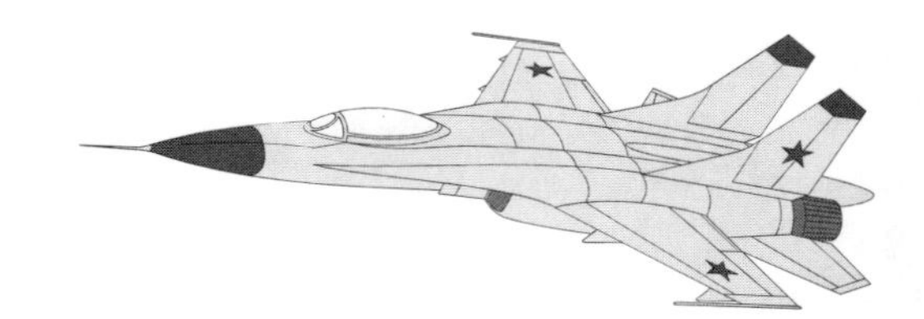

Pick Up Point: ______________

Rescue Mission

Pilot B

You are a fighter pilot and your plane has been shot down behind enemy lines. As you parachute down, you see another plane in your squadron being shot down. You land safely and begin to check your equipment. You have a map with a compass; your radio is working, but if you transmit messages, the enemy will be able to pick up on your location. Therefore, your messages must be short and in code. You also realize that the other pilot will need to be rescued and that it would be safer if the rescue helicopter only had to make one pickup. In a case where two people need to be rescued, your orders are to radio headquarters asking for the direction you should travel to meet the other person. Before proceeding on this course, you must give headquarters your path in the form of a LINEAR code. Headquarters will radio back to you the other pilot's linear code. Once you have both codes, you will determine the POINT where the two paths will meet. This is the place where the rescue helicopter will land. If your linear code is wrong, you and the other pilot may miss the pickup point.

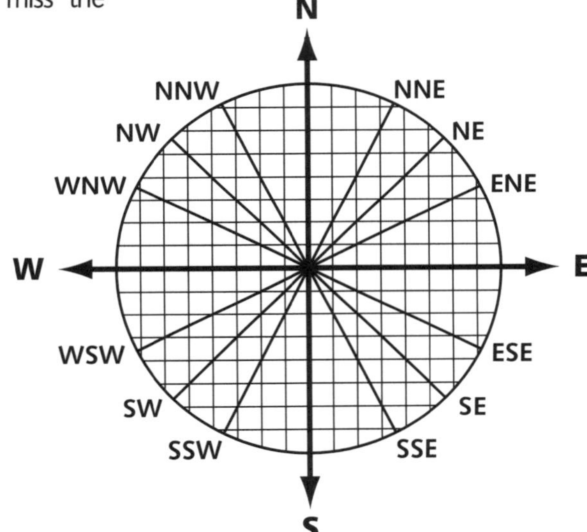

PART ONE

When you land, you are at 56 degrees longitude and 36 degrees latitude. Headquarters determines that you will need to walk in a Northeast (NE) direction. Use your compass and your algebra skills to help you write a LINEAR code for the path you will be walking.

Linear Code for you, Pilot B: ____________________

PART TWO

a) When your linear code is finished, write it on the index card provided and exchange it with Pilot A.

Linear Code of Pilot A: ____________________

b) Now determine the location of the pick-up point.

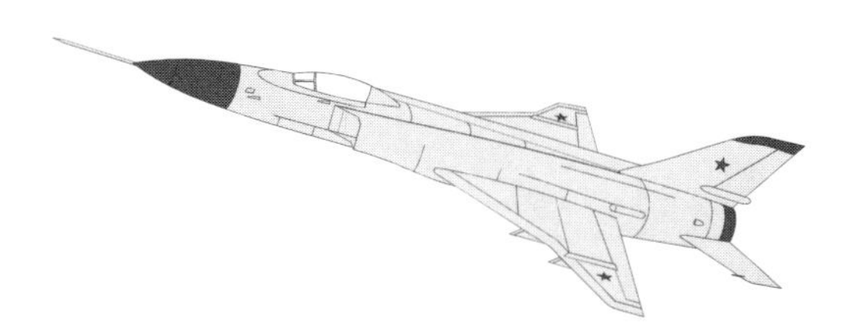

Pick Up Point: __________

Rescue Mission

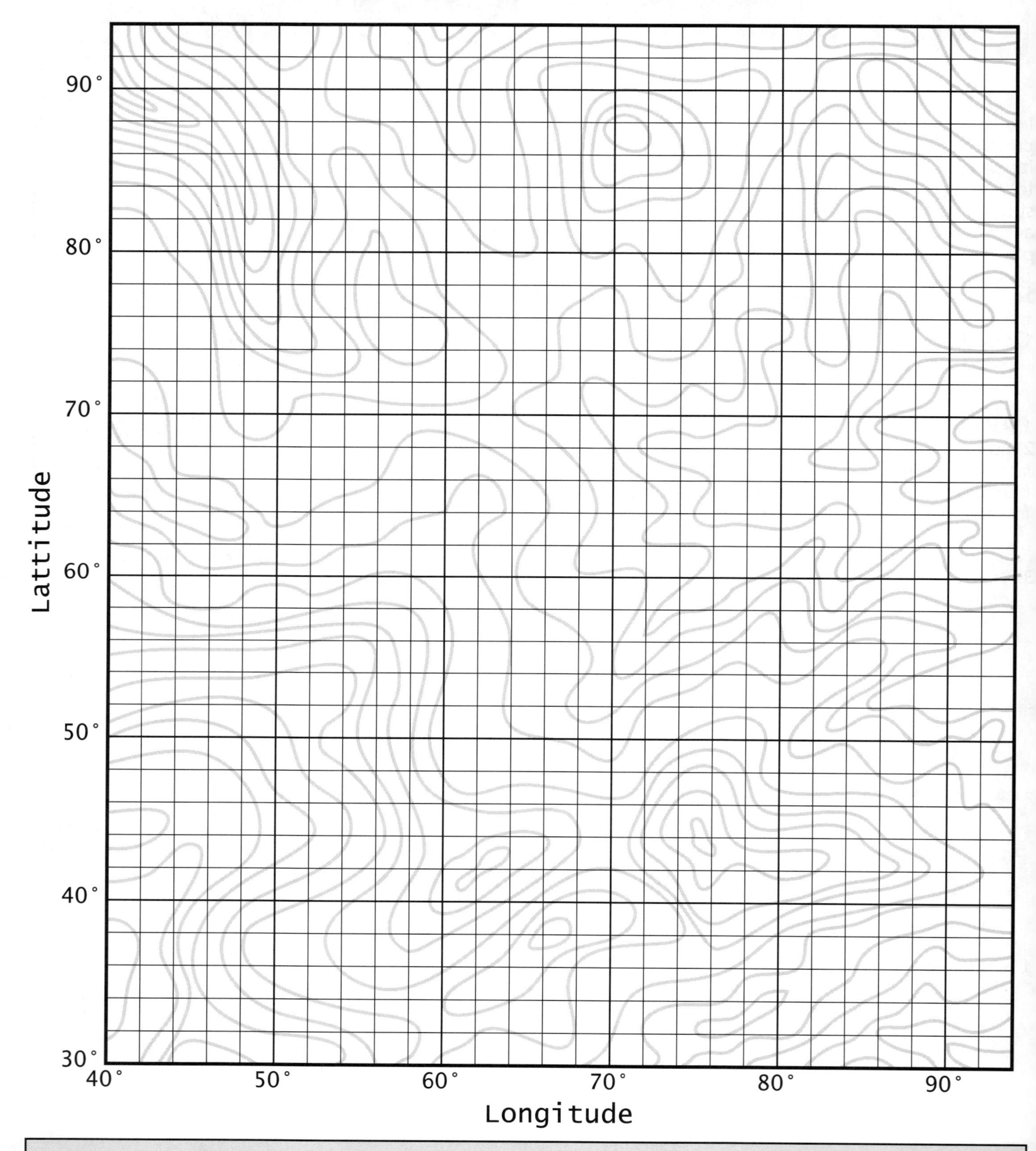

The X-Files

Introduce students to the scenario, and then let them work with the various problems. This activity serves more as an assessment than an instructional tool. It reveals the students understanding of perpendicular equations, the solution to a system and the relationship of equations to their graphs. Allow students to struggle if that is the case. Their mistakes will make for rich points of discussion as much as their correct answers will. Students should establish the coordinate plane on their own graph paper, with each unit representing one mile.

Concepts
Slope, writing and graphing linear equations, slopes of perpendicular lines, solving systems, distance formula, distance-rate-time.

Time: 1hour

Materials
Graph paper, Student handout

Preparation
This is predominantly a review and assessment activity, therefore students should have a strong background in all the concepts listed above.

SOLUTIONS

1. Slope:

Scully

$$m_S = \frac{100-40}{60-0} = \frac{60}{60} = 1$$

$$\mathbf{m_S = 1}$$

Mulder

$$1 \cdot m_M = -1$$

$$\mathbf{m_M = -1}$$

2. Equations:

Scully

$$\mathbf{y_S = x + 40}$$

(FBI headquarters on the same line)

Mulder

$$y_M = -x + b$$
$$-80 = -(20) + b$$
$$-60 = b$$
$$\mathbf{y_M = -x - 60}$$

3. The location of the Alien Space Ship is the intersection of the two equations for Scully and Mulder.

$$x + 40 = -x - 60$$
$$2x = -100$$
$$x = -50$$

A(-50, -10)

$$-50 + 40 = -(-50) + 60$$
$$-10 = -10$$

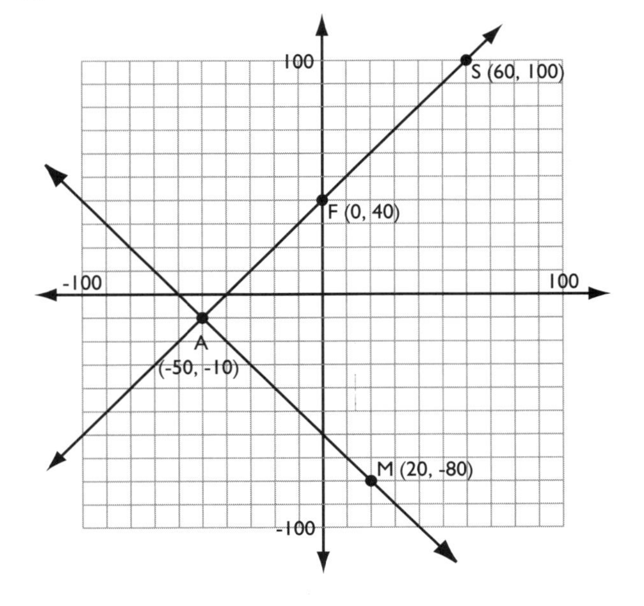

4. Distance:

$$d_S = \sqrt{(60 - (-50))^2 + (100 - (-10))^2}$$
$$= \sqrt{110^2 + 110^2}$$
$$= 110\sqrt{2} \approx 155.5$$

$$d_M = \sqrt{(20 - (-50))^2 + (-80 - (-10))^2}$$
$$= \sqrt{70^2 + (-70)^2}$$
$$= 70\sqrt{2} \approx 99$$

5. Time:

Scully

$$d_S = r_S \cdot t_S$$
$$155.5 = 45t_S$$
$$3.5 = t_S$$

Scully < 4 hrs
OK

Mulder

$$d_M = r_M \cdot t_M$$
$$99 = 45t_M$$
$$2.2 = t_M$$

Mulder > 1 hr
Sorry, Mr. President

The X-Files

THE SCENARIO

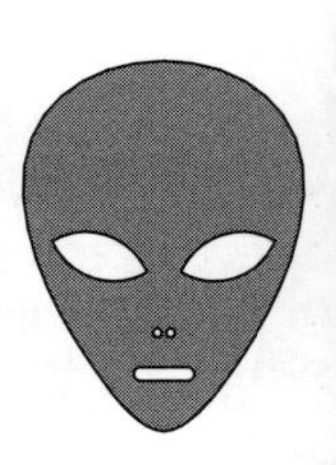

Scully and Mulder are two FBI agents who have dedicated their lives to solving the mysteries behind alien activities in the United States. The FBI headquarters, located at coordinates (0, 40), receives an anonymous phone call at 2:00 p.m. They are told that the President has been kidnapped and is in an alien space ship that has landed somewhere in the third quadrant. They are also told that the space ship will be leaving with the President of the United States as their hostage if the alien that Mulder has captured is not returned to them by 6:00 p.m.

Scully is located at a hospital with coordinates (60, 100). She leaves immediately for the space ship. Three hours later, Mulder is located with the alien at the coordinates (20, -80). Mulder and the alien depart for the space ship as soon as they receive the call from the FBI.

CLUES

1) The alien space ship is on the same line as Scully and the FBI headquarters.
2) Mulder and the alien space ship are on a line that is perpendicular to the line through Scully and the alien space ship.
3) Both Mulder and Scully must drive their automobiles to get to the space ship and because of rough terrain they both drive at a constant speed of 45 mph.

YOUR TASK

Your job is to find the location of the alien space ship and determine if Mulder and Scully have enough time to get there before the ship leaves with the President. All work must be shown along with an explanation of the process that was used and a graph to scale.

	Scully	Mulder
1) Direction/Slope	m_S = ___________	m_M = ___________
2) Equation	y_S = ___________	y_M = ___________
3) Location of the Alien Space Ship	A(_____ , _____)	
4) Distance from the Space Ship	d_S = ___________	d_M = ___________
5) Time to get to the Space Ship	t_S = ___________	t_M = ___________

OLYMPIC SWIM TIMES

OBJECTIVE

Students are to write equations for the lines of best fit and solve a system of linear equations to predict when the fastest woman in the world will swim faster than the fastest man.

> ***Concepts***
> Best fit lines, solving systems of equations, graphing, slope, writing linear equations
>
> ***Time:*** 2-3 hours
>
> ***Materials***
> Graph paper and student handout, including the winning Olympic swim times
>
> ***Preparation***
> Students need a strong understanding of solving systems and writing a linear equation for a line of best fit. This project may be presented at the beginning of a unit as a hook for students, i.e. "We need to learn some things first, before we can solve this."

LESSON PLAN

Provided on the student handout are the Gold Medal swimming times for Men's and Women's 100 meter Freestyle. Point out to the students that the rate of improvement appears faster for the women than for the men, implying that the women will eventually swim faster than the men. For the sake of the project, we will assume that the improvement for each gender is linear. The students are to predict when the women's winning time will equal or surpass (be faster) than the men's winning time. The student is also to predict what that winning time will be, and write an article describing that event.

After much intense instruction in systems of equations, students will be ready to tackle the problem. First, they are to set-up the one-quadrant graph for the problem (no negative values for the domain and range are required). Establishing the scale for each axis is critical. The student needs enough room to plot all current points as well as projected points. Help them understand that the domain may be labeled with the number of years after 1900. Therefore, 81 represents 1981, and 104 represents 2004. It is suggested to label the "x-axis" in increments of four to correspond with the Olympiads. Students may want to create a broken scale for the y-axis. The swim times fall within a given range from approximately 75 seconds to 40 seconds, with the projected winning time still greater than 30 seconds.

For each set of points (one set per gender), the students should plot the points, draw the best fit line and write the equation for that line. Then the two equations should be viewed as a system. Students will most likely solve the system by the substitution method, since it is most likely that both equations will be in slope-intercept form. After the students solve the system of equations, the value found for the year will probably not coincide with an Olympic year. Therefore, the students must be sure to mention this in the article that they will write (how the women and men have tied), or project the winning times in the following Olympiad. Having the students find the times in the year 2000, and the year that the women surpass the 50 second mark, give students the opportunity to solve for y given x and vice-versa. Be sure that for the year 2000, students use 100 (years past 1900) instead of 2000.

The purpose of having the students write the article is to assess their global understanding of the problem — what the significance of the slope and y-intercept are, what a solution for a system truly means, etc. Emphasize to students that the y-intercepts for the equations should correspond with the y-intercepts on their graphs. The improvement that they mention in the article should correlate to the slope shown in the equations.

Enjoy the color commentation of their articles. Students tend to really get into the story writing.

Olympic Swim Times
Solutions

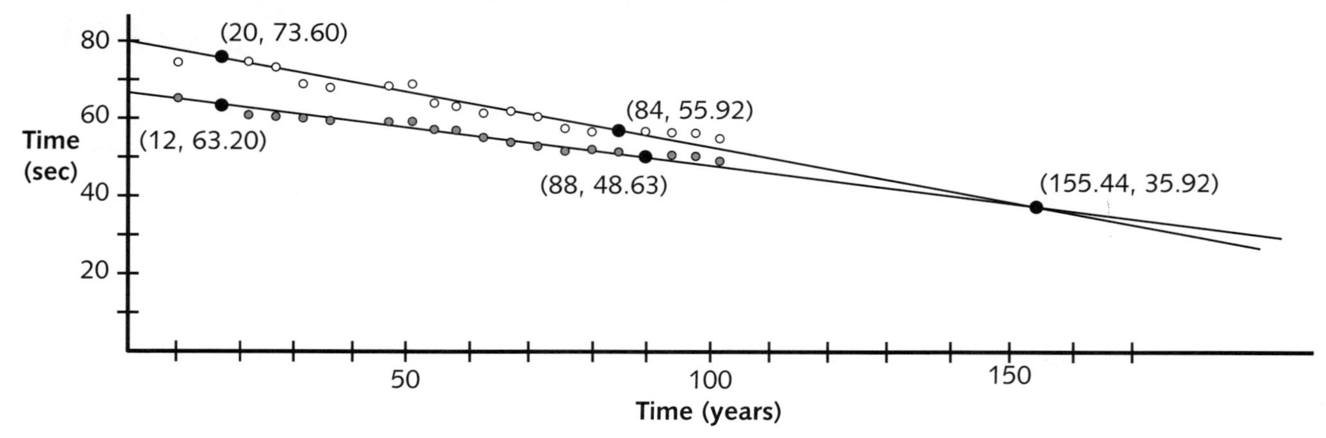

	MEN'S SOLUTIONS	**WOMEN'S SOLUTIONS**
Choosing points	(12, 63.20) & (88, 48.63)	(20, 73.6) & (84, 55.92)
Calculating slopes	m = (48.63 - 63.20)/(88 - 12) m = -14.57/76 **m = -.19**	m = (55.92 - 73.6)/(84 - 20) m = -17.68/64 **m = -.28**
Determining y-intercepts	y = -.19x + b 48.63 = -.19(88) + b 48.63 = -16.72 + b **b = 65.45**	y = -.28x + b 55.92 = -.28(84) + b 55.92 = -23.52 + b **b = 79.44**
Writing equations	**y = -.28x + 79.44**	**y = -.19x + 65.45**

Solving the system

-.19x + 65.45 = -.28x + 79.44
.09x + 65.45 = 79.44
.09x = 13.99
x = 155.44

-.19(155.44) + 65.45 = -.28(155.44) + 79.44
-.29.53 + 65.45 = -43.52 + 79.44
35.92 = 35.92

The women's time will equal the men's time (35.92) in the year 2055. However, since this is not the year of a summer Olympiad, the women will actually exceed the mens' time in the 2056 Olympics.

Projecting gold medal times	y = -.19(156) + 65.45 **= 35.81 sec**	y = -.28(156)+ 79.44 **= 35.76 sec**
Calculating the time in the year 2000	y = -.19(100) + 65.45 **= 46.45 sec**	y = -.28(100)+ 79.44 **= 51.44 sec**
Finding the year women break the 50 second mark		50 = -.28(x)+ 79.44 -29.44 = -.28x **x = 105.14 (year 2005)**

OLYMPIC SWIM TIMES

Below are listed the winning 100 meter Olympic swim times for men and women. The women's times appear to be decreasing at a faster rate than the men's, implying that the women will eventually swim faster than the men. Under this assumption, you are to do the following:

Make a mathematical analysis:

- Plot the points for both men's and women's times on the same axes.
- Graph the best fit line for each set of points.
- Write the equation for each of the best fit lines.
- Show all calculations for the information requested below.

Write a news article reporting the event in which women make history by surpassing the men's winning time. In the article include:

- The date & winning swim time when the women's time meets or exceeds the men's. (You may make this an Olympic or non-Olympic year.)
- The general rate of improvement for each gender over the last century.
- The winning times for both the men & women in the year 1900. (label these as an ordered pair on your graph)
- The winning times for both the men & women in the year 2000. (label these as an ordered pair on your graph)
- When the women's time broke the 50 second barrier. (label this as an ordered pair on your graph)

Hint: Creating names and quotes of key figures at the event will make for a more colorful article.

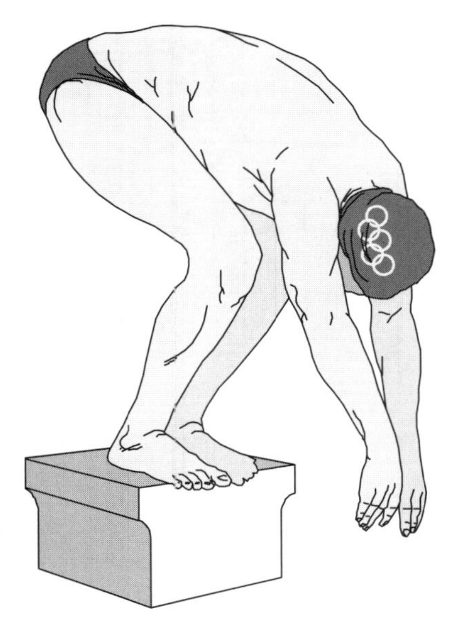

100 Meter Freestyle
Olympic Winning Times (seconds)

Year	Men's	Women's
1912	63.20	72.20
1920	61.40	73.60
1924	59.00	72.40
1928	58.60	71.00
1932	58.20	66.80
1936	57.60	65.90
1948	57.30	66.30
1952	57.40	66.80
1956	55.40	62.00
1960	55.20	61.20
1964	53.40	59.50
1968	52.20	60.00
1972	51.22	58.59
1976	49.99	55.65
1980	50.40	54.79
1984	49.80	55.92
1988	48.63	54.93
1992	49.02	54.64
1996	48.74	54.50
2000	48.30	53.83

Robotic Disney

Miketron, Plutonic, and Digibell

LESSON PLAN

Have students read the scenario and plot the main attraction points and their midpoints (#1 & 2). Be sure that all students have correctly plotted the points before moving on. It will help the students to picture ΔFCS. Most of the struggle for the students comes in writing the equations of the perpendicular bisectors (question #4). Encourage them to graph these equations in order to check their accuracy. The circumcenter (intersection of the bisectors) is found by solving a system of any pair of the bisector equations. Again, their algebraic solution should coincide with the graphic solution. Have students continue to rework the equations until they all validate the graph.

The key to finding the placement of the Digibell control center is recognizing that the intersection of the perpendicular bisectors is the circumcenter of the triangle. Have students use a compass to accurately draw the circle through each of the points F, C, and S with center D. Doing so reinforces that the circumcenter is equidistant from each of the vertices.

Concepts
Distance Formula, midpoint, perpendicular slope, writing & graphing equations, solving systems, perpendicular bisector, circumcenter, and area of circles and triangles.

Time: 1-2 hours

Materials
Student handout, compass, straightedge.

Preparation
Students should have a firm understanding of the algebraic concepts above. They need only minor exposure to the concept of the circumcenter of a triangle (intersection of the perpendicular bisectors).

SOLUTIONS

1. See graph
2. See graph. Midpoint: $T\left(\frac{-20+80}{2}, \frac{200+180}{2}\right) = \mathbf{T(30,\ 190)}$
3. $TA = \sqrt{30^2 + 50^2} = \sqrt{3400} \approx 58.3$
   $LA = \sqrt{50^2 + 10^2} = \sqrt{2600} \approx 51.0$ $\quad TA + AL \approx \mathbf{109.3}$
4. Find the slope of FC ($^{-5}/_3$), a side of the triangle. The perpendicular bisector has a slope that is the opposite reciprocal of the original line ($^3/_5$) and passes through the midpoint:
   $150 = (^3/_5)(10) + b; \quad y = (^3/_5)x + 144$
5. Solve the system of any pair of perpendicular bisectors.
   $5x + 40 = -0.5x + 170 \qquad 5(23.6) + 40 = -0.5(23.6) + 170$
   $5.5x = 130 \qquad 158.1 \approx 158.2$
   $x \approx 23.6 \qquad \mathbf{D\ (23.6,\ 158.1)}$
6. Draw the circle about point D with a compass.
7. $r = \sqrt{(80-23)^2 + (180-158)^2} = \sqrt{57^2 + 22^2} = \sqrt{3733} \approx 61$ paces
   $A = \pi(61)^2 = 3733\pi \approx \mathbf{11{,}722\ paces^2}$

Bonus:
Pick's algorithm (counting squares): $A \approx (29 + ^{31}/_2)(102) \approx \mathbf{4{,}450\ paces^2}$

Triangle area formula: $A = (^1/_2)bh = (^1/_2)(117)(76) = \mathbf{4{,}446\ paces^2}$
(where b and h are found as shown below)
$b = FC = \sqrt{(-20-40)^2 + (200-100)^2} = \sqrt{(-60)^2 + 100^2} \approx 117$ paces
h = the distance of the altitude from point S to point P on FC.
base: $y = (^{-5}/_3)x + ^{500}/_3 \qquad$ altitude: $y = (^3/_5)x + 132$
P(15, 141)
$h = SP = \sqrt{65^2 + 39^2} \approx 76$ paces

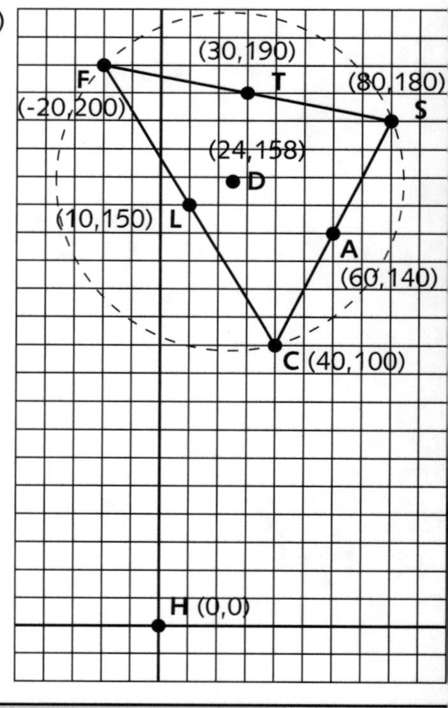

Robotic Disney

Miketron, Plutonic, and Digibell

Disneyland California does not have enough employees to walk the park in their character costumes and take pictures with all the little kids. Therefore, the Imagineers at Disney have created robot characters. These robots each have different guidance systems that allow them to travel the park. For example, Micketron, the robot version of Mickey Mouse, only walks in the directions of true north, south, east and west, and only turns at right angles. His memory chip has the entire park superimposed on a coordinate grid. His home, City Hall, has the coordinates of H(0, 0).

1. Micketron has three main attractions that he must visit. Cinderella's Castle, Festival Arena and It's a Small World. Micketron leaves City Hall (point H) and walks to Cinderella's Castle (point C). To get there, he walks 40 paces East, then 100 paces North. Label points H & C on the grid. Include the coordinates.

   After Cinderella's Castle, Micketron walks to the Festival Arena (point F). To get there from the castle, he walks 60 paces west and 100 paces north. Label point F on the grid. Include the coordinates.

   To visit Small World (point S), Micketron must return to Cinderella's Castle. From Cinderella's Castle, he walks 40 paces east and 80 paces north. Label point S on the grid. Include the coordinates.

Also working the park is Plutonic, the robot version of Pluto. Pluto works three main attractions and is programmed to run in straight-line paths directly to and from these stations. He is currently stationed in Toon Town (point T), which is the midpoint between Festival Arena and Small World. He then races to Alice in Wonderland (point A), the midpoint between Small World and Cinderella's Castle. From there, he bolts to a Lemonade Stand (point L), half-way between Cinderella's Castle and Festival Arena.

2 Label points T, A & L. Include the coordinates.

3. How many total paces (same units as Micketron) does Plutonic run from Toon Town to Alice in Wonderland to the Lemonade Stand?

   TA + AL = ___________ paces

There are spotlights atop each of Pluto's stations (T, A & L), The beams of these three spotlights all converge on the Dumbo attraction (point D). In order to do this, the lights must shine perpendicularly to their corresponding segments. For example, LD ⊥ FC.

4. Write a linear equation (y = mx + b) for each spotlight. Draw the line that represents the beam. Remember that the beam passes through the attraction and is perpendicular to its corresponding segment.

   y_{LD} = ________________ y_{AD} = ________________ y_{TD} = ________________

Robotic Disney

5. Label Dumbo point D, where the light beams converge. Include the coordinates. (______, ______)

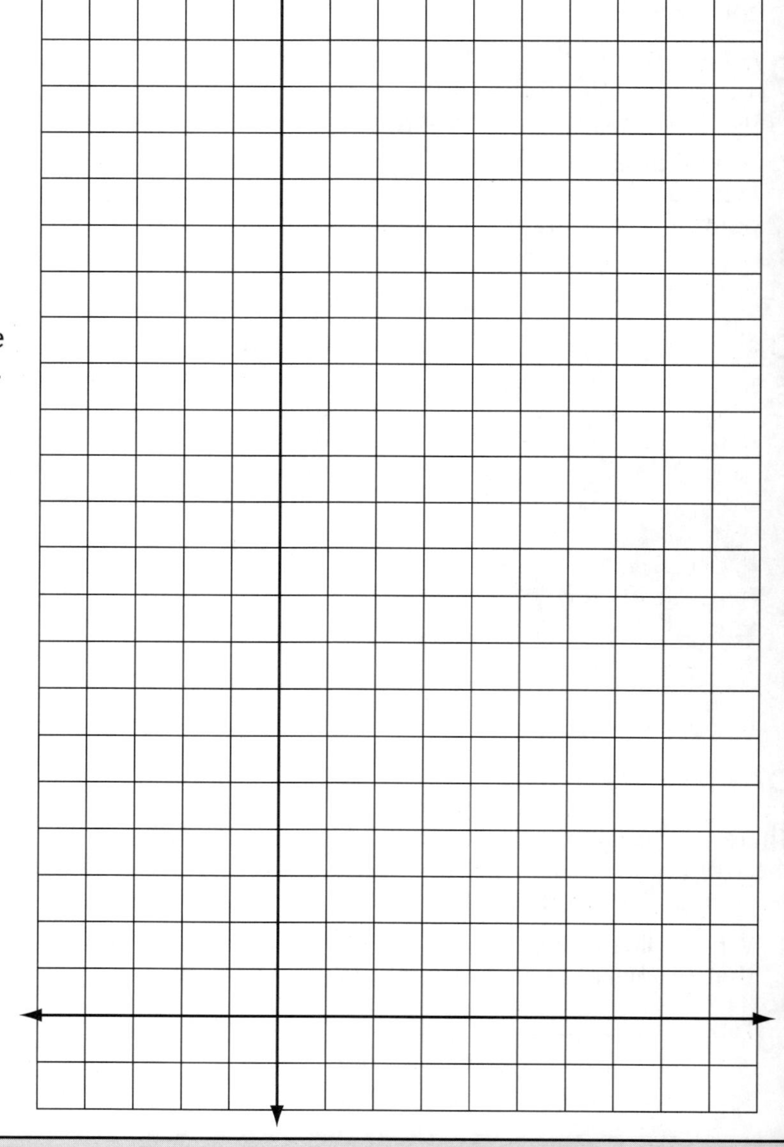

Disney also has created a flying robotic Tinkerbell, called Digibell. Digibell flies in a circle such that she passes over each of Micketron's stations (F, C, & S). Her control mechanism needs to be located at the center of her circle.

6. Using a compass, draw Digibell's flight circle passing through points F, C & S.

7. Find the area within Digibell's circle.
   $A_{\circ}$ = ____________ paces^2

Bonus: Find the area within Micketron's triangle.
   A_{Δ} = ____________ paces^2

PROJECT

The Migration Paths of the Sandhill Cranes and the Snow Geese

Contributed by Loretta Griffy, Assistant Professor, Austin Peay University

DAY ONE (1 hour)

1. Discuss the migration situation and ask several students to briefly share their answers to the initial question.
2. Distribute graph paper and discuss an appropriate domain and range for this situation.
3. Once the students have plotted the paths of the birds on the coordinate plane, discuss how to find the location of the intersection of the paths in reference to home (the origin).
4. Ask the students to answer question #4 by using the graphing calculator to plot the two paths and determine the point of intersection of the two lines.
5. Once the intersection has been located, remind the students how to find the distance between two points and ask them to answer question #5.

DAY TWO (1 hour)

6. Discuss calculating the location of the birds at any particular time. Ask them to complete the charts in #6 and #7.
7. Discuss parametric equations. Point out that the last rows of the charts are the parametric equations themselves. Use your calculator to demonstrate, in the parametric mode, the flight of the birds.
8. At this point, students should be able to mathematically support (or oppose) their answer to questions #1 and #2.

DAY THREE (1 hour)

9. Begin the day by discussing how to determine the distance between the flocks of birds at any given time. Remind the students to look at their charts in questions #6 and #7 to find the distance between the two points at the requested time. (Find the distance between the points using the Distance Formula or Pythagorean Theorem.)
10. Have the students answer questions #9 and #10 using this information.
11. Discuss how, in general, one could determine the distance between the birds at any given time. Encourage the students to look at their charts and to use the distance formula when working on question #11. (Use the ordered pairs for time t to determine the function at a given time.)
12. Once the correct function in question #11 has been developed, instruct the students to answer question #12.

DAY FOUR (1 hour)

13. Graph the function from #12 on the calculator (in function mode).
14. Discuss with the students the function and how it describes the time and the distance between the birds, emphasizing the function's minimum and why it exists. Include discussion on what the minimum value means to the distance between the flocks of birds. Afterwards, ask the students to answer question #13.
15. Finish the project by animating the path of the birds to illustrate that they do not collide. Set the calculator's Mode to Parametric and Simultaneous. Graph the parametric equations, setting the function to "animation path." To do this, start in the Y= screen and press the left arrow key to move the cursor all the way to the left and press enter a few times to scroll through the icons. Do this for each equation and press Graph.

Concepts

Graphing lines, systems of equations, developing parametric equations, function minimums

Time: 3-4 hours (depending on the skill level of the class)

Materials

TI-83 graphing calculator (or its equivalent), graph paper and student handout

Preparation

This lesson is graphing calculator intensive; a working knowledge of the TI calculator is necessary. 1) Read through the lesson to become familiar with the activity. 2) Complete the activity as if you are the student and make note of topics you will need to address prior to beginning the activity. 3) As a homework assignment the day prior to the start of the lesson, distribute the initial question to the students.

Sandhill Cranes and the Snow Geese (Instructor's Solutions)

This solutions page is a combination of the answers to the questions asked on the handout, notes for the instructor, and hints that the instructor can give students to guide them along the way.

1. Yes, the paths do appear to cross when the points are plotted on a coordinate grid with the home location as the origin. Students may include such topics as slope, systems of equations, or the drawing of lines.
2. The flocks of birds will not collide because in two hours the Cranes are past the intersection point of the paths and the Geese have not yet reached the intersection point.
3. Slope for the Geese: $\frac{-270 - 30}{120 - (-80)} = \frac{-300}{200} = -\frac{3}{2}$ Slope for the Cranes: $\frac{-275 - (-25)}{-50 - 50} = \frac{-250}{-100} = \frac{5}{2}$
4. Equation for Geese: $y = (-3/2)x - 90$ Equation for Cranes: $y = (5/2)x - 150$ Intersection: (15, -112.5) NOTES: The instructor should require students to answer all questions algebraically as well as use the graphing calculator. The ideal calculator window settings for this are: $x_{min} = -100$, $x_{max} = 100$, $x_{scale} = 10$ $y_{min} = -150$, $y_{max} = 50$, $y_{scale} = 10$.
5. Home: (0,0) Intersection: (15, -112.5). Using the Distance Formula, you must travel approximately 113.5 miles from home to see the birds directly overhead.
6. Position of the Sandhill Cranes in relation to time.

Time in Hours	Horizontal Position (x) in miles	Vertical Position (y) in miles
0	50	-25
1	50 - 20(1) = 30	-25 - 50(1) = -75
2	50 - 20(2) = 10	-25 - 50(2) = -125
3	50 - 20(3) = -10	-25 - 50(3) = -175
10	50 - 20(10) = -150	-25 - 50(10) = -525
t	50 - 20(t) = 50 - 20t	-25 - 50(t) = -25 - 50t

7. Position of the Snow Geese in relation to time.

Time in Hours	Horizontal Position (x) in miles	Vertical Position (y) in miles
0	-80	30
1	-80 + 40(1) = -40	30 - 60(1) = -30
2	-80 + 40(2) = 0	30 - 60(2) = -90
3	-80 + 40(3) = 40	30 - 60(3) = -150
10	-80 + 40(10) = 320	30 - 60(10) = -570
t	-80 + 40(t) = -80 + 40t	30 - 60(t) = 30 - 60t

8. Graph the parametric equations. The ideal window settings for this are: $T_{min} = 0$, $T_{max} = 5$, $T_{step} = 0.2$, $x_{min} = -100$, $x_{max} = 100$, $x_{scale} = 10$, $y_{min} = -150$, $y_{max} = 50$, $y_{scale} = 10$.
9. At 6:00 am: Cranes: (50, -25), Geese: (-80, 30). $d = \sqrt{(-80 - 50)^2 + (30 - (-25))^2} = 141.16$ miles apart.
10. After 3 hours: Cranes: (-10, -175), Geese: (40, -150). $d = \sqrt{(40 - (-10))^2 + (-150 - (-175))^2} = 56$ miles apart.
11. At time t: Cranes: (50 - 20t, -25 - 50t), Geese: (-80 + 40t, 30 - 60t). $f(t) = \sqrt{[(-80 + 40t) - (50 - 20t)]^2 + [(30 - 60t) - (-25 - 50t)]^2}$ $f(t) = \sqrt{3700t^2 - 16{,}700t + 19{,}925}$
12. At 2:00 pm, t=8: $f(8) = \sqrt{3700(8)^2 - 16{,}700(8) + 19{,}925} = 350.9$ miles
13. Graph the function, f(t), on the calculator and determine the minimum value of the function using the CALC button. The minimum is (2.26, 32.88). The closest the flocks of birds get to each other is 32.88 miles when time is 2.26 hours since first spotted, or approximately 8:15 am.

Contributed by Loretta Griffy, Assistant Professor, Austin Peay University

Using a Graphing Calculator to Analyze

The Migration Paths of the Sandhill Cranes and the Snow Geese

During the winter months many birds migrate south to warmer climates. You are part of a team of observers who spot and record the movement of a flock of Sandhill Cranes and a flock of Snow Geese. After a couple of observations, you are interested in exploring the relationship between the paths of the birds. The locations of the birds are first recorded at 6:00 am. The Cranes are originally spotted at 50 miles east and 25 miles south of your home location and the Geese at 80 miles west and 80 miles north of your home. The second recording of the birds is at 11:00 am. During this spotting, the Cranes are 50 miles west and 275 miles south of your home location and the Geese are 120 miles east and 270 miles south of your home location.

1. Assuming the Sandhill Cranes and the Snow Geese are flying at the same altitude and their paths are straight lines, will the paths of the flocks intersect? Use the coordinate plane to help you answer this question and provide a written explanation of your answer.
2. Assuming the Cranes and Geese are flying at the same altitude and their paths are straight lines, will the flocks of birds collide? Again, use the coordinate system to help you answer this question.
3. The rate of change is the slope of the lines that describe the flying paths. What is the rate of change for the Snow Geese? What is the rate of change for the Sandhill Cranes?
4. At what location will the paths of the birds intersect? Provide algebraic support of your answer.
5. How far from your home must you travel to see both flocks of birds directly above you from one location?
6. Complete the chart below that identifies the position of the Sandhill Cranes in relation to time.

Time in Hours	Horizontal Position (x) in miles	Vertical Position (y) in miles
0		
1		
2		
3		
10		
t		

7. Complete the chart below that identifies the position of the Snow Geese in relation to time.

Time in Hours	Horizontal Position (x) in miles	Vertical Position (y) in miles
0		
1		
2		
3		
10		
t		

8. Using your calculator, graph the parametric equations from the last row on your charts and verify that the flocks do not collide.
9. How far apart are the flock of Snow Geese from the flock of Sandhill Cranes at 6:00 am?
10. How far apart will the Snow Geese be from the Sandhill Cranes three hours later?
11. Write another equation as a function of time, t, to represent the distance between the Geese and the Cranes.
12. Using the equation developed in #11, determine the distance between the Snow Geese and the Sandhill Cranes at 2:00 pm (8 hours after the starting time).
13. Graph the equation to determine the minimum distance between the Snow Geese and the Sandhill Cranes over the course of their migration.

The empty box project is an extension of a popular math problem, maximizing the volume of a box formed from a net that is created by cutting out congruent squares from a rectangular sheet of material. This version offers a unique twist by writing polynomials to analyze the relationships between length, surface area and volume. The resultant graphs serve as a precursor to quadratic functions.

LESSON PLAN

1. Each member of the group should choose a different value for x to represent the side of the square to be cut from each corner. Then each member is to actually cut the "x by x" square out of each corner and fold along the dotted lines as shown in the diagram. Do not allow students to tape the box, since they will fold and unfold repeatedly throughout the project. Have students measure the width and length of the box and write these dimensions directly on the box near each edge.
2. Have each student calculate the area of his/her box.
3. Have the students write the equation for the "area of our box, as a class." This is written as $A(x) = (8 - 2x)(11 - 2x)$. Evaluating this equation with the student's chosen value of x yields the student's value for area.
4. Have the students simplify the equation: $A(x) = 4x^2 - 38x + 88$. Evaluating this equation with the student's chosen value of x again yields the student's value for area. This means that while these two equations look different, they represent the same quantity.
5. Graphing the extra (new) point here helps the students understand the usefulness of the equation that they have written.
6. Thinking of the surface area as the sum of the five rectangles yields the following long, ugly equation:
   $S(x) = (11 - 2x)(8 - 2x) + 2x(11 - 2x) + 2x(8 - 2x)$. Simplifying this yields: $S(x) = 88 - 4x^2$.
7. Leave the students to decide which of the two equations above they want to use to create the graph. The short one is the obvious choice which shows them the value of simplifying equations.
8. Writing the equation for volume is easy if the students realize that it is simply the area formula multiplied by the height, which is x. $V(x) = x(11 - 2x)(8 - 2x) = 4x^3 - 38x^2 + 88x$. The students believe that this graph is also a parabola. That offers the opportunity to compare the graph of quadratic to that of a cubic.

> ***Concepts***
> Writing and simplifying expressions, adding and multiplying polynomials, distributive property
>
> ***Time:*** 2-3 hours
>
> ***Materials***
> Scissors, ruler, 8"x 11" paper, graph paper
>
> ***Preparation***
> Students should possess a basic knowledge of writing and simplifying expressions.

EXTENSION: Introducing Quadratics & Cubics

Now that students have been introduced to the factor and quadratic equation forms in the Empty Box lesson, a connection can be made between the equations and parabolas. Each student should have a graph of four data points (x, A) relating the side "x" of the square and the area of the bottom of the box. These data points were calculated through the factored form $A = (11 - 2x)(8 - 2x)$. Then have the students use the form $y = 2x^2 + 19x + 44$ to find and plot additional data points. They will see that the points still fall on the same parabola.

For Advanced Algebra lessons, these equations may be used to teach solving of quadratics. For instance, given the equation $A = x^2 - 35x + 300$ and an area, A, of the bottom as 126 square inches, what are the dimensions of the original cardboard sheet and what is the value of x? Also, the equation and graph for the volume of the box may be used to explore the properties of cubics.

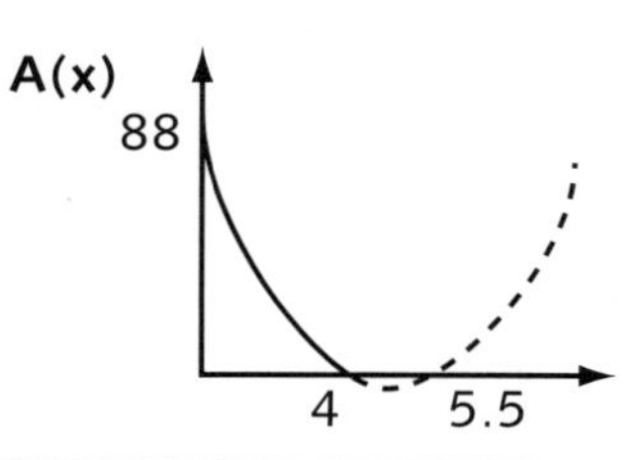

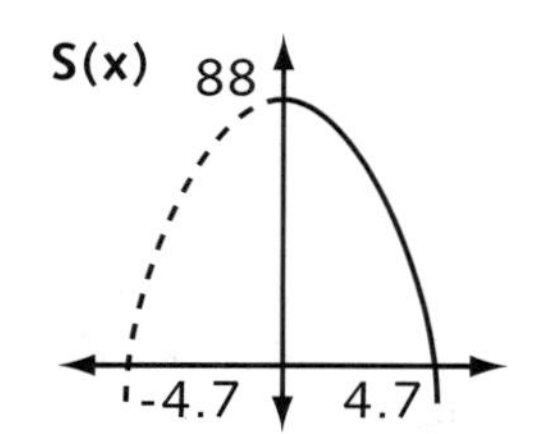

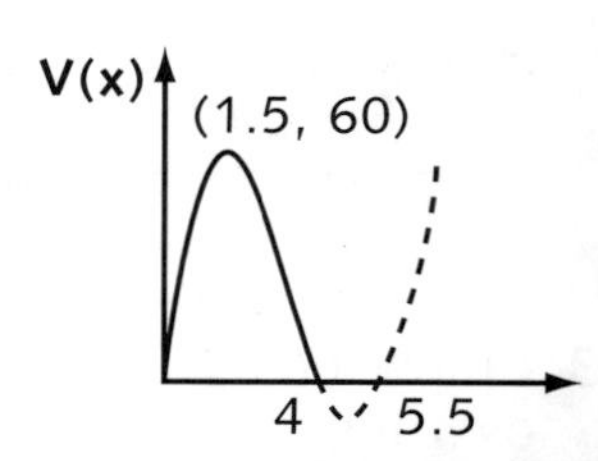

STUDENT HANDOUT

Your assignment is to make a box without a lid from an 8″ x 11" sheet of paper by cutting out a square that is x inches on each side and folding along the dotted lines, as shown below. Answer the following questions with regard to the area of the BOTTOM, the SURFACE AREA, and the VOLUME of the box.

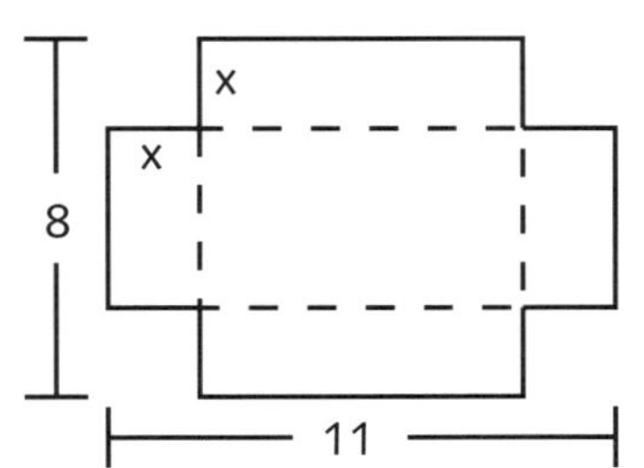

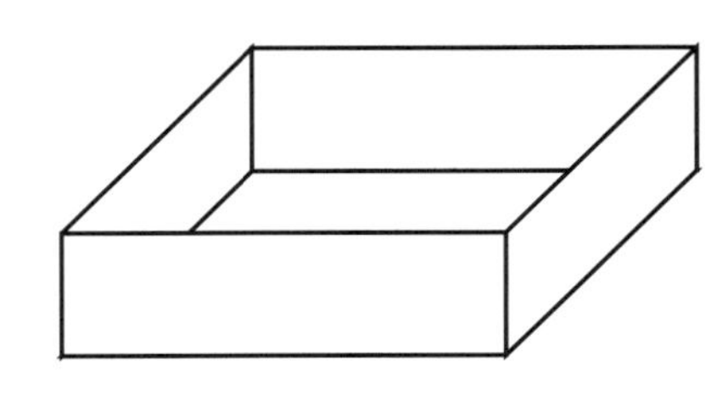

1. Choose a value for x, then actually cut the "x by x" square out of each corner and fold. Each member of your group should have a different value of x. Measure the dimensions of your box, and write these dimension on the box.

   Your x value: _______ Your width: _______ Your length: _______

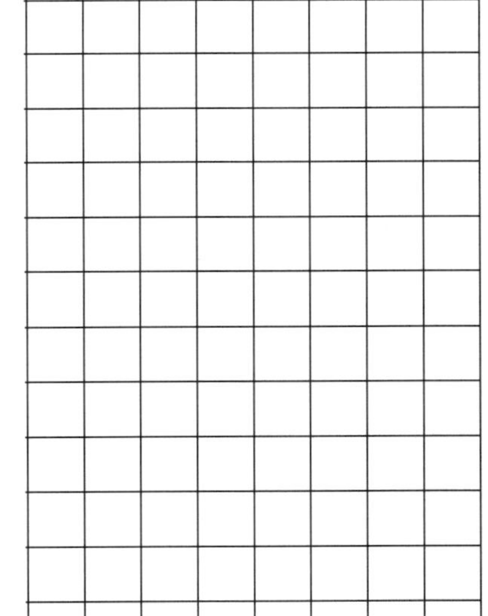

PART ONE: Area of the Bottom

2. Calculate the area of the BOTTOM of your box.

3. a) Write an equation for the area of the bottom of the box in terms of x, as the product of the width and length.

   A(x) = ____________________________

   b) Evaluate your equation with your value for x. Compare your answer to #2.

   A(__) = ____________________________

4. a) Multiply the two binomials in your equation, A(x), to form a new equation.

   A(x) = ____________________________

   b) Evaluate this new equation with your value for x. Compare your answer to #2. What does your comparison say about the two equations.

   A(__) = ____________________________

5. a) With a domain of $0 \le x \le 4$, and the area of the bottom of the box, A(x), as the range, graph each group member's point (x, A) and draw a curve through the points.

   b) Calculate the dimensions of the box for a new value of x and plot its new area on the graph.

   New x value: _______ New width: _______ New length: _______ New area: _______

   c) Use your simplified equation from #4 to calculate the area of the bottom of the box for the new value of x. Compare your answer to part (b).

EMPTY BOX

PART TWO: Surface Area

6. a) Write and simplify an equation for the surface area of the outside of your box in terms of x.

original: S(x) = ______________________________

simplified: S(x) = ____________

b) Evaluate either equation above with your value for x.

S(__) = ________________

7. With a domain of $0 \leq x \leq 4$, and the surface area of the box, S(x), as the range, graph each member's point (x, S) and draw a curve through the points (using the graph on the bottom left).

PART THREE: Volume

8. a) Write and simplify an expression for the volume of your box. Then evaluate for your value of x.

V(x) = ____________________ V(__) = ____________________

b) With a domain of $0 \leq x \leq 4$, and the volume of the box, V(x), as the range, graph each member's point (x, V) and draw a curve through the points (using the graph on the bottom right).

Surface Area

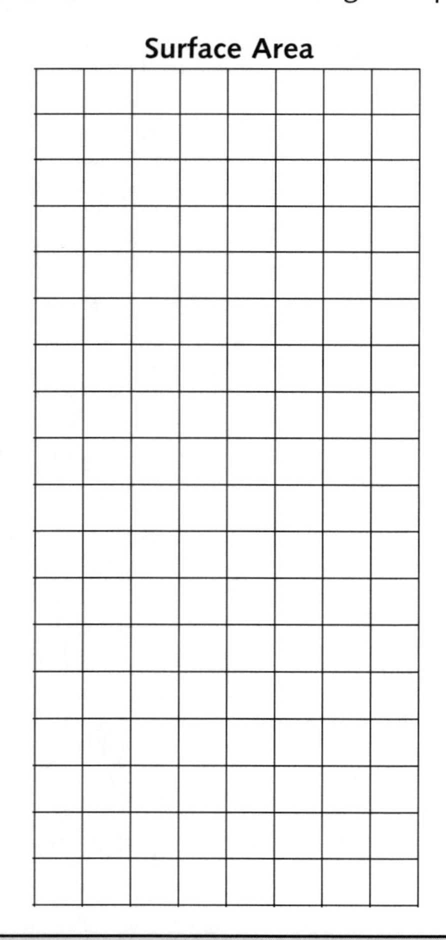

Volume

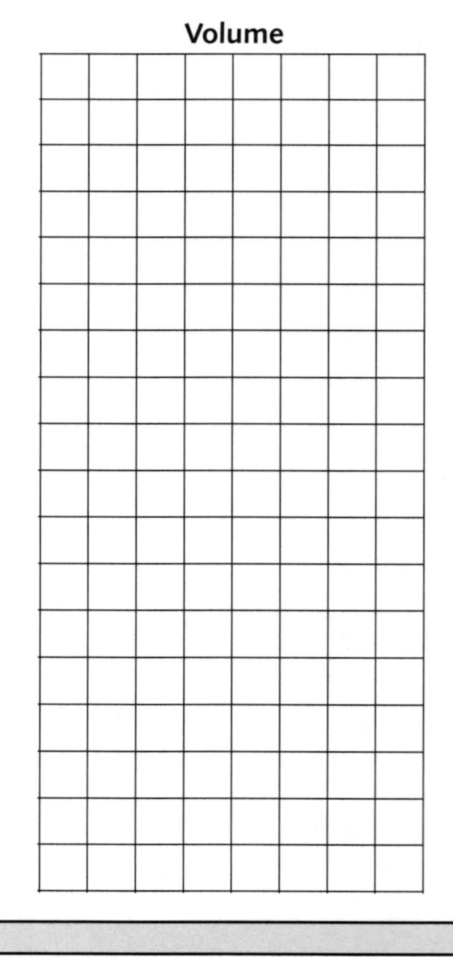

PROJECT

THE PIZZA BOX

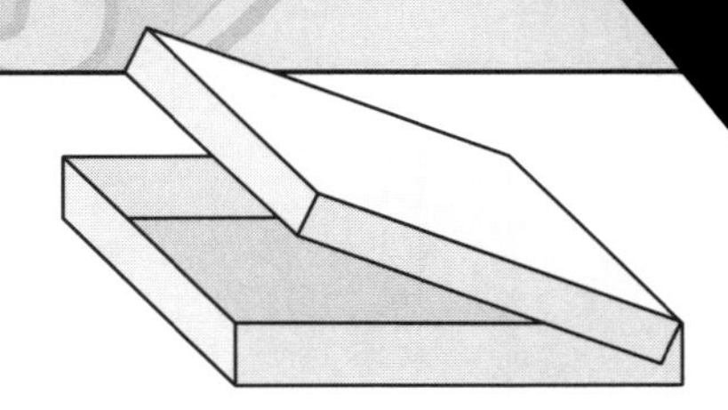

LESSON PLAN

The first side of the handout deals with a concrete investigation of the box by choosing various values for the length of each side of the square that is cut out. The second side deals with the generalization of the problem — thinking of each side of the square in terms of x.

1-4) Be sure that students trim whatever size paper they have to 6″ x 11″. Once this is done, have the students choose a value for their cuts. Be sure each student in the group chooses a different value. When completing the charts and graphs, they may share their values with their group, but will also need to calculate additional values.

5. When the students write the expressions for the Area of the Bottom, Surface Area and Volume, be sure they write the expressions as a product (initial) and as a sum (simplified).

6. It is important that the students verify both the initial and the simplified versions of their expressions. This reinforces the idea that these expressions are equivalent. They should get the same answers for both.

7-8) This chart is an expanded version of that in #4, intended to give students a more complete view of the graphs.

Concepts

Writing and multiplying polynomials, graphing quadratic and cubic equations, solving quadratics by the quadratic formula and completing the square.

Time: 2-3 hours

Materials

Scissors, notebook paper, calculator and student handout

Preparation

Students need a basic understanding of binomial multiplication and of quadratic equations. This project usually follows the Empty Box Project.

EXTENSIONS

i) Find all the values of x that will produce a square bottom for the box. Algebraically, this can be accomplished by setting the width and length equal: $6 - 2x = (11 - 3x) / 2$. Solving for x produces a single solution of one inch.

ii) Find the roots of the quadratic equations for the Area of the Bottom and Surface Area, by using the quadratic formula and by completing the square.

SOLUTIONS

1) Assume a cut of 1.5 inches
2) Length=3.25; Width=3; Height=1.5
Bottom=9.75; Surface area=52.5; Volume=14.625.
3)

Cut	Length	Width	Height	Bottom	S.A.	Vol.
0.0	5.50	6	0.0	33.00	66.0	0.000
0.5	4.75	5	0.5	23.75	64.5	11.875
1.0	4.00	4	1.0	16.00	60.0	16.000
1.5	3.25	3	1.5	9.75	52.5	14.625
2.0	2.50	2	2.0	5.00	42.0	10.000
2.5	1.75	1	2.5	1.75	28.5	4.375
3.0	1.00	0	3.0	0.00	12.0	0.000

4)

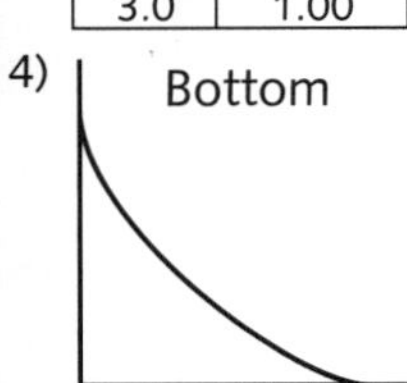

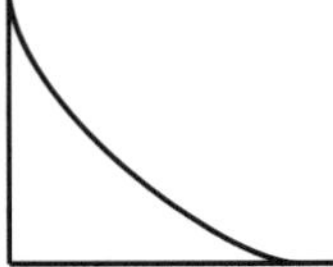

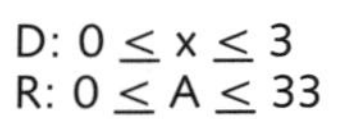

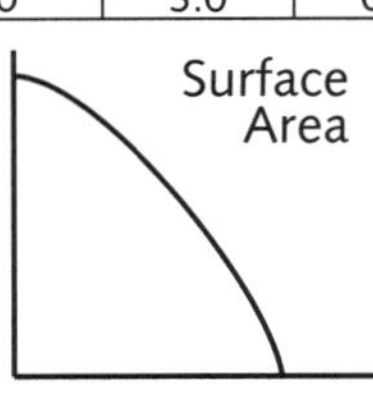

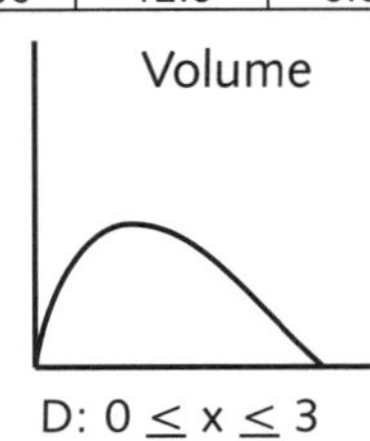

D: $0 \le x \le 3$ R: $0 \le A \le 33$

D: $0 \le x \le 3$ R: $12 \le S \le 66$

D: $0 \le x \le 3$ R: $0 \le V \le 16$

5) Length = $(11 - 3x)/2$; Width = $6 - 2x$; Height = x;
Bottom = $0.5(11 - 3x)(6 - 2x) = 3x^2 - 20x + 33$;
S.A. = $2(6 - 2x)(11 - 3x)/2 + 4x(11 - 3x)/2 + 3x(6 - 2x) = 66 - 6x^2$; Volume = $x(6 - 2x)(11 - 3x)/2 = 3x^3 - 20x^2 + 33x$.

6) Answers will vary. Sample: $x = 1.5$.
Bottom: $0.5(11 - 3(1.5))(6 - 2(1.5)) = 0.5(11 - 4.5)(6 - 3)$
$= 0.5(6.5)(3) = 9.75$ square inches
$3(1.5)^2 - 20(1.5) + 33 = 3(2.25) - 30 + 33$
$= 6.75 + 3 = 9.75$ square inches

7)

Cut	Length	Width	Height	Bottom	S.A.	Vol.
-4	11.5	14	-4	161	-30	
-3	10.0	12	-3	120	12	
-2	8.5	10	-2	85	42	-170
-1	7.0	8	-1	56	60	-56
0	5.5	6	0	33	66	0
1	4.0	4	1	16	60	16
2	2.5	2	2	5	42	10
3	1.0	0	3	0	12	0
4	-0.5	-2	4	1	-30	4
5	-2.0	-4	5	8	-84	40
6	-3.5	-6	6	21	-150	126
7	-5.0	-8	7	40	-228	

8)

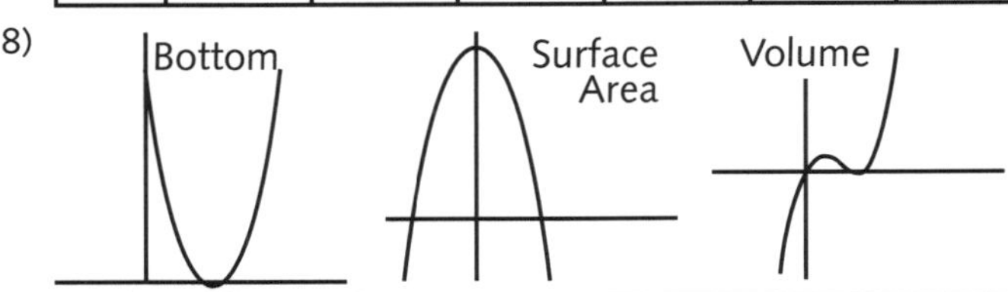

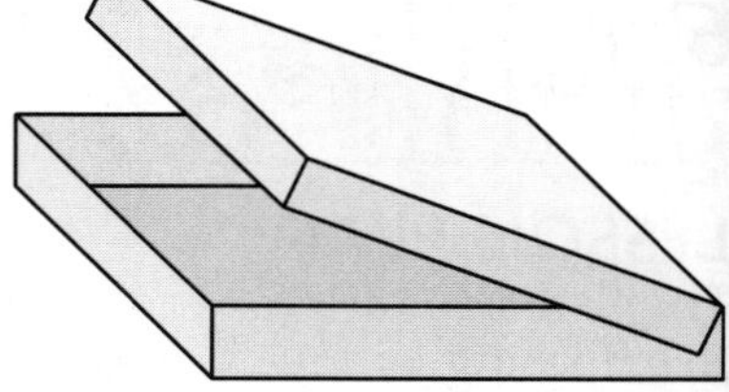

Your challenge is to create a pizza box from a 6" x 11" piece of paper, by cutting six congruent squares and folding along the dotted lines.

1. Trim a sheet of paper down to a 6" x 11" rectangle. Then, from the chart in Step 3 below, choose a value for the length of your cut and actually cut a square out of each of the six positions. Once completed, create a pizza box by folding along the dotted lines shown. Each member of your group should have a different value for your cut. Length of your cut = _______ inches.

6

11

2. For the length of your cut, determine the dimensions of your box and then calculate the AREA OF THE BOTTOM, the SURFACE AREA, and the VOLUME. (show work below)

Length = _______ Area of the Bottom = _______

Width = _______ Surface Area = _______

Height = _______ Volume = _______

3. Complete the chart below for all the given cut lengths:

Length of Cut	Length of Box	Width	Height	Area of Bottom	Surface Area	Volume
0.0						
0.5						
1.0						
1.5						
2.0						
2.5						
3.0						

4. Create a graph for each of the following below, using the lengths of the cuts from the chart above as the domain. List the domain and the range for each (e.g. $0 \leq x \leq 10$):

Area of the Base

Domain: _______________
Range: _______________

Surface Area

Domain: _______________
Range: _______________

Volume

Domain: _______________
Range: _______________

THE PIZZA BOX

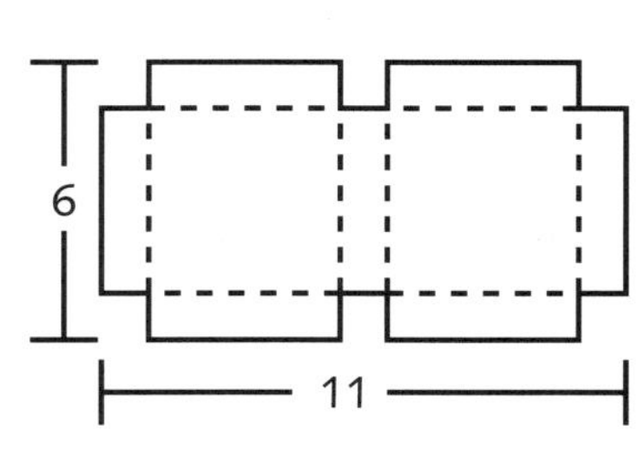

5. Allow x to represent the length of the cut. Complete the algebraic equations below for the dimensions (**Length**, **Width**, and **Height**). Also give your calculations for the **Area of the Bottom**, the **Surface Area**, and the **Volume** in unsimplified and simplified form.

Length: l = ____________ Area of the Bottom: A(x) = ____________________ = _____________

Width: w = ____________ Surface Area: S(x) = ____________________ = _____________

Height: h = ____________ Volume: V(x) = ____________________ = _____________

6. Verify your answers by evaluating the equations using the length of your original cut, x. Check that your answers here match your answers from Step 2.

Length : Width: Height:

Area of the Bottom: Surface Area: Volume:

7. Complete the chart below for all the given cut lengths.

Length of Cut	Length of Box	Width	Height	Area of Bottom	Surface Area	Volume
-4						
-3						
-2						
-1						
0						
1						
2						
3						
4						
5						
6						
7						

8. On a separate sheet of graph paper, create another set of graphs for the Area of the Bottom, Surface Area, and Volume, using the lengths of the cuts from the chart above as the domain. Also, list the domain and the range below each graph (e.g. $-4 \leq x \leq 7$).

This is a 5-part unit on quadratics. The common theme, area of a pool and its surrounding deck is woven throughout the unit to convey all the important principles of working with quadratic equations.

> ***Concepts***
> Writing, graphing, solving quadratic and linear equations; factoring, quadratic formula, finding roots
>
> ***Time:*** 5 days
>
> ***Materials***
> Student handout, graph paper
>
> ***Preparation***
> This is designed to be an introduction to quadratics. Students do not need any prior knowledge.

DAY ONE: Writing Quadratic Equations

Introduce students to the idea of a square pool or spa (**Pool #1**). On graph paper, have each student draw a square that has different dimensions than those of the other members in the group. In other words, one student draws a 1 x 1 square; another draws a 2 x 2 square, etc. Each student calculates the area of their own pool. For example, let us assume Jackie drew a 4 x 4 pool, then she would record A = 16.

The critical part of the lesson now is to have students write the algebraic representation for the area of the pool given the length of its side: "What is the area of your group's pool?" For this one, we get $A(x) = x^2$. The students check their equation by evaluating it with the dimensions of their own pool. Jackie would then record: $A(4) = (4)^2 = 4$.

For **Pool #2**, tell the students that the pool now has a concrete deck to the right of it. This deck is three feet wide and as long as the pool itself. Have each student redraw their original pool, and this time draw in the deck. Then the students are to find the total area represented by the diagram, as well the areas of just the pool and the deck. Jackie would record: A = 28, P = 16, D = 12. The students should compare answers and then generate the functions for these areas for the entire group/class. Those functions are: $A(x) = x^2 + 3x$, $P(x) = x^2$, $D(x) = 3x$. Again students should check the validity of these functions by evaluating for their respective values of x. Jackie would then record: $A(4) = (4)^2 + 3(4) = 28$, $P(4) = (4)^2 = 16$, $D(4) = 3(4) = 12$.

Pool #3 is similar to Pool #2, plus a two-foot square portion of deck (for the towel rack). The students go through the same process as with #2. The functions for #3 are: $A(x) = x^2 + 3x + 4$, $P(x) = x^2$, $D(x) = 3x + 4$.

DAY TWO: Graphing Quadratic Equations

Students generally take a long time to graph the simplest of quadratics, therefore, the entire second day is dedicated to problem #4. Have the students complete the table of values for each pool. The domain of odd integers from -5 to 5 is intentional. They offer a well balanced parabola for Pool #1, and is an easy transition for the students. For Pool #2, the vertex is (-1.5, -2.25), which does not appear in the table. The students will have to rely on their instincts regarding symmetry and roots to find this vertex. Pool #3 also offers challenges. Much more of the right side than the left side of the graph can be seen within the given domain.

We want the students to struggle with these issues. It is this struggle that will motivate them to learn the strategies that make graphing quadratics easy. This is a good time to show the various methods for finding the vertex, and pointing out the y-intercept in the equation. Finding the roots, however, is the topic of the next day.

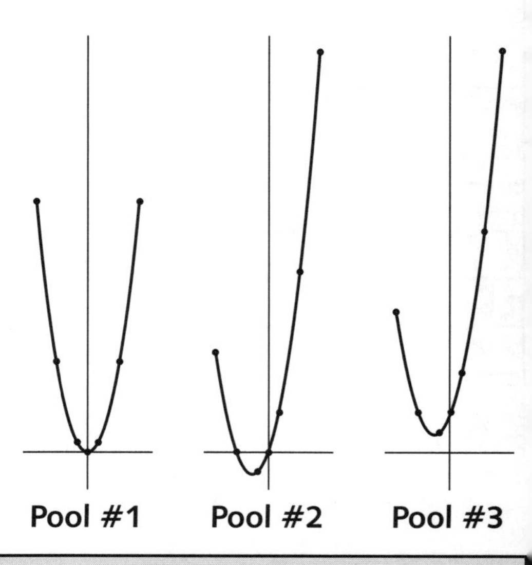

DAY THREE: Factoring Quadratic Equations

The students now get three new pools to study. They should draw the various instances on graph paper just as they did for the first three pools. It is assumed that the students have a basic competency in factoring. Therefore, the objective here is to show them how factoring may be applied to quadratics. Besides the work on factoring, this lesson allows students to practice the skills that they have learned in the previous two days.

Pool #4 is our square pool again, surrounded this time by deck on two sides — 5 feet to the east and 3 feet to the south. (This diagram mimics the use of algebra tiles closely.) For the sake of the factoring portion of this activity, have the students subdivide the rectangle into smaller rectangles. In this case, the sub-rectangles have areas of x^2, 5x, 3x and 15, respectively. The sum of these areas yields the equation below, which then factors as shown. The point of emphasis here is that the factors are the dimensions of the sides of the entire area.

$$A(x) = x^2 + 8x + 15 = (x + 5)(x + 3)$$

Pool #5 is a 3′ by 4′ spa with a deck surrounding the entire spa. The deck may be of any width (x). Again, have the students subdivide the diagram into $4x^2$ (the corners), 4x, 4x, 3x, 3x (the remaining portions of the deck) and 12 (the spa). The sum of these areas yields the equation below, which then factors as shown. The point of emphasis again is that the factors are the dimensions of the sides of the entire area.

$$A(x) = 4x^2 + 14x + 12 = 2(x + 2)(2x + 3)$$

Pool #6 has a deck that is 3 feet wide on all sides of a pool that is twice as long as it is wide. Subdividing the diagram yields the following areas $2x^2$ (the pool), 6x, 6x, 3x, 3x (the side portions of the deck) and 36 (the four 3′ x 3′ corners). The sum of these areas yields the following equation, which then factors as shown.

$$A(x) = 2x^2 + 18x + 36 = 2(x + 3)(x + 6)$$

DAY FOUR: Solving Quadratic Equations

The students have the opportunity to revisit each of the six pools. Now that they have the equations to represent the various areas, a value is assigned to the area of each pool design. The students are then to find the appropriate value for x that will yield the necessary dimensions for that given area. This gives the teacher an opportunity to illustrate two critical points:

1) Solving a quadratic means to find the value of x that yields the given value for y.
2) Factoring and the Quadratic Formula are two different methods to find the same answer.

These are many principles of solving quadratics that may be straightforward to the math teacher, but seem to elude most math students. The context here helps the students overcome this cognitive hurdle. Note, however, that while this lesson offers students the opportunity to apply the Quadratic Formula, it is assumed here that the instructor has taught it to the students by some other means.

DAY FIVE: Assessment

Pools #7 & 8 are intended to evaluate what the students have learned about quadratics from this series of lessons. The four traditional problems give them a chance to practice all the basic skills that a typical algebra student is expected to have when working with quadratics. From there, they should be on their own for the two pools offered.

Pool #1

x
x

Pool #2

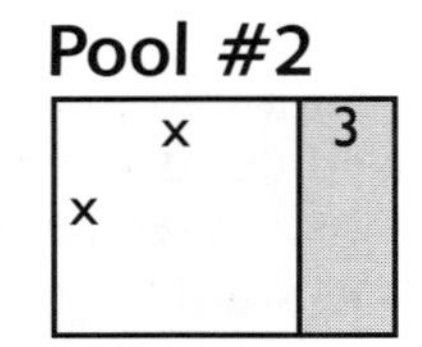

Pool #3

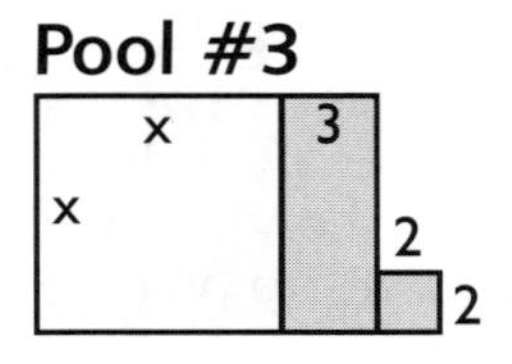

1. a) Draw **Pool #1** on graph paper with a value for *x* that is different than the others' in your group.

   b) Find the area of your pool.
   A = ________

   c) Write the function, A(x), for the area of your group's pool.
   A(x) = ________________

   d) Evaluate A(x) for your value of *x* and confirm the result with your group. A(__) = __________

2. a) Draw **Pool #2** on graph paper with a value for *x* that is different than the others' in your group.

   b) Find the total area, pool area and deck area. A = ________ P = ________ D = ________

   c) Write the functions, A(x), P(x) and D(x), for the various areas of your group's pool.
   A(x) = ______________ P(x) = ______________ D(x) = ______________

   d) Evaluate A(x), P(x) and D(x) for your value of *x* and confirm the result with your group.
   A(__) = ______________ P(__) = ______________ D(__) = ______________

3. a) Draw **Pool #3** on graph paper with a value for *x* that is different than the others' in your group.

   b) Find the total area, pool area and deck area. A = ________ P = ________ D = ________

   c) Write the functions, A(x), P(x) and D(x), for the various areas of your group's pool.
   A(x) = ______________ P(x) = ______________ D(x) = ______________

   d) Evaluate A(x), P(x) and D(x) for your value of *x* and confirm the result with your group.
   A(__) = ______________ P(__) = ______________ D(__) = ______________

4. a) For Pools #1-3, complete the charts below (plus any additional points needed).

Pool #1

x	-5	-3	-1	0	1	3	5
A(x)							

Pool #2

x	-5	-3	-1	0	1	3	5
A(x)							

Pool #3

x	-5	-3	-1	0	1	3	5
A(x)							

   b) Graph each function above on one plane.

   c) For Pools 2 & 3, graph D(x) on one plane.

   d) What would make graphing the quadratics easier?

Pool #4

x x 5 3

Pool #5

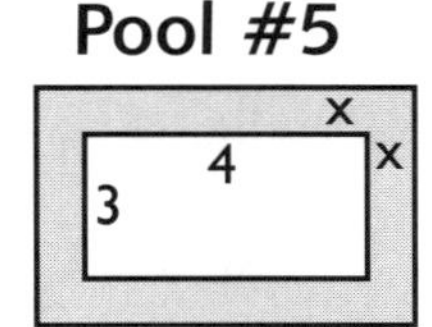

Pool #6

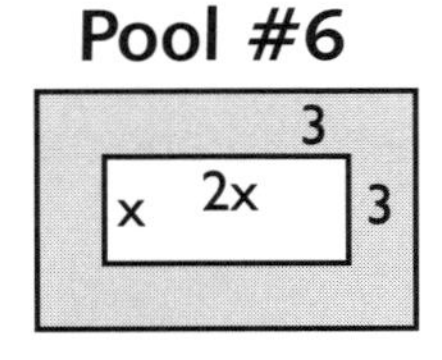

5. a) Draw **Pool #4** on graph paper with a value for *x* that is different than the others' in your group.

   b) Find the total area, pool area and deck area. A = ________ P = ________ D = ________

   c) Write the functions, A(x), P(x) and D(x), for the various areas of your group's pool.
   A(x) = ____________ P(x) = ____________ D(x) = ____________

   d) Evaluate A(x), P(x) and D(x) for your value of *x* and confirm the result with your group.
   A(__) = ____________ P(__) = ____________ D(__) = ____________

   e) Find the vertex and y-intercept. Also, find the roots by factoring. f) Graph each function.

6. a) Draw **Pool #5** on graph paper with a value for *x* that is different than the others' in your group.

   b) Find the total area, pool area and deck area. A = ________ P = ________ D = ________

   c) Write the functions, A(x), P(x) and D(x), for the various areas of your group's pool.
   A(x) = ____________ P(x) = ____________ D(x) = ____________

   d) Evaluate A(x), P(x) and D(x) for your value of *x* and confirm the result with your group.
   A(__) = ____________ P(__) = ____________ D(__) = ____________

   e) Find the vertex and y-intercept. Also, find the roots by factoring. f) Graph each function.

7. a) Draw **Pool #6** on graph paper with a value for *x* that is different than the others' in your group.

   b) Find the total area, pool area and deck area. A = ________ P = ________ D = ________

   c) Write the functions, A(x), P(x) and D(x), for the various areas of your group's pool.
   A(x) = ____________ P(x) = ____________ D(x) = ____________

   d) Evaluate A(x), P(x) and D(x) for your value of *x* and confirm the result with your group.
   A(__) = ____________ P(__) = ____________ D(__) = ____________

   e) Find the vertex and y-intercept. Also, find the roots by factoring. f) Graph each function.

Below are diagrams representing the pools that you have studied thus far. The equations that you generated for each pool still apply. For each of the following, find the value of x that will yield the desired area given. Solve by both factoring and by the quadratic formula. Then draw the dimensions of the diagram to confirm your result.

#1) $A(x) = 16$

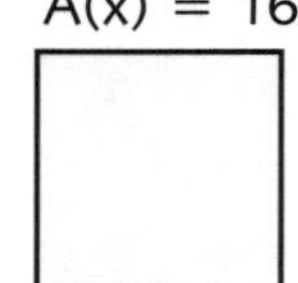

#2) $A(x) = 40$

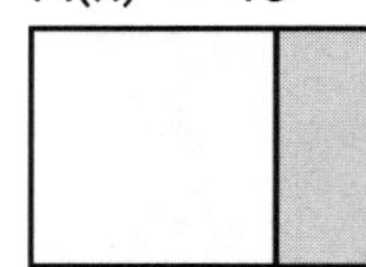

#3) $A(x) = 8$

#4) $A(x) = 48$

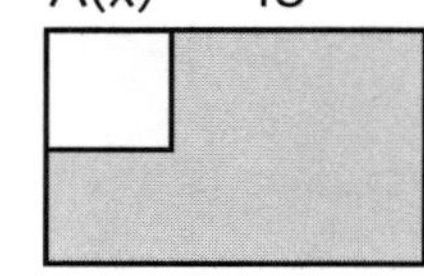

#5) $A(x) = 182$

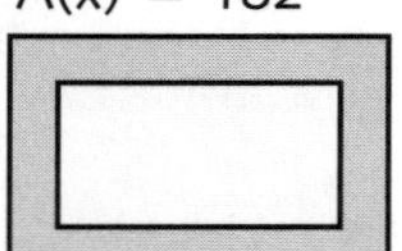

#6) $A(x) = 140$

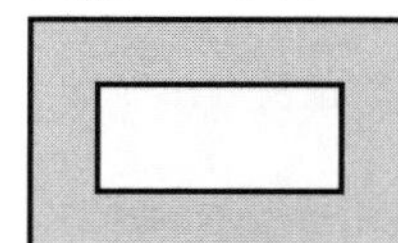

For each equation, calculate the vertex, the roots (by factoring and the quadratic formula) and two other points; then graph (on a separate sheet of paper).

	a) the vertex	b) the roots by factoring	c) the roots by the quadratic formula	d) two other points
1) $y = x^2$	(,)	(,)(,)	(,)(,)	(,)(,)
2) $y = x^2 + 4x$	(,)	(,)(,)	(,)(,)	(,)(,)
3) $y = x^2 + 8x + 12$	(,)	(,)(,)	(,)(,)	(,)(,)
4) $y = -3x^2 - 12x + 15$	(,)	(,)(,)	(,)(,)	(,)(,)

For questions 5 - 8, use **Pool #7.**

5. Write equations for the area of the pool, the deck and the total area.

   P(x) = ______________________________

   D(x) = ______________________________

   A(x) = ______________________________

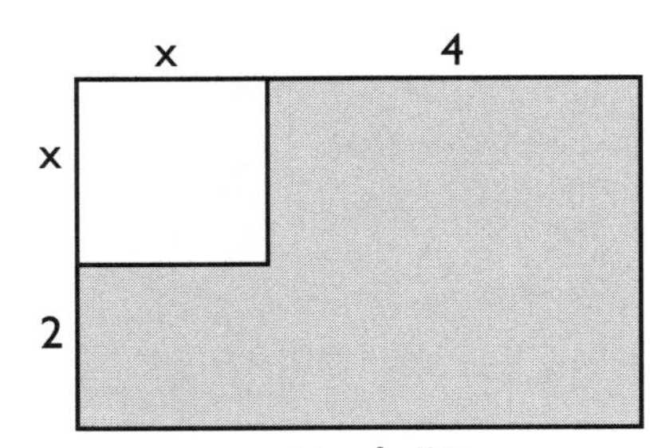

Pool #7

6. For the area of the pool, P(x), find the vertex and then graph.

7. For the area of the deck, D(x), find the slope, the y-intercept, three other points and then graph.

8. For the total area of the pool and deck, A(x), find:
   a) the vertex
   b) the y-intercept
   c) the x-intercepts by factoring
   d) the x-intercepts by the quadratic formula
   e) and graph
   f) and solve A(x) = 35

For questions 9 - 12, use **Pool #8.**

9. Write equations for the area of the pool, the deck and the total area.

   P(x) = ______________________

   D(x) = ______________________

   A(x) = ______________________

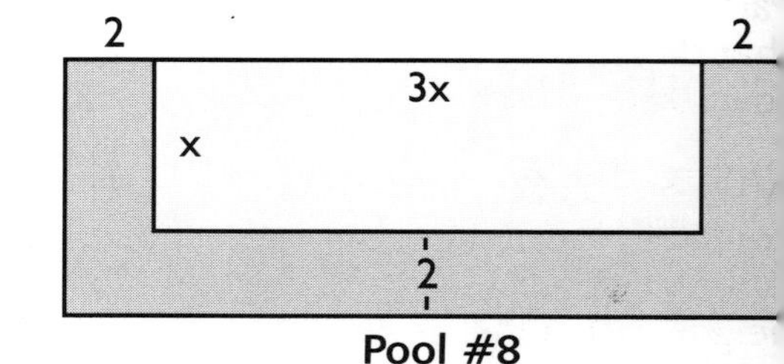

Pool #8

10. For the area of the pool, P(x), find the vertex and then graph.

11. For the area of the deck, D(x), find the slope, the y-intercept, three other points and then graph.

12. For the total area of the pool and deck, A(x), find:
   a) the vertex
   b) the y-intercept
   c) the x-intercepts by factoring
   d) the x-intercepts by the quadratic formula
   e) and graph
   f) and solve A(x) = 133

PROJECT

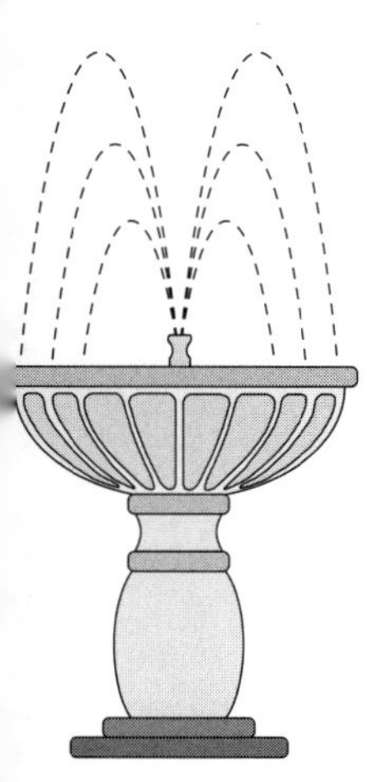

THE COIN FOUNTAIN

The study of parabolic curves through the design of water arcs

LESSON PLAN

The twelve questions on the student handout are listed in ascending order of difficulty. The number of elements that students will be capable of completing depends on the course. We suggest the following:

Beginning Algebra: #1-7
Advanced Algebra: #1-10
Bonus/Revisit: #11-12

The overall purpose of this lesson is to understand the relationship between an equation and its roots. Rather than deconstructing an equation to get to the roots, this lesson constructs the equation from the roots.

Present the students with the project at the beginning of a unit on quadratic equations. Over several days, take the entire class through each phase of the activity using the solution offered below: launch (0, 1), land (0, 17) and vertex (9, 32). These elements offer a very easy value of "a." While, students usually have little difficulty with Part One, the challenges come in Part Two, where the students find the roots by several methods. The fact that the students repeatedly achieve their launch and landing points as solutions reinforces their understanding of the roots.

After completing the assignment with the given values, the students are to complete the assignment again using their own values. They choose the launch and land points of the arc, and write an equation that produces a height within the given range. The purpose for this range of acceptable heights is to force the students to "play" with a variety of values for "a" until they get a correct one. This better helps them understand the influence of "a" on the graph, rather than if the students simply solved for "a."

Concepts

Writing, graphing and solving quadratic equations; finding the roots, vertex, axis of symmetry, directrix, and focus; solving by factoring, completing the square, and the quadratic formula; systems of quadratics

Time: 2-4 hours

Materials

Graph paper, calculator and the student handout

Preparation

Students need to have a basic understanding of each of the methods at each step. This project gives them a conceptual understanding of what the methods produce for them, but it is not an exploratory or investigative activity.

SAMPLE SOLUTIONS

1.

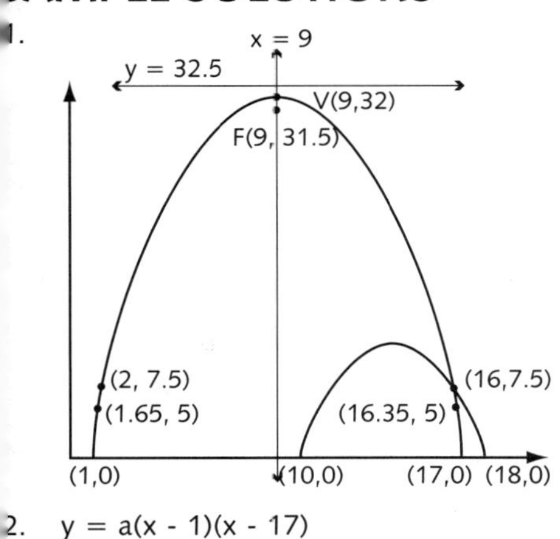

2. $y = a(x - 1)(x - 17)$
   $y = a(9 - 1)(9 - 17)$
   $y = a(-8)(8)$
   $y = -64a$
   i.e. $a = -0.5$ produces height of 32
   $y = -0.5(x - 1)(x - 17)$
   $y = -0.5x^2 + 9x - 8.5$
3. Roots (1,0) & (17,0) for height of 32
4. $y = -0.5(2 - 1)(2 - 17) = 7.5$
   The arc will be 7.5 ft high after 1 ft.
5. vertex: V(9,32)
   axis: $x = 9$
6. $y = -0.5x^2 + 9x - 8.5$
   $y = -0.5(x^2 - 18x + 17)$
   $y = -0.5(x - 1)(x - 17)$
   $x - 1 = 0$ or $x - 17 = 0$
   $x = 1, 17$
7. $(-9 \pm \sqrt{9^2 - 4(-0.5)(-8.5)})/(2(-0.5))$
   $x = 1, 17$
8. $8.5 = -0.5x^2 + 9x$
   $8.5 = -0.5(x^2 - 18x)$
   $8.5 - 40.5 = -0.5(x^2 - 18x + 81)$
   $y = -0.5(x - 9)^2 + 32$
9. $5 = -0.5x^2 + 9x - 8.5$
   $0 = -0.5x^2 + 9x - 13.5$
   $(-9 \pm \sqrt{9^2 - 4(-.05)(-13.5)})/(2(-.05))$
   $x = 1.65, 16.35$
10. $50 = a(x - 1)(x - 17)$
    $50 = a(9 - 1)(9 - 17)$
    $50 = a(-8)(8)$
    $50 = -64a$
    $a = -(25/32)$
11. Second arc: $y = a(x - 10)(x - 18)$
    $10 = a(14 - 10)(14 - 18)$
    $a = -0.625$
    $y = -0.625x^2 + 17.5x - 112.5$
    $-0.5x^2+9x-8.5=-.625x^2+17.5x-112.5$
    $0.125x^2 - 8.5x + 104 = 0$
    $x = 16, 52$; (16, 7.5)
12. $1/(4a) = 1/(4(-0.5)) = -0.5$
    Focus: F(9,32 - 0.5) = F(9,31.5)
    Directrix: $y = 32.5$

THE COIN FOUNTAIN

The study of parabolic curves through the design of water arcs

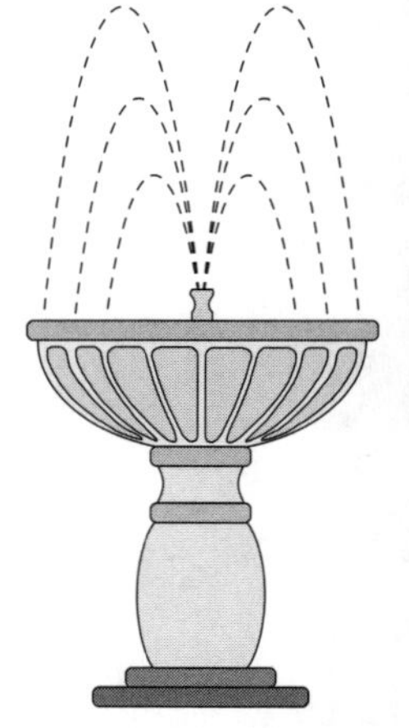

You have been hired to design the water arc of a coin fountain. The pool of the fountain is 20 feet wide, and the water arc is to be greater than 6 feet tall, but less than 50 feet. You will need to determine the locations of the launch point and landing points and the maximum height of the arc. Also, you will need to write an equation that describes the water arc in terms of its height in relation to the horizontal distance along the pool.

PART ONE

1. Place the side view of your fountain in a first quadrant graph. Have the surface of the pool correspond to the x-axis with the left side at the origin. Show the coordinates of the roots and vertex, and include all pertinent data points from the following questions.

2. Using the equation $y = a(x - x_1)(x - x_2)$ where $(x_1, 0)$ & $(x_2, 0)$ are the roots of the parabola, choose a value for "a" that will produce a reasonable arc. Then convert your equation to the form: $y = ax^2 + bx + c$.

3. State your launch and landing points and the height of your water arc.

   Launch (,)
   Landing (,)
   Height ________

4. After one foot of horizontal distance (from your launch point) how high will the water arc be?

5. Find the vertex and the equation for the axis of symmetry.

   Vertex (,)
   Axis: x =

THE COIN FOUNTAIN

PART TWO

6. Factor your equation (in the form $y = ax^2 + bx + c$) to find the roots of the equation and then verify your launch and landing points.

7. Use the quadratic formula to find the roots, again verifying the launch and landing points.

8. Convert your original equation from quadratic form to vertex form, $y = a(x - h)^2 + k$, by completing the square.

9. At what horizontal distance will the arc be 5 feet high?

PART THREE

10. Determine the value of "a" that will produce a 50 foot high arc for your chosen launch and landing points.

11. Create a second arc that intersects the first one and find their point of intersection.

12. Create and graph the coordinates of the focus and the equation for the directrix of your water arc.

LESSON PLAN

The handout deals with six main ideas. We suggest you have the students deal with these main ideas one at a time, discussing each before moving on to the next. This method seems to work better than having the students complete the entire worksheet in one sitting. The six main ideas are as follows:

#1 - 3 Graphing and Writing Linear Equations in Slope-Intercept Form
Fibonacci's Pizza has a constant rate of change for the price, at one dollar per inch (diameter). The students will create a first quadrant graph, so they will not see the y-intercept for this equation (-4.75). You may want to demonstrate it for them.

#4 - 6 Graphing and Writing the Equation for a Best Fit Line
D'Mathematica does not have a constant rate of change for the price, although there is a linear correlation. Have students draw a line that best approximates that correlation. Then have them choose two points on their line to write an equation in slope-intercept form.

#7 - 9 Solving for One Variable Given the Other
These questions help students to solve for x and y, but they also give students a context (price or size) for solving the equations. The power of these problems is in plotting the answers as ordered pairs (size, price), which always lie on the lines graphed. This reinforces the idea that a line represents an infinite number of points — or, in this case, all the combinations of size and price for a pizza.

 # 10-11 The X- and Y-Intercepts
What is the price of a pizza with a diameter of zero? What is the diameter when the price is zero? These solutions do not make sense: A pizza with a diameter of zero, still costs the customer money; and a pizza that is free would have a negative diameter. These nonsense values grab the students attention and actually help them understand the concept of intercepts.

#12-13 Graphs of Quadratic Equations
Once the students have completed the table and graph, they will see that the relationship between diameter and area is not linear. This is due to the fact that the radius squared implies a quadratic relationship.

 # 14 Analyzing Graphs and Data
If the profit for each pizza were consistent, then the price would also be a quadratic relationship. When the students calculate the price per square inch of pizza, it becomes obvious that less profit is made on the larger pizzas than on the smaller (assuming the cost per square inch is consistent).

Concepts
Linear and quadratic equations, best fit lines, slope, x- and y-intercepts, area of a circle, graphing

Time: 2-3 hours

Materials
Straight edge, calculator and student handout

Preparation
Students need to have a basic understanding of writing and graphing linear equations and best fit lines. They should also know the area of a circle.

SOLUTIONS

1.
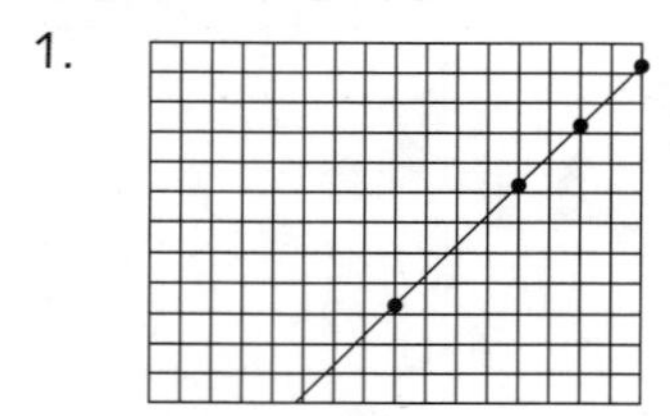

2. One dollar per inch, or $1/in
3. P = d - 4.7
4.
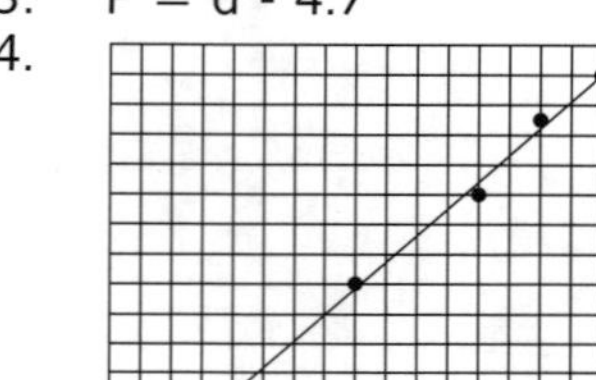

5. Answers vary, e.g. 89¢ per inch
6. Answers vary. e.g. P=0.89d - 3.29
7. $5.25, $5.64
8. 19.75 inches, 20.5 inches
9. They are on the line that represents the equation.
10. (4.75, 0), (3.7, 0); Fibonacci's would give away a 4.75" pizza for free, D'Mathematica would give away a 3.7" pizza. They would lose money for pizzas less than 4.75 and 3.7 inches, respectively.
11. (0, -4.75), (0, -3.29); For a pizza of zero diameter, Fibonacci's and D'Mathematica would pay $4.75 and $3.29, respectively.
12.

Diameter	Area
2	3.14
5	19.6
8	50.2
12	113
14	154
16	201

13. The graph is shaped as it is because the area formula for a circle is a quadratic equation.
14.

Diameter	Fibonacci's	D'Mathematica
8	6.5	8.0
12	6.4	6.2
14	6.0	6.2
16	5.6	5.5

There is a definite relationship between the diameter of a pizza and its area, just as there is with any circle. However, is there a relationship between the diameter and the price of a pizza? The following activity offers pizza prices from two different pizzerias. Your task is to explore the relationship between the size of pizzas and their prices.

Diameter	Fibonacci's Pizza	Pizzeria D'Mathematica
8″	\$3.25	\$4.00
12″	\$7.25	\$7.00
14″	\$9.25	\$9.50
16″	\$11.25	\$11.00

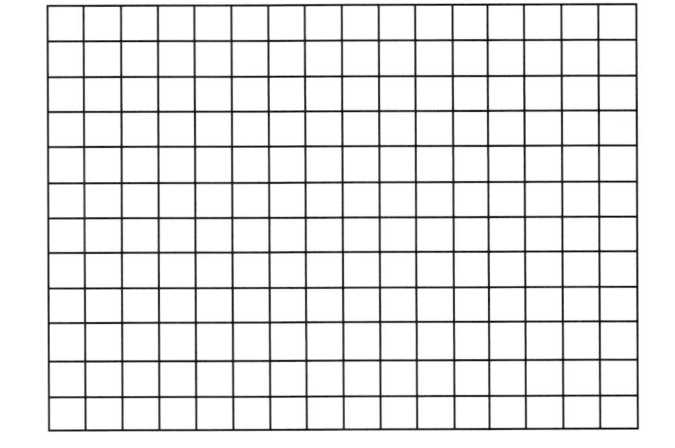

1. Begin with **Fibonacci's Pizza**. Graph the data, using the diameter of the pizzas as the domain, and the prices as the range.

2. What is the rate of increase in the price of the pizza in relation to the increase in the diameter of the pizza.

3. Write a linear equation to represent the relationship of the price of the pizza, P, according to the diameter, d.

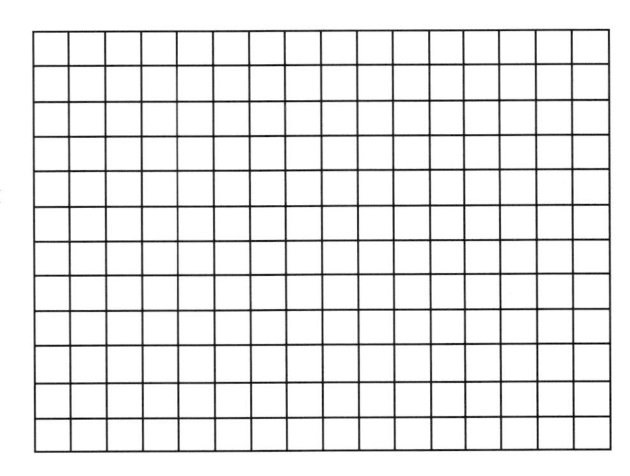

4. Now use the data for **Pizzeria D'Mathematica**. Graph the data, using the diameter of the pizzas as the domain, and the prices as the range.

5. What is the rate of increase in the price of the pizza in relation to the increase in the diameter of the pizza. If it is not consistent (like Fibonacci's Pizza), approximate and explain how you determined the rate of increase.

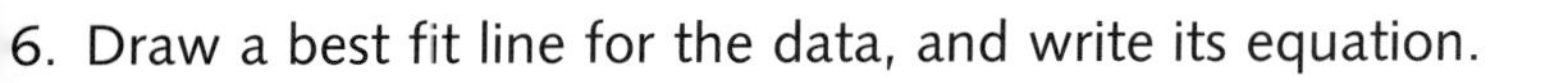

6. Draw a best fit line for the data, and write its equation.

7. For each restaurant, use its respective equation to determine the cost of a 10″ pizza, and then plot the data.

8) For each restaurant, use its respective equation to determine the size of a pizza that would cost \$15, and then plot the data.

PARABOLIC PIZZA

9. How do your answers to #7 and #8 relate to the graphs?

10. Find the x-intercepts of both equations. According to these points, how big would a free pizza be? For which sizes of pizza would the restaurants lose money?

11. Find the y-intercepts of both equations. What do these points say about the cost and size of the pizzas?

12. For the sizes listed in the chart, graph the the relationship of the diameter, d (domain), to the area, A (range).

Diameter	Area
2	
5	
8	
12	
14	
16	

13. Explain why the graph is shaped like it is.

14. a) If there is a linear relationship between the diameter and the price of the pizza, and a quadratic relationship between the diameter and the area, then what do you predict about the profit made on a pizza (profit = price minus cost) in relationship to the size of the pizza.

b) Verify your conjecture by determining the price per square inch of each pizza size for both pizza parlors.

	Price per square inch (¢)	
Diameter	Fibonacci's	D'Mathematica
8		
12		
14		
16		

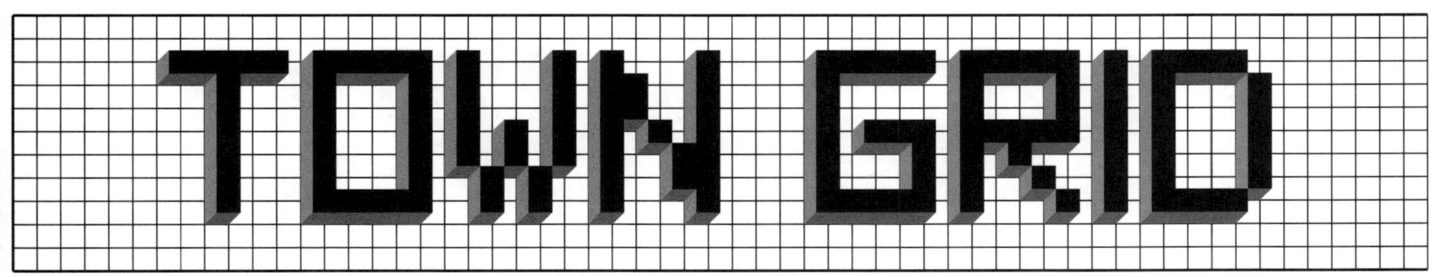

Submitted by Jan Vance, Temecula Valley High School, California

LESSON PLAN

The lesson will work well if you use the map provided on the student handout. However, it may make the activity more meaningful, if you can find a map of your town and mimic this activity with your own locally popular landmarks.

Demonstrate each segment of the lesson (units, inches and miles) for the first pair of landmarks (T & W). Then allow the students to work with the other landmarks on their own. Calculating with the distance formula will be the easiest portion of the assignment. In the second step, it is most important to have the students both *calculate* the distance in inches using the unit scale on the axes and to actually *measure* using the ruler. The fact that the answers are so close makes a distinct impression in the students' minds. The most difficult portion of the lesson is calculating the scale factor. "How many miles are represented on the legend? How many inches long is the legend?" The concept of converting 1.5 miles per 2.5 inches to 0.6 miles per one inch seems to elude most students. Take it slow here.

Concepts
Distance Formula, measurement, scale factor, unit conversion.

Time: 1 hour

Materials
Student handout, ruler.

Preparation:
Students should have exposure, but not necessarily mastery of the distance formula.

SOLUTIONS

1.
T (3, -6) & W (14.5, 2) $TW = \sqrt{(14.5 - 3)^2 + (2 - (-6))^2}$ = **14.1 units**
M (12, -6.5) & G (-7, 21) $MG = \sqrt{(-7 - 12)^2 + (21 - (-6.5))^2}$ = **33.4 units**
C (-6.5, -11) & P (7, 10) $CP = \sqrt{(7 - (-6.5))^2 + (10 - (-11))^2}$ = **25 units**
R (1, -7.5) & L (7, -7) $RL = \sqrt{(7 - 1)^2 + (-7 - (-7.5))^2}$ = **6 units**

2.

Calculated Distance	Measured Distance
TW = 14.1($^1/_4$") = **3.5 inches**	TW = **3.4 inches**
MG = 33.4($^1/_4$") = **8.4 inches**	MG = **8.5 inches**
CP = 25($^1/_4$") = **6.25 inches**	CP = **6.25 inches**
RL = 6($^1/_4$") = **1.5 inches**	RL = **1.75 inches**

3. Scale
1.5 miles : 2.5 inches
0.6 miles : 1 inch

TW = 3.4(0.6) = **2.04 miles**
MG = 8.5(0.6) = **5.10 miles**
CP = 6.25(0.6) = **3.75 miles**
RL = 1.75(0.6) = **1.05 miles**

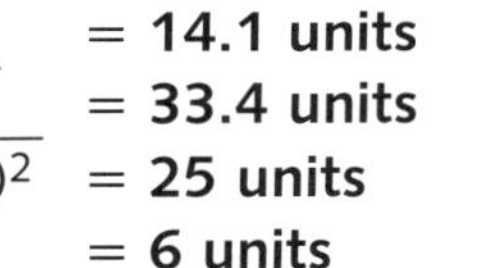

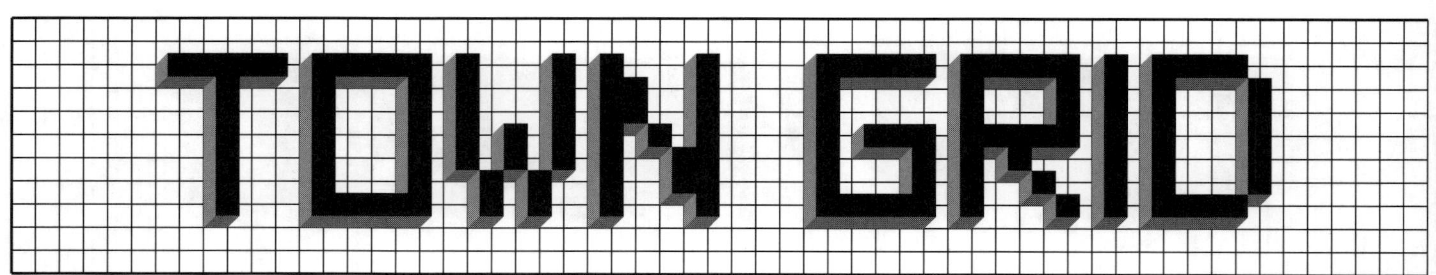

1. Using the distance formula, find the distance (crow's flight) between.

   a) Temecula Valley High School & Thornton Winery. T(,) & W(,)

      TW = __________ units

   b) Temecula Middle School & SCGA Clubhouse (California Golf Club). M(,) & G(,)

      MG = __________ units

   c) Temecula Cemetery & Riverton Park. C(,) & P(,)

      CP = __________ units

   d) Temecula Community Rec Center & Linfield School. R(,) & L(,)

      RL = __________ units

2. Convert your units above to inches, using the scale of the axis (one unit = ¼"). Then use a ruler to measure the actual distance between the two points of each pair above and compare your calculated distances to your measured distances.

<u>Calculated Distance</u>	<u>Measured Distance</u>
TW = __________inches	TW = __________inches
MG = __________inches	MG = __________inches
CP = __________inches	CP = __________inches
RL = __________inches	RL = __________inches

3. Use the legend at the top of the map to determine the true distance (miles) between the points.

Scale: ______ miles = ______ inches **Unit Scale:** ______ miles = 1 inch

TW = __________ miles CP = __________ miles

MG = __________ miles RL = __________ miles

TOWN GRID

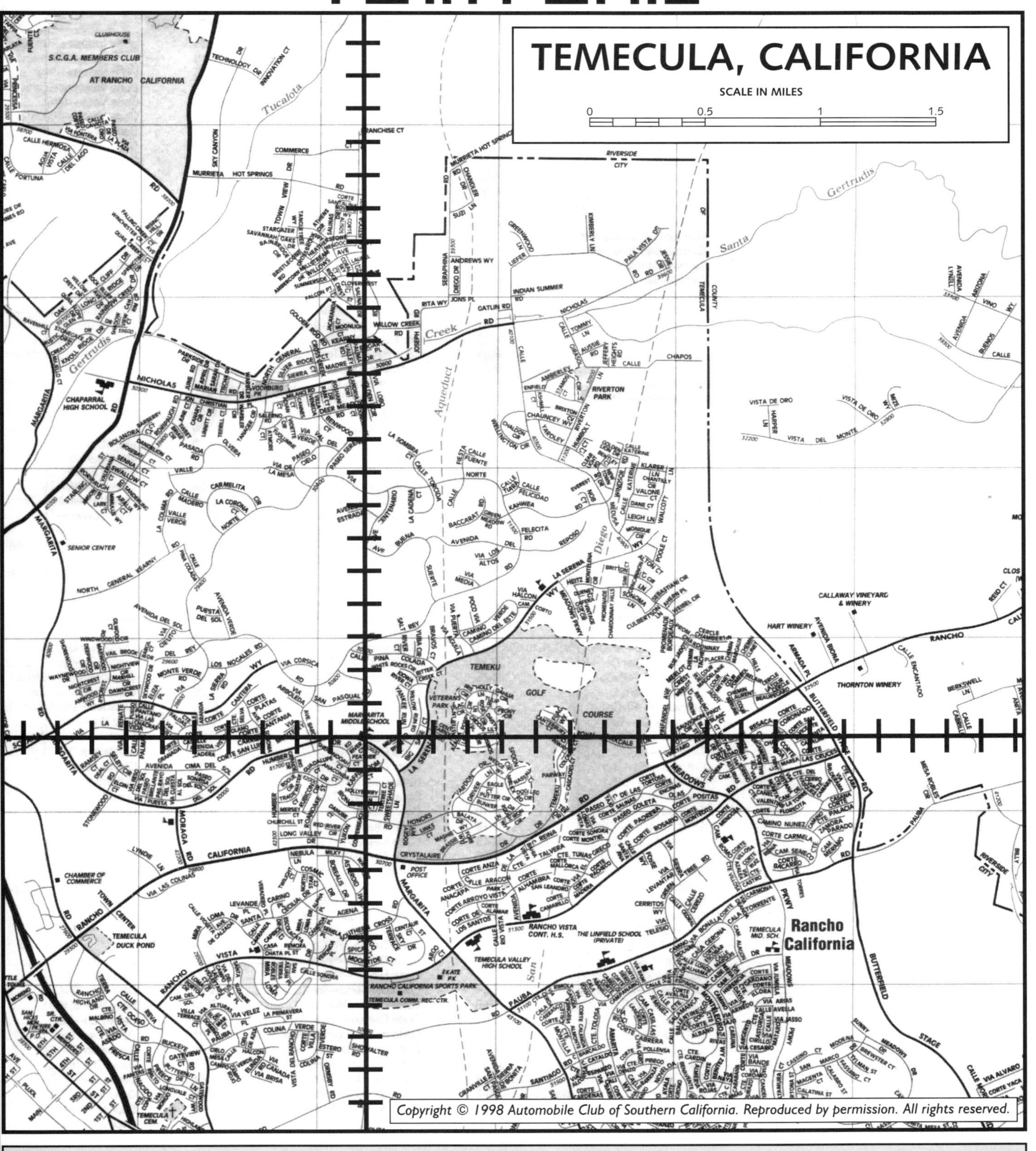
TEMECULA, CALIFORNIA
SCALE IN MILES
0
0.5
1
1.5
TOWN GRID
Rancho
California
CHAPARRAL HIGH SCHOOL
TEMEKU
GOLF
COURSE
RIVERSIDE CITY
TEMECULA VALLEY HIGH SCHOOL
CHAMBER OF COMMERCE
SENIOR CENTER
CALLAWAY VINEYARD & WINERY
THORNTON WINERY
HART WINERY
RANCHO CALIFORNIA SPORTS PARK
TEMECULA DUCK POND
POST OFFICE
MARGARITA MIDDLE SCHOOL
RIVERTON PARK
S.C.G.A. MEMBERS CLUB AT RANCHO CALIFORNIA

THE BILLABONG & THE BUSH

Analyzing Animal Feeding Grounds

Submitted by Michael Wheal, Mathematics Education Consultant, South Australia

LESSON PLAN

This is a complex project involving advanced manipulation of matrices, which is appropriate for most Advanced Algebra students. You can walk students through Part 1, use Part 2 as guided practice, and leave Part 3 for homework or future assessment. In order to complete Part 4, students need experience with a matrix applications, most likely found in Honors or Advanced Placement classes.

Concepts
Writing & Multiplying Matrices

Time: 5-7 days

Materials
Student handout, graphing calculator (optional)

Preparation
Students should have a firm understanding of matrices.

Part One

Discuss the meaning of the numbers in Matrix T — that each number represents the portion of the animals from the feeding ground listed at the row header that move to the feeding ground listed in the column header. For example, Row B states that 60% of the animals in Feeding Ground B will move to Ground A, while 30% will go to Ground C, and 10% will remain in B. Note that the sum of each row is 1 (100%), but that the sum of each column is not. This is because the animals do not move uniformly to each feeding ground. T^2 will be the transition matrix for two consecutive days. For instance, after the second day, 40% of the animals in Ground A will still be in Ground A after the second day. "How is this possible, if only 20% remain in A after the first day?" (Because a large percentage of the animals from the other two grounds return to A.)

Before students multiply matrices D & T, it is good to discuss why matrix multiplication is not commutative (we can have DT, but not TD). Interestingly, the product of a 1 by 3 matrix and a 3 by 3 matrix is another 1 by 3, which is what we need to determine future distributions. Walk students through calculating D_2 & D_3. Emphasize their interpretation. What do these say about the distribution of the animals in each feeding ground? How does the original matrix suggest that the majority of the animals will be found in Feeding Ground A?

Part Two

While the dimensions of the matrices are larger now, encourage the students that the process for multiplying is still the same. Again focus on the interpretation of the data. Allow the students to commit to their answers before discussing it with the class. Many of the same principles from the smaller matrices in Part 1 apply here. Part 3 simply asks the students to create their own data and calculations for a similar scenario.

AUTHOR'S NOTE

I would normally use this project at the end of a seven week topic on applications of matrices. The first part of the investigation enables all students to start with something familiar and sets the scene for the real investigation. The second part introduces the idea of an absorbing state and directs students toward a particular result. The third part begins the investigation gently at first with an explained variation on the theme and then moves to a challenge in which the students are required to devise and investigate their own scenario with an absorbing state (e.g. human emigration, debt payment, radioactive decay).

We are required to give grades (A, B, C, D and E) and performance in a piece such as this is a very good guide. Part 1 is worth one-quarter of the marks, Part 2 is also worth one-quarter and Part 3 is worth half of the marks. A student who makes only some progress with Part 1 will be awarded a D, whilst one who completes it successfully will be awarded a C. A student who also completes Part 2 successfully will be awarded a B, whilst one who completes everything successfully will be awarded an A. The term completes successfully is used deliberately because students will make calculation errors and we allow a little leeway in this regard. Moreover, if students omit written explanations and commentary or treat them superficially, the grade awarded will reflect this.

THE BILLABONG & THE BUSH

SOLUTIONS

Part 1

$$T = \begin{matrix} & A & B & C \\ A & 0.2 & 0.4 & 0.4 \\ B & 0.6 & 0.1 & 0.3 \\ C & 0.3 & 0.5 & 0.2 \end{matrix} \qquad T^2 = \begin{bmatrix} 0.40 & 0.32 & 0.28 \\ 0.27 & 0.40 & 0.33 \\ 0.42 & 0.27 & 0.31 \end{bmatrix}$$

(i) The 0.3 means that 30% of the animals in Feeding Ground C will move to Feeding Ground A. Similarly, 30% of the animals in Feeding Ground B will move to Feeding Ground C.

(ii) See above

(iii) $DT = [\ 120\ \ 240\ \ 240\ ]$

(iv) $DT^2 = [\ 240\ \ 192\ \ 168\ ]$

(v) $D_2 = [\ 120\ \ 240\ \ 240\ ]$
$D_3 = [\ 240\ \ 192\ \ 168\ ]$
$D_4 = [\ 213.6\ \ 199.2\ \ 187.2\ ]$
$D_5 = [\ 218.4\ \ 199\ \ 182.6\ ]$
$D_6 = [\ 217.8\ \ 198.6\ \ 183.6\ ]$
$D_7 = [\ 217.8\ \ 198.8\ \ 183.4\ ]$

(vi) The number of animals in any given feeding ground seem to stabilize. Interestingly, though, there are still animals moving in and out of the each of the feeding grounds. While the numbers stabilize, the model is not static.

(vii) $\frac{218}{600} = 36\%$ $\frac{199}{600} = 33\%$ $\frac{183}{600} = 31\%$

(viii) Two weaknesses of the model are that the model assumes that the daily rates are consistent, and that the animals do not choose a favorite feeding ground.

Part 2

Another model of feeding behaviour proposes that animals will eventually settle in a favourite area and remain

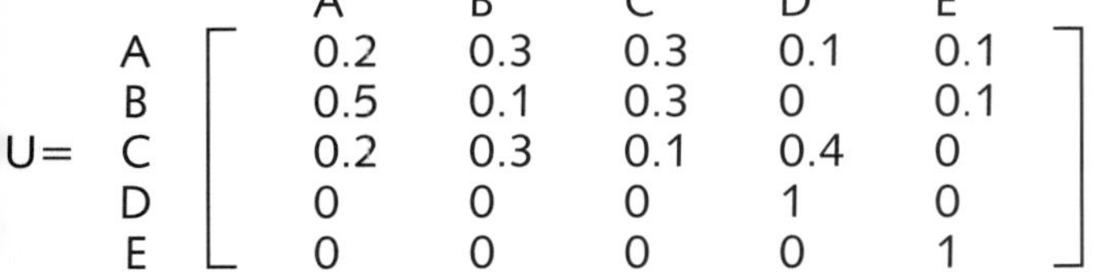

$$U = \begin{matrix} & A & B & C & D & E \\ A & 0.2 & 0.3 & 0.3 & 0.1 & 0.1 \\ B & 0.5 & 0.1 & 0.3 & 0 & 0.1 \\ C & 0.2 & 0.3 & 0.1 & 0.4 & 0 \\ D & 0 & 0 & 0 & 1 & 0 \\ E & 0 & 0 & 0 & 0 & 1 \end{matrix}$$

(ix) D and E are the preferred feeding grounds, as designated by the 1 in each of those corresponding rows and column. The 1 implies that one hundred percent of the animals in that feeding ground remain there.

(x) Determine the distribution on each of the next six days.
$D_2 = [\ 120\ \ 180\ \ 180\ \ 60\ \ 60\ ]$
$D_3 = [\ 150\ \ 108\ \ 108\ \ 144\ \ 90\ ]$
$D_4 = [\ 105.6\ \ 88.2\ \ 88.2\ \ 202.2\ \ 115.8\ ]$
$D_5 = [\ 82.9\ \ 67.0\ \ 67.0\ \ 248.0\ \ 135.1\ ]$
$D_6 = [\ 63.4\ \ 51.6\ \ 51.6\ \ 283.1\ \ 150.2\ ]$
$D_7 = [\ 48.8\ \ 39.7\ \ 39.7\ \ 310.1\ \ 161.7\ ]$

(xi) After Day 4, half of the animals are in the preferred feeding areas.

(xii) D appears to be more popular. By reading down columns D & E in the transition matrix U, one sees that D has a greater total weight. This shows to be true as the distribution matrices for the next six days are analyzed.

THE BILLABONG & THE BUSH

Analyzing Animal Feeding Grounds

INTRODUCTION

Many animals alternate between a number of feeding grounds. In this investigation their feeding behaviour is modeled using transition matrices.

PART ONE

Consider the case in which animals move randomly between three feeding grounds A, B and C. Suppose that T is the transition matrix showing the change in feeding ground from day to day where:

$$T = \begin{matrix} & A & B & C \\ A \\ B \\ C \end{matrix} \begin{bmatrix} 0.2 & 0.4 & 0.4 \\ 0.6 & 0.1 & 0.3 \\ 0.3 & 0.5 & 0.2 \end{bmatrix} \qquad T^2 = \begin{bmatrix} \_\_\_ & \_\_\_ & \_\_\_ \\ \_\_\_ & \_\_\_ & \_\_\_ \\ \_\_\_ & \_\_\_ & \_\_\_ \end{bmatrix}$$

(i) What do the two entries 0.3 tell you about the animals feeding habits?

(ii) Calculate T^2 and explain what the entries in the second row tell you.

Suppose that the row D = [100 200 300] describes the distribution of the animals on the first day of observations. Calculate each of the following and explain what the entries tell you.

(iii) DT = [____ ____ ____] (iv) DT^2 = [____ ____ ____]

Suppose that the row D = [600 0 0] describes the distribution on the first day of observations.

(v) Calculate the distribution on each of the next six days.

D_2 = [____ ____ ____] D_3 = [____ ____ ____] D_4 = [____ ____ ____]

D_5 = [____ ____ ____] D_6 = [____ ____ ____] D_7 = [____ ____ ____]

(vi) What appears to happen to these animals?

(vii) Over time, according to this model, what proportion of the animals' feeding time is spent in each area?

(viii) Suggest two weaknesses of this model and the likely effects they would have on the description of the distribution.

THE BILLABONG & THE BUSH

PART TWO

Another model of feeding behaviour proposes that animals will eventually settle in a favourite area and remain there. Suppose that there are five areas, A, B, C, D and E, that the transition matrix U shows the change from day to day and that the row D = [600 0 0 0 0] gives the distribution on the first day of observations.

U =

	A	B	C	D	E
A	0.2	0.3	0.3	0.1	0.1
B	0.5	0.1	0.3	0	0.1
C	0.2	0.3	0.1	0.4	0
D	0	0	0	1	0
E	0	0	0	0	1

ix) Identify the preferred feeding grounds.

x) Determine the distribution on each of the next six days.

D_2 = [____ ____ ____ ____ ____] D_5 = [____ ____ ____ ____ ____]

D_3 = [____ ____ ____ ____ ____] D_6 = [____ ____ ____ ____ ____]

D_4 = [____ ____ ____ ____ ____] D_7 = [____ ____ ____ ____ ____]

xi) After how many days are half of the animals in the preferred feeding areas?

xii) Explain which of the two preferred feeding grounds appears to be more popular.

PART THREE

xiii) Set up your own initial distribution and 5 x 5 transition matrix V with two preferred feeding grounds.

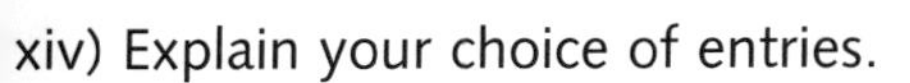
xiv) Explain your choice of entries.

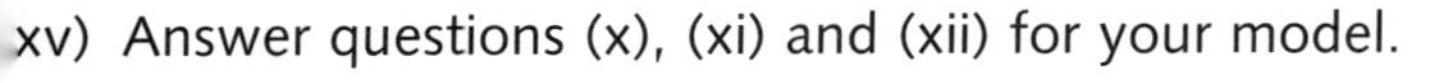
xv) Answer questions (x), (xi) and (xii) for your model.

xvi) Suggest two weaknesses of this model and the likely effects they would have on the description of the distribution.

PART FOUR - Extra Credit

xvii) Describe another situation which could be modeled usefully by this type of transition matrix.

xviii) Set up a transition matrix and initial distribution for the situation you have described and explain your choice of entries.

xix) Determine the numbers in each state over the next few periods. Relate the results of your calculations to the situation you described in parts (xvii) and (xviii).

LESSON PLAN

Begin with the story of Zeno's Paradox. Zeno was a Greek philosopher who imagined an arrow being shot at a target. Before the arrow can reach the target it must first cover half the distance, i.e. reach the midpoint of the distance. From there, it must then cover half the remaining distance, or the next midpoint. This leaves one quarter of the original distance. This concept of continually covering half the distance perplexed Zeno. If you continually cut something in half, when do you have nothing left to cut? In other words, if we start with a value of 1 (the whole distance) and repeatedly multiply by one-half, when do we reach zero? We never will. This implies that motion is impossible, which is contrary to our reality. So how do we reconcile Zeno's Paradox?

Simple. Motion is a matter of addition, not multiplication. Rather than multiplying by half each time, we add each successive segment that the arrow travels. In Zeno's example, the arrow would travel 1/2 + 1/4 + 1/8 + 1/16 + . . . In other words, the distance that the arrow covers is the sum of an infinite series. This is still perplexing for most students, because intuition says that if you add forever, you'll never reach a final sum.

This is where we apply the concepts of convergence and divergence. If the sequence of numbers increases without bound, no limit exists. Since the fractions decrease at a constant ratio, we have a convergent sequence, with a limit that the students can easily see as zero.

Concepts:
Sum of infinite series, limits

Time: 1 Hour

Materials:
Student Handout, 2 or more bouncing balls (tennis & racquet), 2 or more yardsticks or tape measures.

Preparation:
The students will have an easier time reading the heights of bounces if the balls are bounced on a table top. This may be done in the classroom or on the lunch tables.

To test this, we will release a ball from a given height and analyze its bounce height until the ball comes to rest. The sum of the differences between all successive pairs of heights should be equal to the initial height. Have one half the class use a racquet ball while the other half uses a tennis ball. Begin both at 36 inches. Students should record each successive bounce height on the diagram provided. Column "a" contains the difference of each successive pair of heights. Since the students are doing rough estimates of the bounce heights, these numbers will not make an exact geometric series, so have them also use the "fudge" column, in which the numbers are slightly altered to fit the apparent geometric sequence. Once they have this pattern, they may apply the formula for the sum of the infinite series. If prompted prior to this calculation, the students will almost unanimously say that the sum should be the initial height of the ball. The calculations should hold true to this conjecture.

Have the groups switch balls and repeat the process. Then have the students try it a third time with a different ball and/or a different height. The persistence of this pattern will confirm for students this anti-intuitive principle of the sum of an infinite series.

SAMPLE SOLUTION

A tennis ball bounces to about 55% of its previous height. Assuming an initial height of 36 inches, the values are: $b_1 = 16$, $b_n = 0.55b_{n-1}$, after fudging the numbers. Therefore, the sequence will be $b_1 = 16$, $b_2 = 9$, $b_3 = 4$, $b_4 = 2.2$... The sum of the infinite series, 16 + 9 + 4 + 2.2 + ..., is 16 / (1 - .55) which equals 35.6, or approximately 36 inches.

STUDENT HANDOUT

Have you ever examined a bouncing ball? The height of each successive bounce obviously gets shorter. But, how does it change? Does it decrease arithmetically or geometrically? Let's find out?

1. a) Hold a ball next to a yardstick so that the bottom of the ball is exactly 36 inches off the ground. Drop it and estimate the height of the bottom of the ball as it bounces back up. Make this estimation for each successive bounce. You may repeat this process several times until you have at least four bounces recorded. Each person in your group can be responsible for watching a different bounce. Also, you may restart the ball at the height of the previous bounce in order to accurately measure each successive bounce.

36"

Preliminary	Fudged
a_1 ______	b_1 ______
a_2 ______	b_2 ______
a_3 ______	b_3 ______
a_4 ______	b_4 ______
a_5 ______	b_5 ______

If the height of the ball decreases arithmetically, then there should be a constant difference between the heights of the successive bounces. If the height of the ball decreases geometrically, then there should be a constant ratio between the heights of the successive bounces.

b) Determine whether there exists a constant difference or constant ratio. What is the approximate difference or ratio between two consecutive heights?

2. What is the approximate difference between each bounce height and the previous bounce height. This is a_n, and should be recorded in the preliminary column.

3. Time to fudge the numbers. The preliminary column will be very close to some kind of sequence. Any error is likely due to both human approximation and friction between the ball and the ground. Adjust the numbers slightly so they more closely approximate a sequence. List these numbers in the fudged column.

4. Find the sum of the infinite series: $b_1 + b_2 + b_3 + b_4 + b_5 + \ldots =$ ____________

Bonus: Find the distance that the ball travels, both up and down paths, until it comes to rest.

Zeno's Bouncing Ball

Repeat this process two more times by using either different balls or different starting heights or both.

5. a) For the new ball and/or height, complete the diagram below and record the successive bounce heights.

Ball Type __________ Initial Height __________

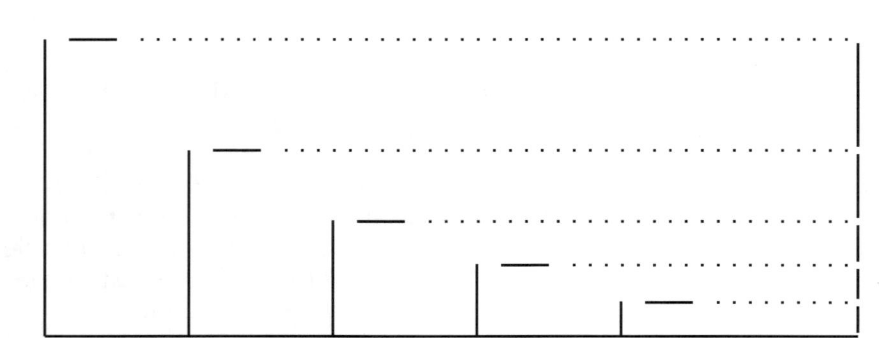

Preliminary	Fudged
a_1 ______	b_1 ______
a_2 ______	b_2 ______
a_3 ______	b_3 ______
a_4 ______	b_4 ______
a_5 ______	b_5 ______

b) What is the approximate difference between each bounce height and the previous bounce height. Record this data in the preliminary column.

c) Fudge the numbers, if necessary.

d) Find the sum of the infinite series: $b_1 + b_2 + b_3 + b_4 + b_5 + \ldots =$ __________

6. a) For another new ball and/or height, complete the diagram below and record the successive bounce heights.

Ball Type __________ Initial Height __________

Preliminary	Fudged
a_1 ______	b_1 ______
a_2 ______	b_2 ______
a_3 ______	b_3 ______
a_4 ______	b_4 ______
a_5 ______	b_5 ______

b) What is the approximate difference between each bounce height and the previous bounce height. Record this data in the preliminary column.

c) Fudge the numbers, if necessary.

d) Find the sum of the infinite series: $b_1 + b_2 + b_3 + b_4 + b_5 + \ldots =$ __________

CALCULATOR LOGOS III

CONTRIBUTED BY DIANE KARST
BISHOP LUERS HIGH SCHOOL, FT. WAYNE, IN

A logo is a design or symbol that prompts a person to think about a particular company, product, or activity. Companies spend thousands of dollars just to develop the perfect logo. This is exactly what students are asked to do in this assignment. Students are required to design a logo for a real or fictitious company, using only mathematical equations programmed into a graphing calculator.

> ***Concepts***
> Advanced Algebra through Calculus. Functions (polynomial, trigonometric, logarithmic), conic sections, derivatives, graphing calculators, programming, graphic transformations
>
> ***Time:*** 1-2 weeks (out of class)
>
> ***Materials***
> Graphing calculators (required for each student)
>
> ***Preparation***
> There is a high degree of algebraic manipulation and calculator proficiency required to accomplish this assignment.

A NOTE ON LEVEL OF DIFFICULTY

This project can be used with a variety of upper level math classes, depending on the difficulty of the requirements you establish. By changing the type of functions involved in the logo, you can cater this assignment to the level of your students. For Advanced Algebra students, you might only require parabolics, logarithms, and conic sections. With Pre-Calculus students you might add in trigonometric functions, and with Calculus students you can even include derivatives and antiderivatives. The assignment listed below is designed for Calculus students.

STUDENT ASSIGNMENT

Your assignment is to create a logo for a real or fictitious company. You should begin by making a rough sketch of one or more ideas and decide how to draw them with mathematical equations. You will design a program for your calculator to draw the logo. To make the picture precise, you must give an exact list of instructions for the calculator, a computer printout of these instructions, and a computer printout of the picture produced.

You must use at least one of each of the following functions: polynomials, trigonometric, logarithmic, functions defining conic sections (circles, parabolas, ellipses, hyperbolas), an antiderivative, and a derivative.

HELPFUL TIPS FOR CREATING PICTURES ON A GRAPHING CALCULATOR

It is recommended that you begin the program with `ClrDraw` and `Dot Mode` and set the `x-min`, `x-max`, etc. inside the program. Remember, if you want it "square", the x and y must have a 3:2 ratio. A good setting might be x: 1 to 10.5 and y: 1 to 7.3.

To restrict the domain of a function: (The program must be in `Dot Mode` if you want to restrict domains.)

TO GRAPH: $y = -x^2 + 5$ for $2 < x < 4$
ENTER: `DrawF ((-x²+5)*(x>2)*(x<4))`

To shade just part of the area between two functions:

TO GRAPH: The shaded region above $y=x + 2$ and below $y = (x - 3)^2 + 5$ between $x=2$ and $x=4$
ENTER: `Shade (x+2,(x-3)²+5,2,4,1,1)`

To write the equation of an ellipse: (You must first solve the equation for y)

TO GRAPH: $y = (x - 2)^2 / 16 + (y - 3)^2 / 9 = 1$; center (2, 3); width = 4 + 4; height = 3 + 3
ENTER: `DrawF 3-√(9(1-(x-2)²/16))` and `DrawF 3+√(9(1-(x-2)²/16))`

See the calculator manual for more details on graphing these and many other functions.

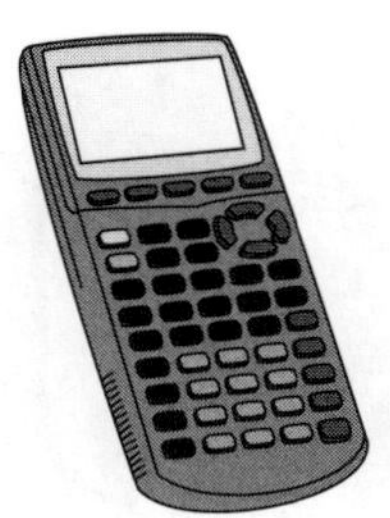

CALCULATOR LOGOS

STUDENT WORK CONTRIBUTED BY
RACHEL FOX & RONE DAVIS, CALCULUS STUDENTS
BISHOP LUERS HIGH SCHOOL, FT. WAYNE, IN

The Taz Vacuum Company was founded in 1974 by a group of Looney Tunes! Mr. Bugs was the head of the creative department that came up with the slogan for the company. Everybody wants a vacuum that can sweep up anything. Well, the Taz Vacuum can sweep up anything from a crumb to a tree. It is a high voltage, intense vacuum, whose growls and snarls make up the amazing product that has made it a household name!!

Editors Note: Due to space restrictions, we were unable to include the students' entire project. The sample listed here was only one of four programs, which were integrated to create the title page, the TAZ drawing and an animated hurricane.

For more information on this assignment you can contact Diane Karst (DKarst@aol.com).

```
ClrDraw
Line(7.861,2.829,7.635,2.829)
Line(7.937,2.727,7.635,2.321)
Line(8.163,2.727,7.861,2.321)
Line(8.388,2.727,8.087,2.321)
Line(3.111,2.829,3.262,2.829)
Line(3.036,2.727,3.337,2.32)
Line(2.81,2.727,3.111,2.321)
Line(2.583,2.727,2.885,2.321)
Line(5.144,5.471,5.75,5.471)
Line(4.537,5.065,4.941,5.065)
Line(5.043,4.963,5.75,4.963)
Line(5.851,5.065,6.35,5.065)
Line(5.851,5.369,5.952,5.065)
Line(4.739,5.979,5.346,5.979)
Line(5.952,6.385,6.255,6.182)
Line(4.915,6.182,5.027,6.182)
Line(5.697,6.182,5.808,6.182)
Pt-On(5.448,6.69)
Pt-On(5.448,6.589)
Pt-On(5.245,6.589)
Pt-On(5.245,6.69)
Pt-On(5.524,6.792)
Pt-On(5.599,6.894)
Pt-On(5.043,5.573)
DrawF (((X-5.5)^2+2.25)*(X≥4)*(X≤7))
DrawF (-(X-5.5)^2+1.75)
DrawF (((.07ln(X-5.5))+6.5)*(X≤6.2))
DrawF (((-(.08ln(X-5.5)))+5.8)*(X≤6.2))
DrawF ((.5(X-6)^3+6.5)*(X≥5.5)*(X≤7))
DrawF ((.35/(-(X-6.1)^2)+6.7)*(X≤5.43)*(X≥4.7))
DrawF ((-.5(X-5)^3+6.5)*(X≥4.06)*(X≤5.5))
DrawF ((-√(.2(X-4.5))+6.2)*(X≤4.8))
DrawF ((√(.2(X-4.5))+6.2)*(X≤4.8))
DrawF ((sin(.5(X-2)+5.5)*(X≥5.9)*(X≤6.4))
DrawF ((5(X-6.1)^2+6)*(X≥6.1)*(X≤6.3))
DrawF ((sin(.5(X-1.9)+5.5)*(X≥5.9)*(X≤6.4))
DrawF ((-(X-5.5)^2+2.5)*(X≤4.8))
DrawF ((-(X-5.5)^2+2.5)*(X≥6.2))
DrawF ((((X-5.5)^2+1.75))*(X≥3.5)*(X≤4.8))
DrawF ((((X-5.5)^2+1.75))*(X≤7.45)*(X≥6.2))
DrawF ((-(X-5.5)^2+9)*(X≥3)*(X≤3.563))
DrawF ((-(X-5.5)^2+9)*(X≥7.409)*(X≤8))
DrawF ((-(X-5.5)^2+11.25)*(X≥2.55)*(X≤3.15))
DrawF ((-(X-5.5)^2+11.25)*(X≥7.83)*(X≤8.46))
DrawF ((√(1-(((X-5.44)^2)/.3))+4.8)*(X≥5.06)*(X≤5.83))
DrawF (((√(1-((-(X-5.44)^2)/.3)))+29)*(X≥5.06)*(X≤5.83))
DrawF ((X-.5)*(X≥5.45)*(X≤5.9))
DrawF (((.5X^2-.5X)-9.75)*(X≥5.95)*(X≤6))
DrawF (((-√(1-(((X-5.45)^2)/7)))-6)*(X≥4.5)*(X≤6.3))
DrawF (((X-5.5)^2+4.85)*(X≥4.065)*(X≤4.35))
DrawF (((X-5.5)^2+4.85)*(X≥6.654)*(X≤6.937))
DrawF ((√((X-3))+5.2)*(X≤4.282))
DrawF ((-.4(X-6.7)^2+6.25)*(X≥6.654)*(X≤7.861))
```

TAZ VACUUM COMPANY

EATS UP EVERYTHING IN ITS PATH

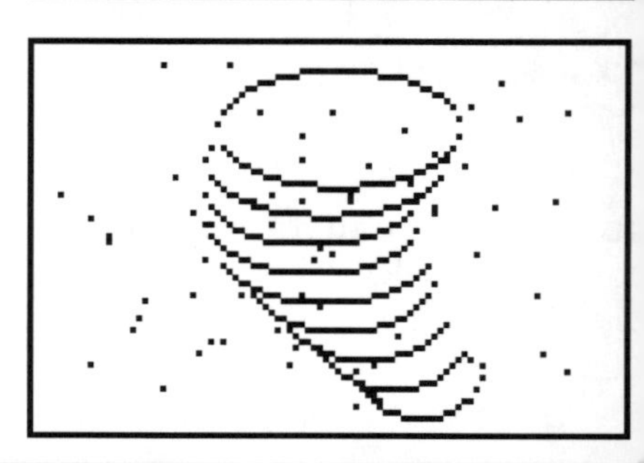

ORIGINAL WORKS
A SERENDIPITOUS PROJECT TREASURE

Select a topic that has been studied in the course, and create an original application or presentation on it.

These are the instructions for an assignment called "Original Works." At the completion of each semester grading period, students make a connection between a concept that they have studied and something that interests them. The assignment is intentionally unstructured to allow for creativity and ingenuity. We do, however, offer examples of past works, most of which are displayed on the classroom walls.

Most students simply recreate a problem that has been done in class, but within a different context. Occasionally, hey create a poster board presentation of an actual or hypothetical problem they solved. We have also had tudents create videos, make wood and clay sculptures and draw instructional comic strips. To show the quality of work that we have received on these projects, here are three samples from the Original Works Hall of Fame.

Our first sample is an area calculation of the United States. Chris, a geometry tudent, had photocopied a map of the continental United States, and traced it onto graph paper. He then applied Pick's algorithm for approximating the area of an irregular region. He compared the map's legend to the width of each grid quare on the graph paper. He then estimated the number of grid squares contained within the map of the United States. With these pieces of information he arrived at an answer that was within 137 square miles of the actual area. On he day the assignment was due, Chris was distressed at this disparity. The instructor old him to relax and explained how to calculate his margin of error. Chris was within four-housandths of one percent. (Then they discovered that the tip of Texas was not on the page.)

Hall of Fame entry number two is Jimmy's Golf Ball. Jimmy, a sophomore geometry student, chose to calculate the volume of a golf ball. The highlight of this story is Jimmy's method of measurement. He measured the circumference with a tape measure to determine the radius and thus the volume of the ball. He then wished to subtract the volumes of the dimples which he assumed to be hemispheres. He first counted all the dimples on the ball by marking each with a black pen. In order to calculate their volume, he filled one with his mother's nail polish, dipped a needle into it, and measured the polish mark on the needle to obtain the dimple's radius. He presented his calculations on a very eye-catching poster board display.

Our third Original Work involves Chad, an unmotivated senior in a freshman Algebra class. A few weeks shy of graduation, Chad was failing because he had not completed his final Original Work assignment. His instructor assured him that he would pass the class and graduate if Chad would submit an acceptable project. Chad then offered an idea. He wanted to increase the diameter of the cylinders of his car engine, and needed to know how much greater its fuel capacity would be. He had to learn quite a bit in a relatively short time in order to answer this question. The true impact of this assignment was evident in his project write-up. Chad wrote, " I have called several machine shops in the past and have asked them how many cubic inches are in my engine with the bore. Not a single one could answer my question even though they specialized in that field. Perhaps if they took the time to understand how to calculate a problem like this one, they could be more educated in their field."

As the anecdotes show, students can, through their Original Works, demonstrate what they know and what they can do in ways that they might not be able to on a traditional test. The project also reveals the impact that a course has made on the student. The most rewarding characteristic of the Original Works is that it creates dialogue between the student and teacher that would not have occurred otherwise. When the assignments are due, we show all of them to the class, and display the more exciting ones on the walls. The days that we share the Original Works are some of the year's best.

Geometry Lessons

HOW HIGH? A Comprehensive Geometry Unit Measuring the Heights of Everyday Objects

Submitted by Greg Rhodes
Trabuco Hills High Shool
Mission Viejo, CA

"How High?" is an extensive, long-term unit centered around determining the heights of tall objects using geometric principles and a variety of measurement tools. Some of the major themes of the unit are: similar triangles, special right triangles, trigonometry, ratio and proportion, measurement, and error analysis. This unit is composed of five projects. Although it is designed to be completed over several weeks, the lessons can be used separately as curriculum or time demands. However, when done together, they can lead to a much deeper understanding of the geometric principles involved. The five lessons in this unit are:

1. **Beyond a Shadow of a Doubt:** Calculating Height through Shadows and Similar Triangles
2. **The "Right" Stuff:** Determining Height with 45°-45°-90° and 30°-60°-90° Right Triangles
3. **Surveyor's Trig Trick:** Determining Height with the Tangent Function
4. **Do You Have the Inclination?:** The Inclinometer: A Tool for Measuring Height using Similar Triangles
5. **The Chinese Difference:** Double-Difference Method for Calculating the Height of an Inaccessible Object

Each of these projects lasts approximately 1-2 hours. They usually require some introduction and demonstration in the classroom, followed by the actual measuring of objects outside the classroom. Your individual schedule and available class time will dictate how much time you are able to spend on each lesson. This unit is designed to develop continuity among the lessons: each has a similar procedure, the students can be in the same group for each project, and the same object can be measured with each different method.

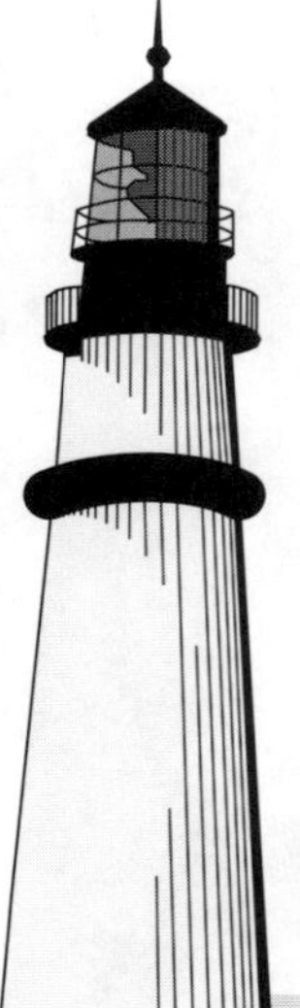

FOLLOW-UP ACTIVITIES & ASSESSMENTS

Option #1 - Portfolio: Have each student turn in a portfolio of the How High unit, including all labs. Included should be a 1-2 page essay covering the following topics: (1) comparison and contrast of the five methods of height measurement; (2) description of the student's growth as a mathematician through the project; and (3) description of the student's best and worst project.
Option #2 - Final Project: Have students, with a partner, find an object and measure it using at least two of the methods. Students should include charts, calculations and explanations.
Option #3 - Practicum Test: Choose an object for the students to measure (the instructor must know the exact height). With a partner, the students are required to measure the object using any method(s) they wish. They will be graded on accuracy of their calculated measurement. Set an appropriate time limit for each phase of the assignment (for example,15 minutes for measurement and 35 minutes for calculation).

HOW HIGH? Beyond a Shadow of a Doubt

Concepts
Similar triangles, ratio and proportion, measurement

Time: 1- 2 hours

Materials
Tape measures, student handout.

Preparation
Choose objects on campus to measure.

The Shadow Project utilizes SIMILAR TRIANGLES to calculate the height of a given object. The ratio of the object's height to the length of the object's shadow is equivalent to the ratio of the student's height to the length of the student's shadow. Being able to measure three of these components, the student can use a proportion to solve for the height of the object. After modeling the procedure for the students, assign them three objects for which they are to find the height (i.e. tree, lamppost, building). Also, have the students calculate the height of a short object, such as a trash can, and then measure it to verify the accuracy of their method. You may also have the students prove that these triangles are indeed similar by the Angle-Angle Similarity Theorem.

HOW HIGH? Beyond a Shadow of a Doubt

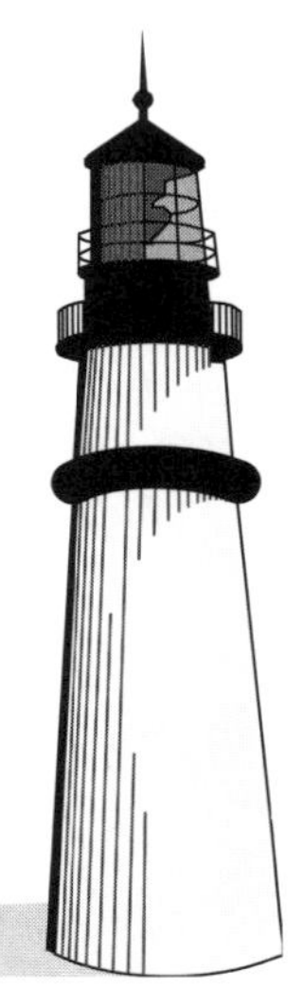

There are several methods to mathematically determine the height of an object. One of those methods involves measuring the object's shadow. It is assumed that the sun's rays strike the earth in parallel, forming a triangle with the ray, the object's height and it's shadow. Any such triangle will be similar to that of another object in the same location, as shown in the diagram below.

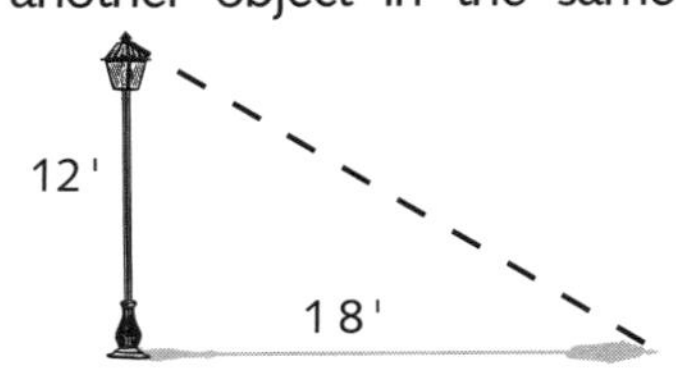

$$\frac{6}{9} = \frac{12}{18}$$

$$\frac{2}{3} = \frac{2}{3}$$

Test this principle of similar triangles among objects with an object that has a height that you can actually measure (short object). Compare the ratio of your height to your shadow (or partner in the group) to the ratio of the object's height to the object's shadow. If these ratios are equal, then measure the shadow of three other objects and estimate their heights by using your new found ratio. Record your measurements and calculations to the chart below.

	Object	Your Height	Your Shadow	Object's Shadow	Object's Height
1					
2					
3					
4					

In the space provided for each object, draw and label the pair of similar triangles and corresponding proportion that you used to calculate the height of the object. Show your calculations.

Object #1
(small object)

Object #2

Object #3

Object #4

Prove geometrically that these triangles will indeed be similar, given that the sun's rays are parallel.

Given: $BC \parallel EF$
$BA \perp AF$, $ED \perp AF$
Prove: $\triangle ABC \sim \triangle DEF$

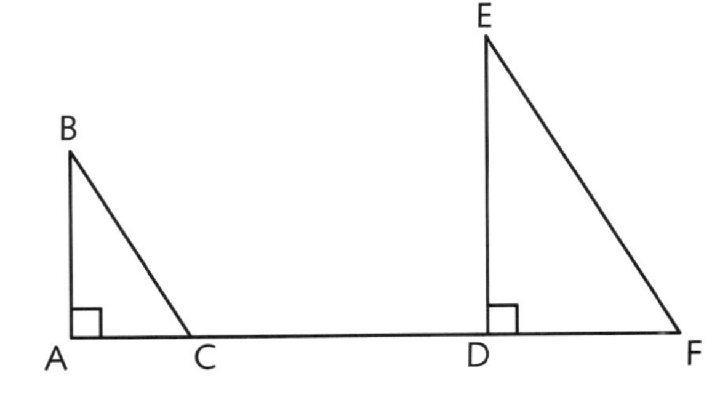

HOW HIGH? The Right Stuff

Determining Height Using 45°-45°-90° and 30°-60°-90° Triangles

In this lesson, students will measure the heights of various objects with 45°-45°-90° and 30°-60°-90° special right triangles. By utilizing the ratios of the sides of the special right triangles, students can take three different measurements that can all be used to separately calculate the height of an object. For example, in the diagram below, assume that a student with an eye height of 65″ is measuring the same object from the three different angles.

45° - 45°
$d = 100''$
$P = 100''$
$T = 100 + 65 = 165$

60° - 30°
$d = 58''$
$P = 58\sqrt{3} \approx 100.4$
$T = 100.4 + 65 = 165.4$

30° - 60°
$d = 173''$
$P = 173\sqrt{3} / 3 \approx 100$
$T = 100 + 65 = 165$

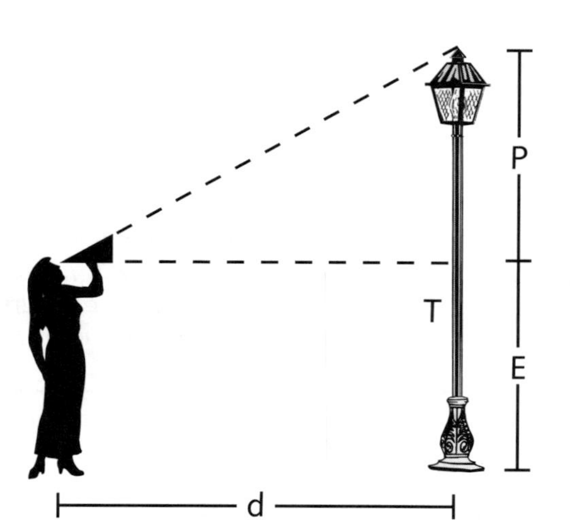

Concepts
Special right triangles, radicals, measurement.

Time: 1- 2 hours

Materials
45°-45°-90° and 30°-60°-90° cardboard triangles (no smaller than 6 inches along a leg), tape measures, straws

Preparation
Students should cut out the triangles ahead of time. Each group needs one of each type of triangle. Choose 2-3 objects around the school to measure (flagpoles, prominent trees, buildings, etc.)

LESSON PLAN

1. Demonstrate the triangle tools used in this project. The instructor may assign the creation of the three cardboard triangles for homework. This is an excellent exercise in helping the students to understand these special ratios, but their lack of accuracy will be an issue when it comes to actually conducting the lesson. Therefore, it is advisable to have a class set available.
2. There are three types of height measurements which the students will perform:
   a) Looking along the hypotenuse from the 45° angle, measure the distance to the object. This distance corresponds to one of the legs of the triangle. This is equal to the length of the other leg, since the two legs of a 45°-45°-90° triangle are equal.
   c) Looking from the 60° angle, measure the distance to the object. This provides the short leg, from which one can calculate the long leg by multiplying by $\sqrt{3}$.
   b) Looking from the 30° angle, measure the distance to the object. This provides the long leg of the triangle, from which one can calculate the short leg by dividing by $\sqrt{3}$.
3. Demonstrate an example of a height measurement in class. Measure the height of a door or wall with one of the tools. Remind the students to add the eye height at the end.
4. Go to the first object and measure according to the instructions. Once all the students have taken their measurements, move on to the next object. You should spend about 10 minutes per object. Spend enough time for the students to take their measurements, but not start their calculations.

TEACHER COMMENTS

- The most common mistake students will make is not having the triangle level with the ground, so monitor them closely.
- To avoid crowding with a large class, split the groups between two nearby objects and switch when they finish the measurements.

HOW HIGH? The Right Stuff

Determining Height Using
45°-45°-90° and 30°-60°-90° Triangles

Since the special right triangle ratios are based on similar triangles (e.g. all 45°-45°-90° triangles are similar), these triangles may be used to determine unmeasureable heights. Use the cardboard triangles to determine the height of three objects by following the instructions below.

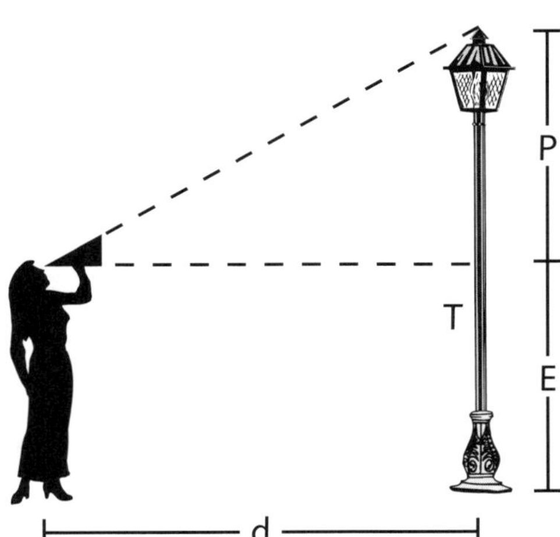

1. Measure the height of your eyes or one person in your group.
2. Look along the hypotenuse of your triangle and move forward or backward to line up the top of your object. REMEMBER: It is very important to keep the bottom edge of the tool parallel with the ground. Have your partner make sure the triangle is level.
3. Once this is achieved, measure the distance from where you are standing to the base of the object.
4. Take this measurement and perform the appropriate calculation; this gives you the preliminary height. Finally, add the eye height to determine the total height of your object.

45° - 45°

Object	Distance to Object	Preliminary Height	Eye Height	Actual Height

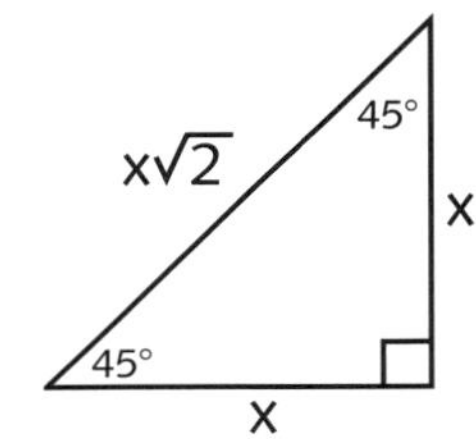

60° - 30°

Object	Distance to Object	Preliminary Height	Eye Height	Actual Height

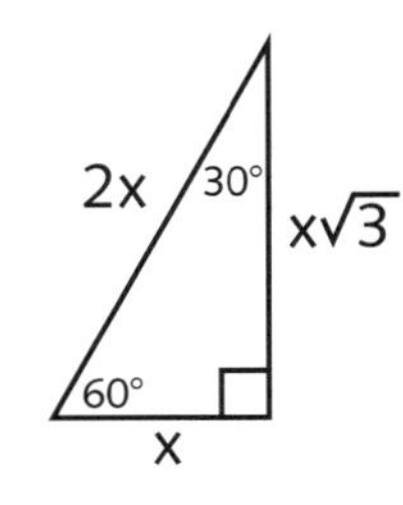

30° - 60°

Object	Distance to Object	Preliminary Height	Eye Height	Actual Height

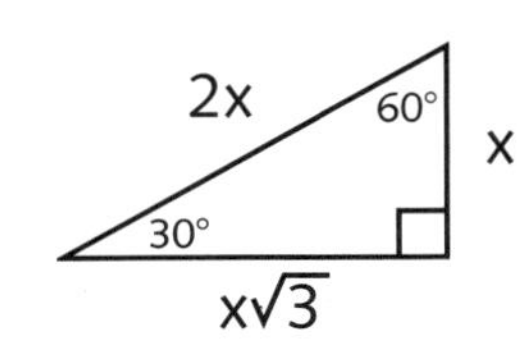

For each object above, on the backside of this sheet, draw and label the appropriate special right triangle diagram and show the corresponding calculation used to determine the height of the object.

HOW HIGH? Surveyor's Trig Trick

Determining Height Using the Tangent Function

In this lesson, students will measure the heights of various objects using angle measurement and trigonometry. If students can measure how far they are from an object and the angle of inclination of their line of sight, they can use the tangent function to determine the height of the object at which they are looking. For example, assume the student has an eye height of 65″ and is standing 17 feet (204″) from a lamp post looking up at the top of the lamppost at a 25 degree angle. The total height, T, is the sum of the preliminary height, P, (from the eye up) and the eye height, E, (from the eye down).

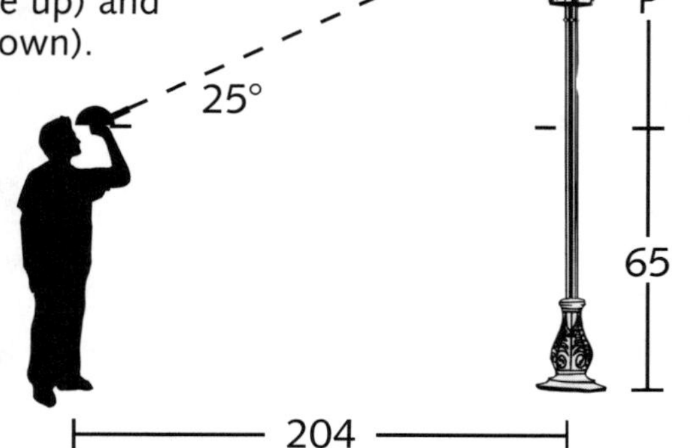

$\tan(25) = {}^{P}/_{204}$
$0.46 = {}^{P}/_{204}$
$P = 0.46(204)$
$P = 95$

$T = P + E$
$T = 95 + 65$
$T = 150''$ (12.5 feet)

Concepts
Trigonometry, tangent function, angles, angle measurement, right triangles, measurement

Time: 1- 2 hours

Materials
Protractors, straws and tape measures

Preparation
Choose 3-5 objects around the school to measure (flagpoles, prominent trees, buildings, etc.)

LESSON PLAN

This is the instrument used for this lesson:

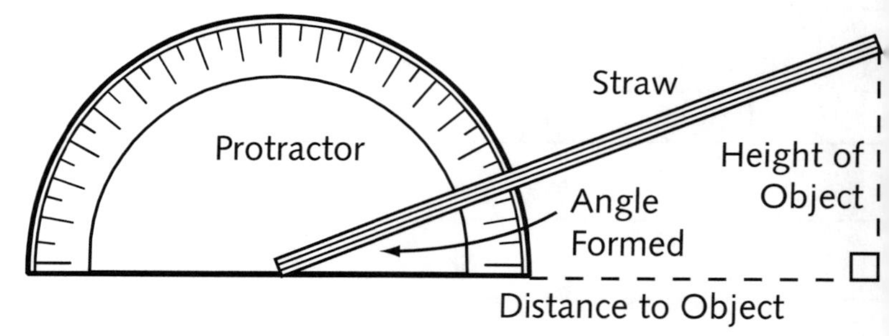

1. Introduce the project and describe the protractor/straw tool used. It is best to have an example to show the students. This is a crude model of an actual instrument, thus the measurements are approximations only.
2. Review the tangent function with a diagram on the board. Emphasize that the tangent of an angle is the ratio of the opposite side and the adjacent side. Show how you can solve for the height (the opposite side) if you know the horizontal distance (adjacent side) and the angle. They will be doing this for the lesson.
3. It is helpful to do a brief example of a height measurement in class to show the students. Measure the height of a door or wall with one of the tools. Show the calculations necessary to determine the height.
4. They are now ready to measure the objects. Go to the first object and let the students begin measuring. Remind them that they can measure from any distance. You can even ask the students to measure the same object from several distances to show that the height does not change.
5. While they measure, suggest to the students that one of them should look through the straw while the other keeps the protractor level, and others in the group may then measure the distance.
6. During the debriefing portion of the activity, emphasize to students that the smaller angle generated a smaller ratio, because the horizontal distance (the denominator) is larger (further away from the wall) at the smaller angles.
7. Also stress the meaning of the decimal version of the ratio. For example, 0.75 means that the vertical height is 75% of the horizontal distance.

HOW HIGH? Surveyor's Trig Trick

Determining Height Using the Tangent Function

Surveyors (those hard-working people in the orange vests along the roadside or at a construction site) often use trigonometry to measure unreachable distances. Similarly, you are to use the tangent function to calculate unreachable heights.

1. Measure the height of your eyes.
2. Look through the straw towards the top of the object.
3. Have your partner hold the protractor against the side of the straw. Be sure the straw passes through the center of the protractor. Be careful to keep the protractor level with the ground. Read and record the angle formed by the straw.
4. Measure the distance from you to the object. Use this distance and the tangent ratio for the measured angle to determine the height of the object.

A typical situation looks like this...

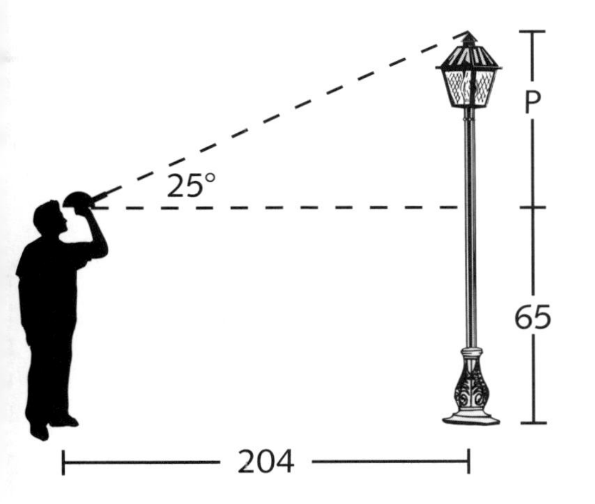

The instrument for this lesson will look like this...

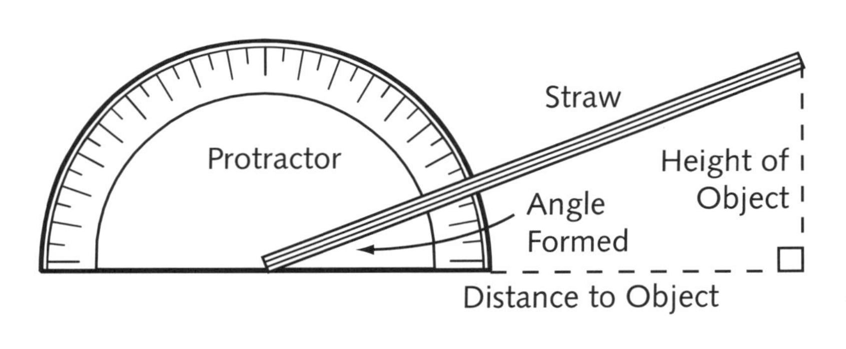

Object	Angle Formed	Tangent of Angle	Distance to Object	Preliminary Height	Eye Height	Actual Height

For each object above, on the backside of this sheet, draw and label the appropriate triangle diagram and show the corresponding calculation used to determine the height of the object. Be sure to show the measured angle and its tangent ratio.

HOW HIGH? Do You Have the Inclination?

The Inclinometer: Measuring Heights Using Similar Triangles

In this lesson, students will measure the heights of various objects through the use of a "mechanical" tool called an inclinometer. An inclinometer is a cardboard measuring device that allows the user to calculate the height of an object using similar triangle proportions.

Concepts
Similar triangles, proofs, ratio and proportion, measurement.

Time:
1- 2 hours (depending on student preparation time)

Materials
Class set of Inclinometers (may be constructed by students) and tape measures

Preparation
The students should construct the inclinometer (according to the handout) ahead of time. Choose 2 objects around the school to measure (e.g. flagpoles, prominent trees, buildings).

HOW TO MAKE AN INCLINOMETER

Materials Needed: Cardboard, ruler, scissors, 12 inch string, straw, small weight (nut, bolt, etc.), tape. The instructor may construct these or have each group of students make one.

1. Cut out a cardboard rectangle at least 8" x 6".
2. Tape a straw along one edge (herein referred to as top) of the cardboard.
3. Mark the bottom and right edges of the cardboard every 1 cm. Begin with the bottom-right side as (0,0) and count up and to the left.
4. Attach one end of the string to the upper right corner of the cardboard. Cut enough string so that it reaches the lower left corner. Tie the weight onto the bottom of the string. See the diagram on the student handout.

LESSON PLAN

1. Walk through the diagrams on the worksheet (The Tool & The Process) and discuss why the two triangles are similar. Since this is the crux of the whole lesson, it is important that the students understand why the inclinometer works (i.e. the triangles created are similar). If your class includes geometric proofs, go through the proof listed below with your students.
2. Next, do a brief example of a height measurement in class to show the students how it's done. Measure the height of a door or wall with an inclinometer. Show the calculations necessary to determine the height, emphasizing the similarity proportion.
3. Go to the first object and mark off 2 or 3 distances from the object. For example, if the students are measuring a flag pole, start at the flag pole and measure out 30, 35, and 40 feet. Then have the students choose one of the measurements from which they will measure. During the actual measuring, suggest to the students that while one of them is looking through the inclinometer, another one should keep the string from swinging and take the reading from the inclinometer.
4. Once all the students have taken their measurements, move on to the next object, spending just enough time for the students to take down their measurements, but not start their calculations. When all the objects have been measured, return to class and use the remaining time to let the students work on their calculations.

PROOF OF THE INCLINOMETER'S SIMILAR TRIANGLES

	Statement	Reason
1.	$CA \parallel XY$	Given in diagram
2.	$\angle XZC \cong \angle X$	Alternate Interior Angles Congruent
3.	$XZ \parallel BA$	Given in diagram
4.	$\angle XZC \cong \angle A$	Corresponding Angles Congruent
5.	$\angle B \cong \angle Y$	All Right Angles are Congruent
6.	$\angle X \cong \angle A$	Transitive (2, 4)
7.	$\Delta XYZ \sim \Delta ABC$	AA Similarity Theorem

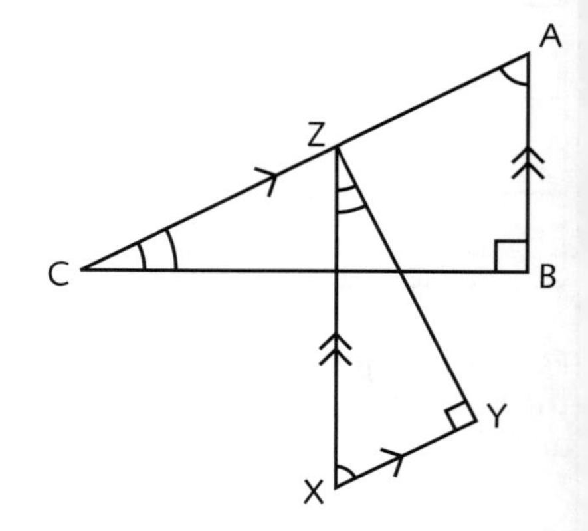

HOW HIGH? Do You Have the Inclination?

The Inclinometer: Measuring Heights Using Similar Triangles

The Tool

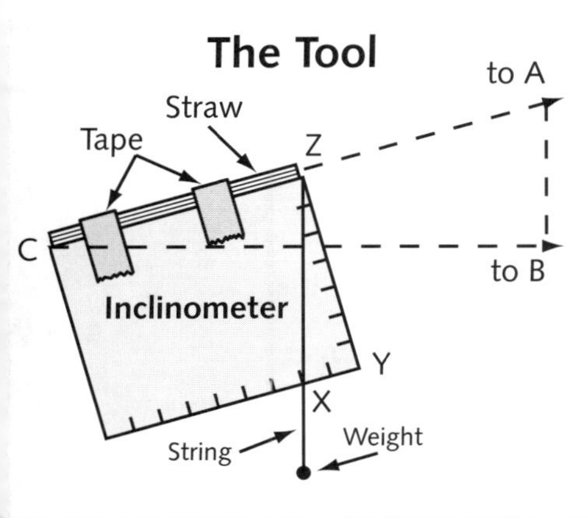

The Process

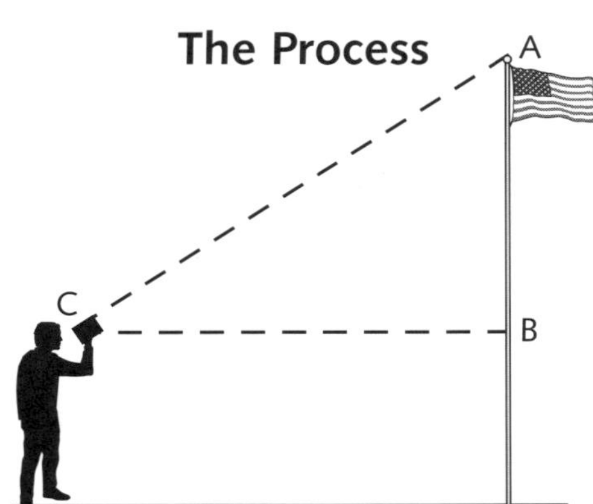

The Explanation

1. $\angle B$ & $\angle Y$ are right angles
2. $\angle X \cong \angle A$ (to be proven)

Thus, because of the Angle-Angle Similarity Theorem:

$\Delta XYZ \sim \Delta ABC$

The Equation

$$\frac{XY}{ZY} = \frac{AB}{CB}$$

$$AB = \frac{XY \bullet CB}{ZY}$$

$$\frac{\text{Distance Marked by String on Bottom of Inclinometer}}{\text{Height of Inclinometer}} = \frac{\text{Preliminary Height of Object}}{\text{Distance to Object}}$$

$$\textbf{Preliminary Height of Object} = \frac{\text{Distance Marked by String}}{\text{Height of Inclinometer}} \cdot \text{Distance to Object}$$

Finally, to find the TOTAL HEIGHT, add the preliminary height to the individual's eye height.

How to Use an Inclinometer

1. Measure the height of your eyes. You will use this piece of information later in your calculations.
2. Look through the straw and line up the top of your object through your straw. The string should be hanging straight down. Have your partner record where the string is crossing the bottom of the inclinometer.
3. Mark where you were standing and measure the distance from that spot to the base of the object.
4. Plug in all the appropriate numbers into the Data Collection Table, set up your proportions, solve for the preliminary height, and finally add the eye height.

Object	Height of Inclinometer (YZ)	Distance Marked by String (XY)	Distance to Object (BC)	Similarity Proportion	Preliminary Height (AB)	Eye Height	Actual Height

HOW HIGH? The Chinese Difference

The Double-Difference Method of Calculating the Height and Distance of an Inaccesible Object

Lesson adapted from: Multicultural Mathematics: Teaching Mathematics from a Global Perspective. *Nelson, David, et al. New York: Oxford Press. 1993*

In this lesson, students will measure the height of a very tall, prominent object through the use of an ancient Chinese method of measuring heights. They will be repeating the original experiment by taking a variety of measurements and then plugging them into two ingenious formulas to find the height and distance of the object.

Concepts
Similar triangles, measurement, evaluating formulas

Time: 2 hours

Materials
Large poles, or boards (at least 4-5' long), towels, tape measures

Preparation
Choose a large, prominent object visible from school (skyscraper, church steeple, mountain, etc.), and a location on campus where the lesson is to be conducted.

LESSON PLAN

The process consists of setting a pole or large stick at a certain location, lying down on the ground, lining up the top of the pole with the top of the object, and then measuring that distance. This is done twice, at two different spots. These numbers are then plugged into two formulas to find the height and distance of the object. See the student handout for more details.

DAY ONE: In Class

1. Distribute the handout and introduce the students to the Double-Difference Method and its background. Inform the students that you will be using this method to measure the height of your specific object. Walk the students through the process involved in this experiment (found on student handout). Explain all the steps in detail.

2. Introduce the two formulas used in this experiment. It is not necessary for the students to understand the origin of the proportions or their consequent derivations, but they should know that the proportions are formed using similar triangles and then algebraically manipulated to solve for a certain variable. More advanced classes can spend more time on the symbolic algebraic manipulation.

3. Have the students form groups of four and trade phone numbers with one another. Each group is responsible for bringing its own pole. Students can use long broom handles, 2 x 4's, or anything else that can be carried easily. Also ask for volunteers to bring in tape measures from home (the more the better). Lastly, warn the students that one person from each group will have to be willing to lie on the ground during the experiment.

DAY TWO: In The Field

5. You are now ready to measure the object. Have the groups collect all the equipment and proceed to the desired location. Follow the instructions given on the student handout. The actual measuring will probably take some time. For the first five minutes, the students will probably be very confused; that is expected. Eventually they will figure things out and start measuring. During this time, walk among the various groups answering questions or looking for errors in calculation or procedure.

6. Once all the groups have finished their measurements, return to class and spend the remaining time working on the calculations.

HOW HIGH? The Chinese Difference

The Double-Difference Method of Calculating the Height and Distance of an Inaccesible Object

Throughout history, mathematicians have used SIMILAR TRIANGLES to calculate the heights of tall objects. However, most of these methods require that the observer know the distance from the object to the point of measurement. In third-century China, a mathematician named Liu Hui developed a means by which he could calculate the height of an inaccessible object, in other words, an object to which you could not measure the distance. His example was a mountain that was on the other side of a river. Hui could not measure the width of the river, but he could still calculate the height of the mountain (in addition to the distance across the water). The process consisted of setting a pole at a certain location, lying down on the ground, lining up the top of the pole with the top of the object, and then measuring the distance to the pole. This is done twice, at two different spots. These numbers are then plugged into the formulas to find the height and distance of the object.

INSTRUCTIONS FOR USING THE DOUBLE-DIFFERENCE METHOD

1. Measure the height of your pole. This distance is labeled h.
2. Choose an initial point at which to place your pole. Place a marker at the base of your pole. You will use the location of your pole in Step 6.
3. Lie down on the ground and move until you line up the top of the pole with the top of your object. It is important that you maintain a straight line between your eye, the pole, and the object.
4. Once you have the top of the object lined up with the pole, mark that spot and measure the distance from it to the pole. This distance is labeled x_1.
5. Move back about 30 - 40 feet behind your initial point. Repeat the process from Step 2. This time the distance is labeled x_2. Notice from the diagram that x_2 should be longer than x_1. If it is not, then there is an error somewhere in your measurements. Double-check x_2 and re-measure x_1 if necessary.
6. Measure the distance between the placement of your first pole and that of your second pole. This distance is labeled d.
7. Plug into the appropriate formulas to calculate the height of the object and the distance to the object.

EXPLANATION

Here is an explanation of how Liu Hui arrived at his formulas. It is based on proportions of similar triangles...

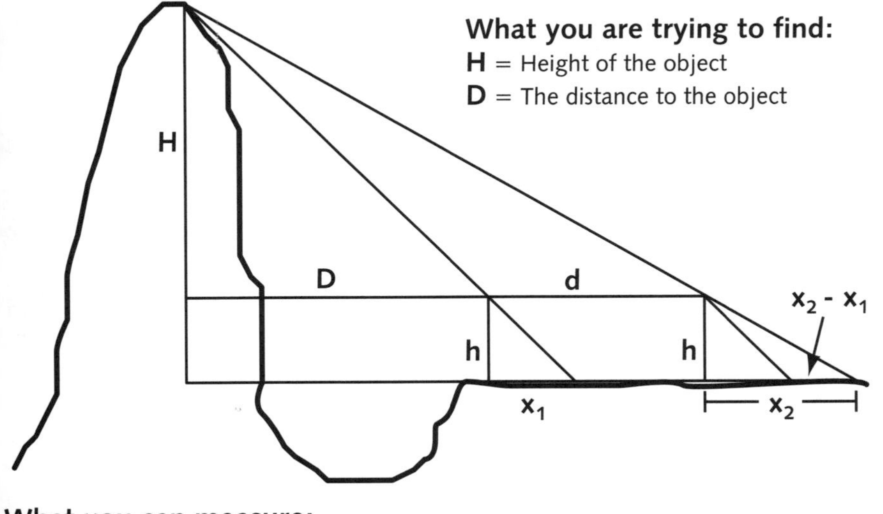

Height of Object

$$\frac{H - h}{h} = \frac{d}{x_2 - x_1}$$

$$h \cdot \frac{H - h}{h} = \frac{d}{x_2 - x_1} \cdot h$$

$$H - h = \frac{d \cdot h}{x_2 - x_1}$$

$$h + H - h = \frac{d \cdot h}{x_2 - x_1} + h$$

$$\boxed{H = \frac{d \cdot h}{x_2 - x_1} + h}$$

What you can measure:

h = Height of your pole
d = Distance between the 1st measurement of your pole and the 2nd
x_1 = The distance between your eyes and the pole at the 1st location
x_2 = The distance between your eyes and the pole at the 2nd location
$x_2 - x_1$ = The difference between these two distances.

Distance from Object

$$\frac{D}{d} = \frac{x_1}{x_2 - x_1}$$

$$d \cdot \frac{D}{d} = \frac{x_1}{x_2 - x_1} \cdot d$$

$$\boxed{D = \frac{x_1 \cdot d}{x_2 - x_1}}$$

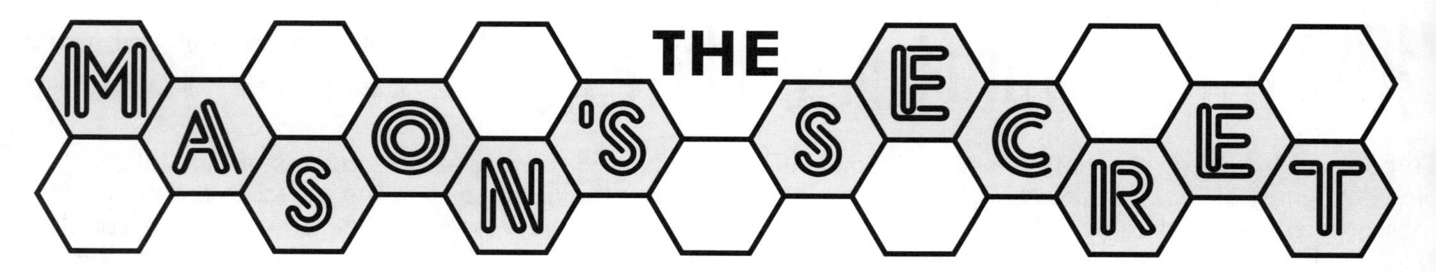

MISUNDERSTANDINGS

Students often have trouble understanding the meaning of the Polygon Sum Theorem (also known as the Interior Angle Formula); usually they simply know how to plug the number of sides into the equation and evaluate. Students also encounter difficulties when they are asked to measure angles with a protractor, especially those greater than 90 and 180 degrees. Another persistent problem is that students often do not see formulas as abstract representations of pervasive patterns in the natural world. This activity helps counter all of these misunderstandings.

Concepts
Polygon Sum Theorem, tessellations and mosaics, measuring angles

Time: 1- 2 hours

Materials
Protractor, student handout.

Preparation
Students should have a familiarity with the Polygon Sum Theorem:
$S = (n - 2) \cdot 180$

DEFINITIONS

Tessellation: A repeated pattern of one figure, also known as the fundamental region. Congruent copies of the fundamental region may be reflected, translated, or rotated (arranged in any position), yet no gaps or overlaps exist in the pattern.
Mosaic: Another type of repeated pattern, similar to the tessellation, except that two or more fundamental regions are perpetuated throughout the pattern.

DEMONSTRATIONS (front side of the student handout)

Compare and contrast the definitions between a tessellation and mosaic (each have a repeating pattern, tessellation contains one fundamental region, mosaic has two). Have students measure the angles in the first hexagon (hourglass). If they have trouble measuring the two angles that are greater than 180 degrees, remind them to measure the acute angle on the exterior of the polygon and subtract it from 360. The sum of the six angles should be close to 720 degrees, as demonstrated by the formula: $(6 - 2) \cdot 180 = 4 \cdot 180 = 720$. The four acute angles are each 60 degrees, the other two are each 240: $4 \cdot 60 + 2 \cdot 240 = 720$.

Once this is complete, show how the sum of all the angles surrounding a single point is 360 degrees. Focus the students on the intersection of the figures in the tessellation, and ask them to measure each angle. In the given example, the angles are 60, 60 and 240, totalling a sum of 360 degrees.

Have the students practice on their own with the second example, mosaic of hexagon hourglasses (the same as the ones in the first example) and regular hexagons. The first shape is the same as the one above and the second will have a sum of 720 also. Each intersection will have a sum of 360.

ASSIGNMENT (back side of the student handout)

Students now have to create their own tessellations and mosaics (one of each) in the spaces provided. For each design, they are to then draw a congruent copy of the fundamental region, measure and record each of the interior angles on the diagram and write the sum in the space provided. They should show the use of the Polygon Sum formula, $(n - 2) \cdot 180$, to calculate the sum measure of all the interior angles. The students should also draw a sketch of each type of intersection in the pattern, showing the measurements of each angle, and their sum (the sum be approximately 360 degrees). Be sure to point out that not all figures tessellate around a point. For example, a regular pentagon will not work because each angle is 108 degrees and 360 is not a multiple of 108.

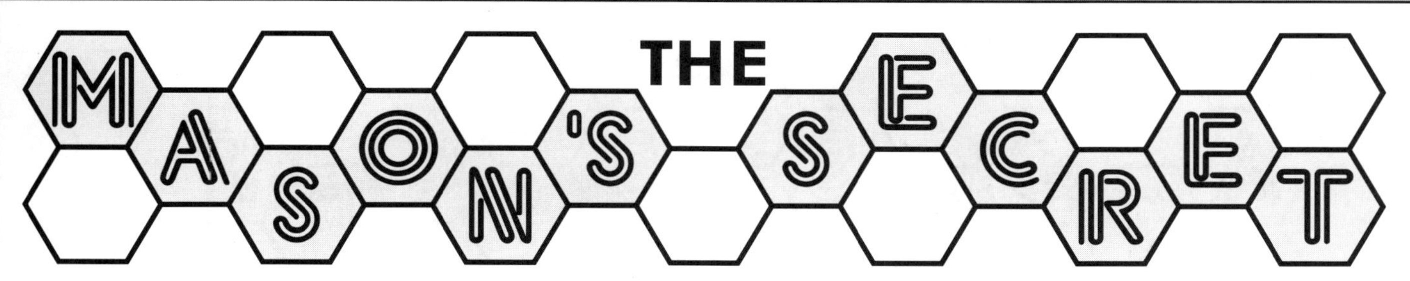

TESSELLATION

Sum of interior angles: ____________
(measured)
Sum of interior angles: ____________
(calculated)

Sum of angles around a point: ____________
(measured)

MOSAIC

Sum of interior angles: ________ ________
(measured)
Sum of interior angles: ________ ________
(calculated)

Sum of angles around a point: ____________
(measured)

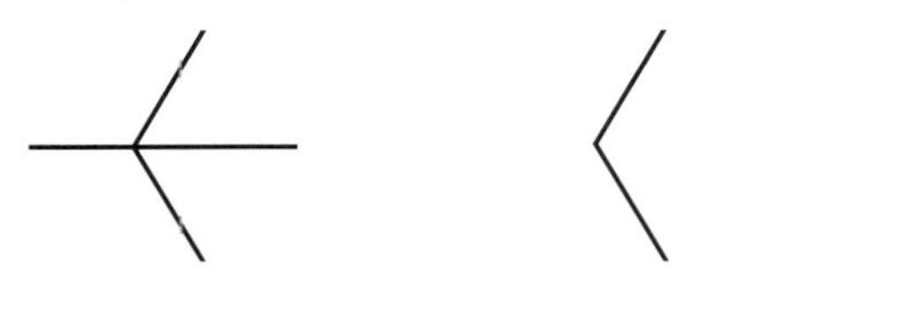

STUDENT HANDOUT

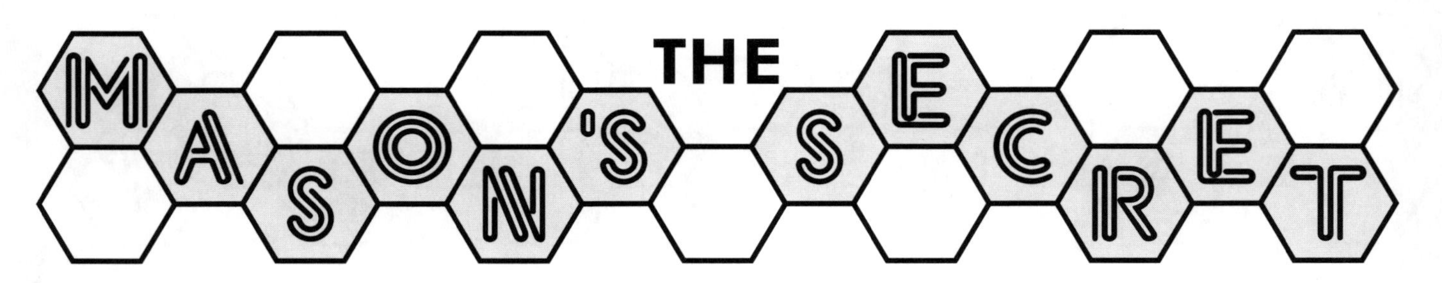

Create your own tessellation and mosaic (one of each) in the space provided. For each design, draw a congruent copy of the fundamental region, measure and record each of the interior angles on the diagram and write the sum in the space provided. Also, draw a sketch of each type of intersection in the pattern, showing the measurements of each angle, and their sum.

TESSELLATION

Sum of interior angles: ____________________
(measured)
Sum of interior angles: ____________________
(calculated)

Sum of angles around a point: ______ ______
(measured)

MOSAIC

Sum of interior angles: ________ ________
(measured)
Sum of interior angles: ________ ________
(calculated)

Sum of angles around a point: ______ ______
(measured)

The America's Cup Sail

Submitted by Jon Hansen, Karen Vega, Jan Vance, and Jeremy Buchholtz of Temecula Valley USD, Temecula, CA

THE TASK

From the information given on the handout, the students are to find the length of each of the seams (14 segments), and each angle (19 of them).

Concepts

Midsegment of trapezoids, Pythagorean Theorem, Polygon Sum Theorem, parallel lines & corresponding angles, triangle congruency proofs. Extensions include similar triangles, areas of trapezoids, and trigonometric functions.

Time: 1hour

Materials

Student Handout.

Preparation

Students should know that the length of the midsegment of a trapezoid is half of the sum of the bases, and that when two parallel lines are cut by a transversal, the corresponding angles are equal.

LESSON PLAN

1. Disseminate the handout and read aloud the given information. Explain the need to find all sides and angles of each of the six panels. (5 stripes and 1 triangle filled with stars)
2. Have the students brainstorm in groups about which mathematical concepts they may use to find these segment and angle measures. Be sure at this point that they make no calculations. "We do NOT want any numbers, yet."
3. Share out. List the student ideas on the board.
4. Students calculate the lengths and measures. The midsegments can be found by averaging the two bases of the trapezoid. The third seam down is $(9 + 1.8)/2 = 5.4$ feet long. The other two are 3.6 & 7.2. Within one trapezoid, the consecutive interior angles will be supplementary and the sum of all four angles will equal 360°. The triangle at the top will have an angle sum of 180°. Therefore, the four angles of each trapezoid will be 53°, 127°, 107° & 73°.

ASSESSMENT: The Proof

Most of the students assume that since the height of each stripe is constant, then all the oblique sides of the panels, along one edge of the sail, are also congruent. This instinct is correct. The challenge for students is to prove it. In the proof, it is given that the heights are congruent (and perpendicular to the seams) and that the seams are parallel.

The proof should show that since the lines are parallel then the corresponding angles should be congruent, $\angle ACB \cong \angle CED$. The right angles are also congruent, therefore, the two triangles are congruent by **A**ngle-**A**ngle-**S**ide, and thus, the corresponding parts will be congruent. Therefore, $\overline{AC} \cong \overline{CE}$.

EXTENSIONS: Revisit The Sail when teaching...

Similar Triangles: The panels can be viewed as overlapping similar triangles rather than trapezoids. The ratio of two consecutive triangles is 1:2. Are the trapezoids also similar?

Area of Trapezoids: Find the area of each trapezoid. Once the length of each base is found, all that is needed is the height. Since each panel is the same height, for a total of 20 feet, the height of each trapezoid is 4 feet.

Trigonometric Functions: Do not offer any of the angle measures (as opposed to giving two as explained earlier) and have students use right triangle trigonometry to find the missing angles.

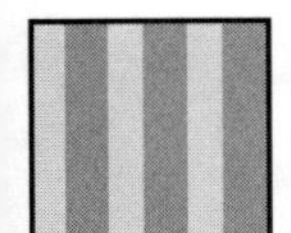

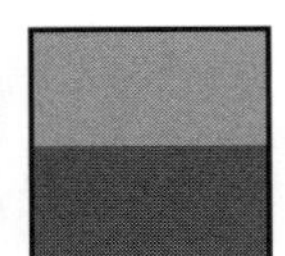

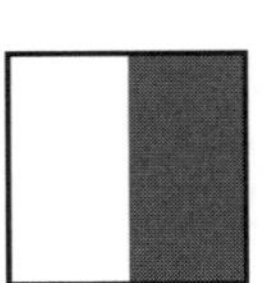

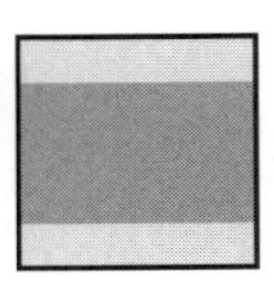

The America's Cup Sail

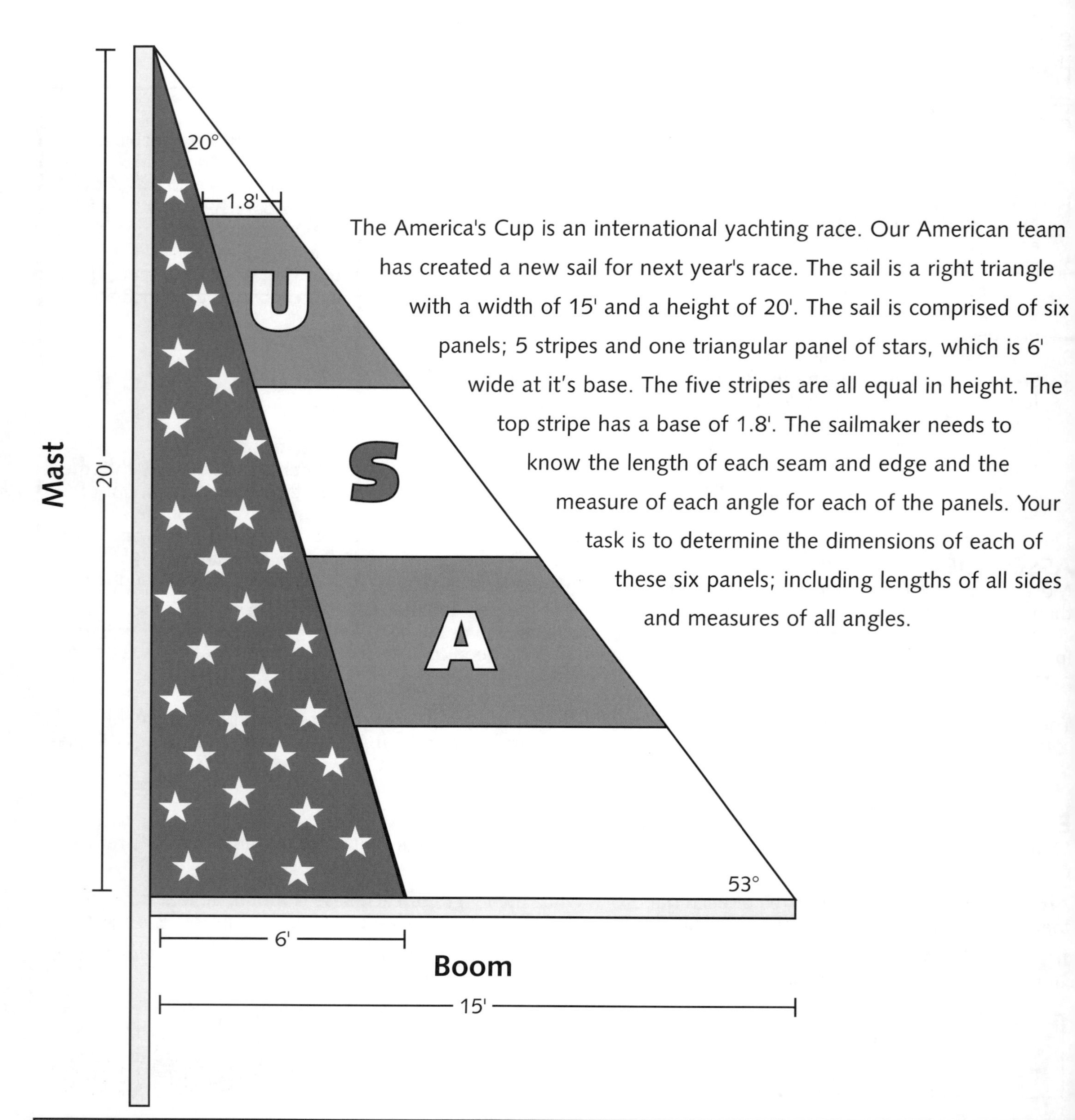

The America's Cup is an international yachting race. Our American team has created a new sail for next year's race. The sail is a right triangle with a width of 15' and a height of 20'. The sail is comprised of six panels; 5 stripes and one triangular panel of stars, which is 6' wide at it's base. The five stripes are all equal in height. The top stripe has a base of 1.8'. The sailmaker needs to know the length of each seam and edge and the measure of each angle for each of the panels. Your task is to determine the dimensions of each of these six panels; including lengths of all sides and measures of all angles.

The America's Cup Sail

Prove that the oblique segments along the side of the sail are congruent.

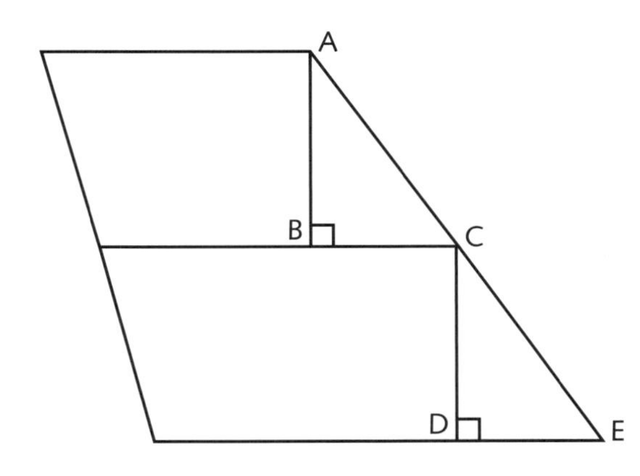

Given: $\overline{AB} \cong \overline{CD}$, $\overline{BC} \parallel \overline{DE}$
$\angle ABC$ & $\angle CDE$ are right angles

Prove: $\overline{AC} \cong \overline{CE}$

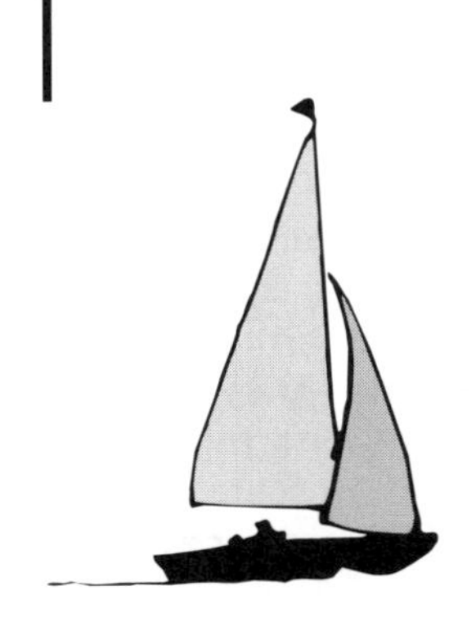

THE FIRE-FISH STORY

Submitted by Don Stoll, El Toro High School, Lake Forest, CA

One of the primary benefits that students receive from learning to write geometric proofs is the development of deductive reasoning. Unfortunately, the inherent lack of this ability to follow a logical argument often precludes students from grasping the underlying concepts behind the proofs. A simple and fun project to help students understand this type of reasoning is "The Fire-Fish Story." The project requires students to prove the following premise:

"If there is a fire, then a fish dies."

Concepts
Two-column geometric proofs.

Time: 1hour

Materials: None

Preparation: None

They are to do this by writing a creative story consisting entirely of conditional statements. The first statement should be of the form: "If there is a fire, then A."The second statement then should be of the form: "If A, then B." The hypothesis of each statement must be the conclusion of the previous statement. The story must be of at least 5 conditional statements, ending with "If D, then a fish dies." Each student's task is to create the conditions of the story, so that it may be logically assumed that "If there is a fire, then a fish dies."The stories should be written in the following format.

If *there is a fire*, then A.
If A, then B
If B, then C
If C, then D
If D, then *a fish dies*.

Let them have fun. The students may suspend realism, but not logic. In other words, a fish may leap from a bowl and out a ten story window; however, it may not leap out of the bowl and end up on the moon — unless, the student can create enough conditional statements to logically explain how the fish got from the bowl to the moon. Have volunteers read their stories aloud in class. It will be quite a treat. Here are couple of student samples...

From a student who understands,
If there is a fire, then the house is burning.
If the house is burning, then the family runs out.
If the family runs out, then they knock over the fish bowl
If they knock over the fish bowl,
then the fish suffocates on the floor.
If the fish suffocates on the floor, then the fish dies.

Conclusion: *If there is a fire, then a fish dies.*

From a student who doesn't understand,
If there is a fire, then a fish dies.
If a fish dies, then the bears have nothing to eat.
If the bears go looking for food,
then they will lose their way.
If the bears lose their way, then they will be sad.
If the bears are sad, then they will attack people.

Conclusion: *This is why some bears attack people.*

The second student's response shows the value of this assignment as an assessment tool. If the student has trouble constructing a logical story about fish, how is the student expected to create a formal proof regarding a geometric theorem? It would be best to have this student rewrite this assignment several times to understand the general flow of logic that deductive reasoning requires. For the first student who possesses a sense of logic, this assignment may be referred to when the teaching in the formal proofs begin. In the proofs, the "Given" is equivalent to "If there is a fire," and the "Prove" is the equivalent to "then a fish dies. It is the student's job to complete the mathematical story that links these two ideas.

BASEBALL CONGRUENCY

LESSON PLAN

PART ONE: Lake Holattawatta

The purpose for this portion of the project is to give the students practice at mapping out the segments, and for exploring the principles behind the application. Discuss with the students why the diagram on the student handout, as given, will NOT work. Most of them will tell you that the segments on either side of the marker are not congruent. Model an example in which the marker is the midpoint of each longer segment, and thus producing two congruent triangles, which in turn produces a distance equal to that of the length of the lake. Then have the students produce their own examples, one for the length and one for the width. Display some student solutions, then have them do it again on the second drawing of the lake, with the marker in a different position.

Use the diagram below to reinforce the idea of congruent triangles. The vertical angles will always be equal; $m\angle CMD = m\angle BMA$. We establish the marker as the midpoint of the longer segments; CM = BM and AM = DM. Therefore, we have congruent triangles by the Side-Angle-Side Theorem of Congruency; $\Delta CMD \cong \Delta BMA$. Since "corresponding parts of congruent figures are congruent" (CPCFC), then the representative distance of the lake is equal to its true measure; CD = BA.

Concepts
Triangle congruency, specifically the "Side-Angle-Side" Theorem

Time: 1hour

Materials
Student handout, class set of rulers, baseball field

Preparation
Students should already be familiar with congruent triangles and the Vertical Angle Theorem (all vertical angles are congruent). Get permission from the baseball coaching staff to use the field. Experience says to be sure that sprinklers are not scheduled to go on during your class time.

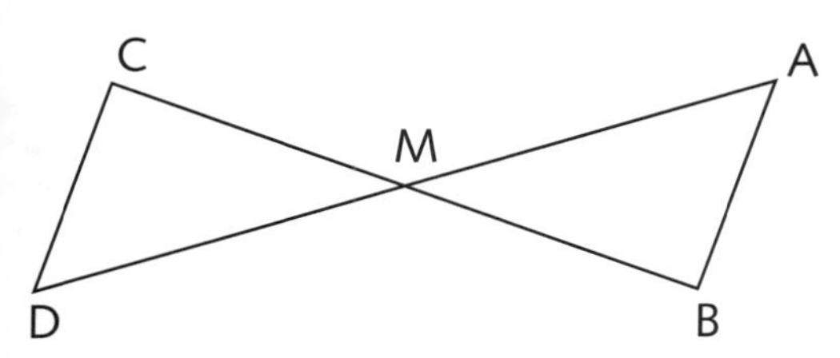

PART TWO: The Throw to First

Have students assign the designated roles (Pacer, Marker, First Baseman, Third Baseman). A group that has only three students may use an inanimate object as the marker. The Marker stands somewhere in the shallow outfield. Each Baseman walks from the assigned base to the Marker, counting the paces. The length of each step is irrelevant as long as the Baseman keeps each stride consistent. Once the Basemen reach the Marker, they should continue walking in a linear fashion past the Marker. The number of paces that a Baseman walks past the Marker should equal the number of paces that it took to reach the Marker. In other words, once a Baseman finally stops, the Marker represents the midpoint between the Baseman and the assigned base. Stress to the Basemen the need to check that the final destination, the Marker and the assigned base are collinear. It is very easy to veer off course.

While the Baseman are walking, the Pacer counts how many paces it takes to walk from third to first — in a straight line. The Pacer then counts the number of paces between the two Basemen (now standing in the outfield). Ideally, these distances should be the same. Have the students repeat this process two more times, for a total of three. Each time the Marker should stand in a different location. Caution the Marker not to stand too close to the perimeter of the field, as this will require the Basemen to walk off the field (through the home run wall).

BASEBALL CONGRUENCY

LAKE HOLATTAWATTA

You are a surveyor and need to calculate the length and width of Lake Holattawatta. The endpoints of your distances are marked by stakes at opposite ends of the lake. Your assistant, Larry Laymo, has already started the job, he shows you a drawing of his method. He placed an extra stake some distance from the lake, and visualized two intersecting lines (shown by the dotted lines). He claims that the solid line in the drawing represents a distance that will be equal to that of the length of the lake.

You, in your infinite wisdom, see that Larry has the right idea, but that his method will only work with congruent triangles. Show Larry on his drawing below, how to use congruent triangles to get the true length and width of the lake. Write your measurements of the segments and angles formed by your new dotted lines to justify your answer. Also, actually measure the lake in the drawing and your solid lines to verify your calculations.

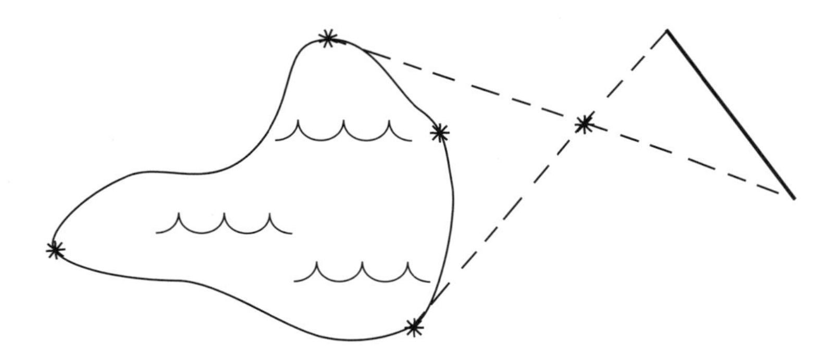

Repeat the process again, in order to find the width of the lake above.

To show Larry that this method will work no matter where you place the intersection of the lines, choose a new location for the marker and determine the length and width of the lake below.

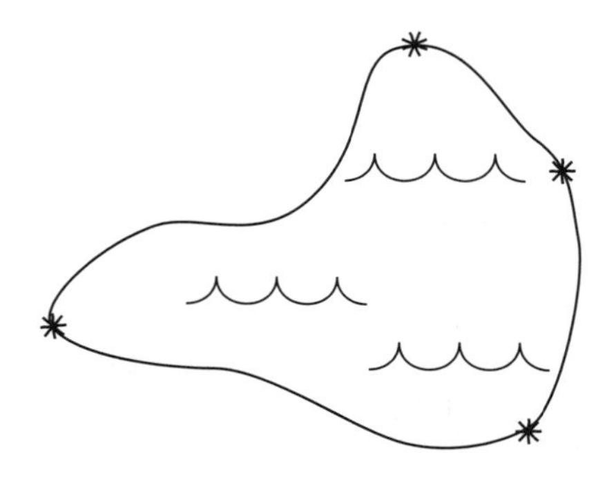

Explain the geometric principle that supports your method.

BASEBALL CONGRUENCY
THE THROW TO FIRST

Apply your method of measuring Lake Holattawatta to the school's baseball diamond. Calculate the distance from third base to first base, in paces. Your group is to designate a person for each of the following roles:

The Marker: Stand somewhere in the outfield. If you have less than four people, any inanimate object can serve as a marker.

First Baseman: Count how many paces it takes you to walk from first base to the Marker; then walk an equal number of paces past the Marker, continuing on a straight line path.

Third Baseman: Do the same as the First Baseman, only from third base. Both Basemen should hold their positions at the end of their line.

The Pacer: Count the number of paces it takes you to walk from third to first base. Then while the Basemen hold their final positions, count the number of paces that it takes you to walk between them.

Complete the process stated above for three different marker positions in the outfield. Give a diagram of each below, and, along each corresponding segment, record the number of paces that each person walked.

First Trial

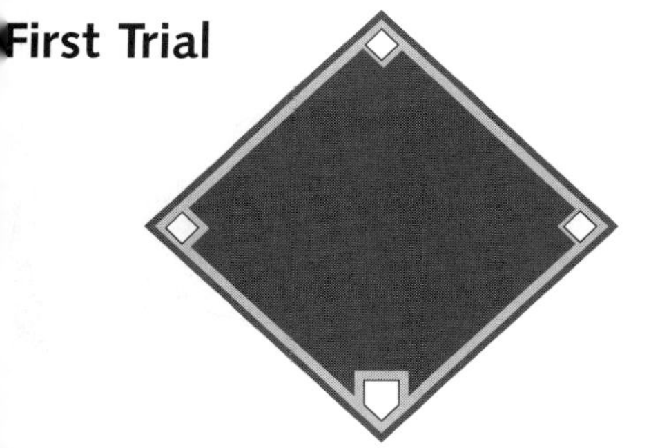

Distance ______

Second Trial

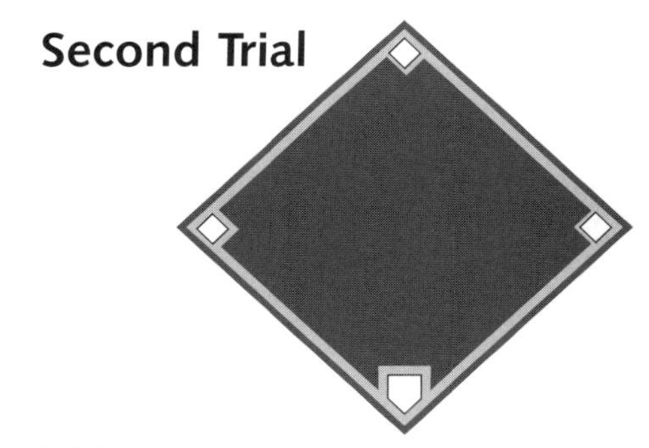

Distance ______

Third Trial

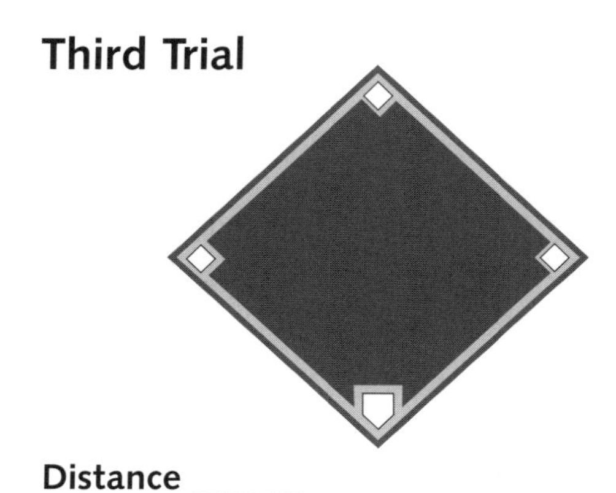

Distance ______

Shipping Conundrum: The Pythagorean Theorem in 3-D

Submitted by Greg Rhodes, Trabuco Hills HS, Mission Viejo, CA

LESSON PLAN

This activity offers students the opportunity to see how the Pythagorean Thoerem can be used to derive the formula for the length of the diagonal of any rectangular box, given the three dimensions of the box. This activity also has students test and prove a mathematical conjecture.

Have students read the Aunt Matilda scenario. The problem basically boils down to determining which of the four boxes offered meets both criteria. The first criterion is that the sum of the lengths of the two shortest sides must be less than 108 inches, which involves unit conversion. The second criterion is that the diagonal of the box must be at least 96 inches long, which involves applying the Pythagorean Theorem.

Have the students read the conjecture regarding the diagonal of a box, and point out to them that the conjecture is very much like the Pythagorean Theorem, except that it involves four variables instead of three. The proof is based on applying the Pythagorean Theorem to two separate triangles within the box, one along the bottom, the other through the center of the box. (See solution below).

Concepts
Pythagorean Theorem, Algebraic Proof

Time: 1hour

Materials
Student handout

Preparation
Students need a sound understanding of the Pythagorean Theorem. They will apply it in a manner that they more than likely have not seen before.

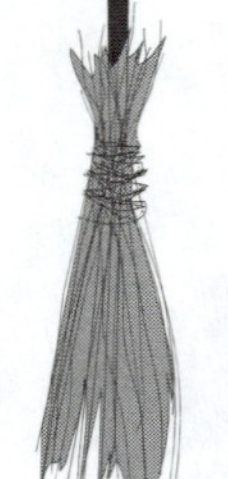

Once the students have proven the new formula, have them verify it by providing two instances. This exercise also gets them familiar with using their new formula.

Once the formula has been verified, have the students apply it to find the length of the diagonal of each of the four boxes. It may be easier to first eliminate the boxes that do not meet the post office's criteria, but since the Pythagorean Theorem is the focus of the lesson, it is best that the students figure the diagonal of all four boxes first.

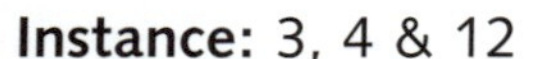

SOLUTIONS

Proof: Drawing the hypotenuse of x & y labeled c, we have $x^2 + y^2 = c^2$. Then c serves as a leg of another triangle with z as the other leg and the diagonal of the box d as the hypotenuse. Therefore, $c^2 + z^2 = d^2$. Substitution yields $x^2 + y^2 + z^2 = d^2$.

Instance: 3, 4 & 12

$3^2 + 4^2 = c^2$, $c = 5$ $\qquad$ $5^2 + 12^2 = d^2$, $d = 13$ $\qquad$ $3^2 + 4^2 + 12^2 = d^2$, $d = 13$

Length of Diagonals of the Boxes (converted to inches):

#1: $50^2 + 60^2 + 70^2 \approx 104.9$

#2: $40^2 + 40^2 + 70^2 = 90$

#3: $30^2 + 70^2 + 70^2 \approx 103.4$

#4: $90^2 + 10^2 + 10^2 \approx 91.1$

Solution: Two of these boxes (#1 & 3) have diagonals that are longer than 96 inches (the length of the broom). Of these two, only box #3 meets the postal criteria of the sum of the two shortest sides being less than 108 inches. Therefore **box #3** is the only one in which you could ship the broom.

Shipping Conundrum: The Pythagorean Theorem in 3-D

You have been asked to ship a large broom to your scary Aunt Matilda in Transylvania. The only problem is that the broom is 8 feet long. (Why does she need an eight foot broom, anyway?!?)

The post office will ship any box as long as the length and the width (the two shortest sides) do not add up to more than 108 inches. The third side can be any length. After talking to the people at the postal annex, you find out that there are four different boxes that might be large enough to work. Their measurements are listed below:

#1: 4'2" x 5' x 5'10" **#2:** 3'4" x 5'10" x 3'4" **#3:** 2'6" x 5'10" x 5'10" **#4:** 7'6" x 10" x 10"

Before determining which box is best suited to ship Auntie M's broom, prove the conjecture below. Once you are confident that the conjecture is true, use it to show which boxes will be big enough to hold the broom:

Box #1 **Box #2** **Box #3** **Box #4**

Of the boxes that are big enough to hold the broom, which is small enough to ship? Why?

Conjecture: Given a rectangular box with dimensions of x, y & z, the square of the length of the diagonal of the box will be equal to the sum of the squares of the dimensions of the box. In other words, $d^2 = x^2 + y^2 + z^2$.

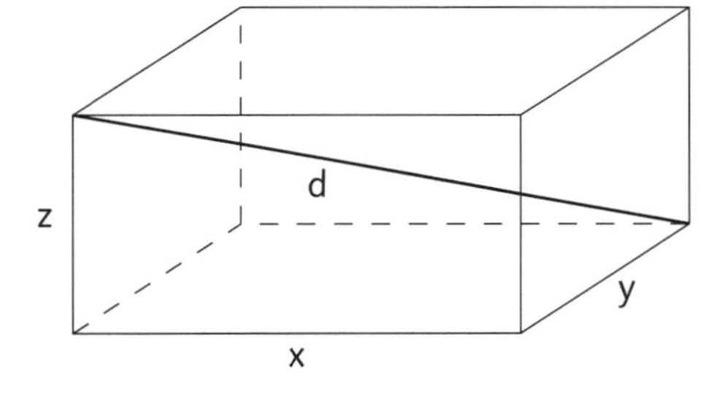

Prove the conjecture:

Verify the conjecture: Use two different instances to support your proof. (Choose two sets of three numbers and show that the formula is true.)

Instance #1 Instance #2

The Gum Drop
Watch Your Step!

LESSON PLAN
The objective of the lesson is two-fold. The primary objective is to give students a general understanding of area. The use of a grid reinforces the idea that area is measured in square units. The size and number of these units may change, but the area of a given figure stays constant. The secondary objective is to introduce students to Pick's Algorithm.

PICK'S ALGORITHM (for finding the area of an irregular figure)
Set the figure on a grid. Coordinates are not necessary. The area of the figure will be the area of one grid unit times the sum of the interior grid squares (units contained entirely within the figure) and half the number of boundary grid squares (units that are only partially contained within the figure). This is based on the assumption that some of the grid squares will only have a small portion within the figure, while others will have a large portion, averaging approximately half of a grid square. The number of boundary squares can be increased by decreasing the size of the grid units. As the number of boundary squares increases, approaching infinity, the average gets infinitely closer to half. If we let N equal the number of interior squares, B equal the number of boundary squares and u equal one linear unit, then the area can be represented by the following formula:
$A = (N + B/2)u^2$.

Concepts
General concept of area, Area of Irregular figures (Pick's Algorithm), Fundamental Theorem of Similarity.

Time: 1-2 Hours

Materials
Student handout, paper, scissors.

Preparation
Find an area on campus that will contains a lot of gum drops (irregular black spots on the ground).

1) Have students find the area of the following first three figures by the traditional formulaic methods. Then ask them to count the number of grid squares to double-check. Students may show an intuitive sense of Pick's Algorithm by counting every two boundary squares as one unit.

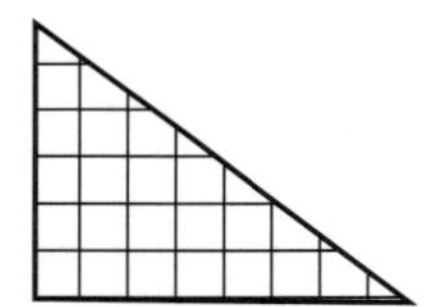
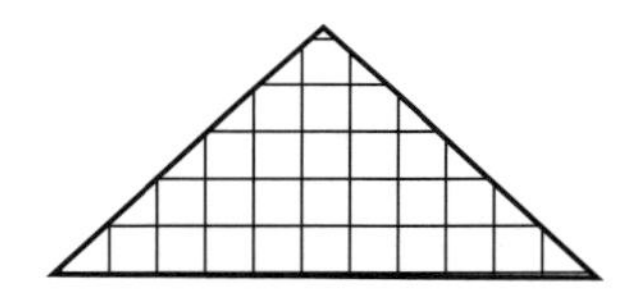
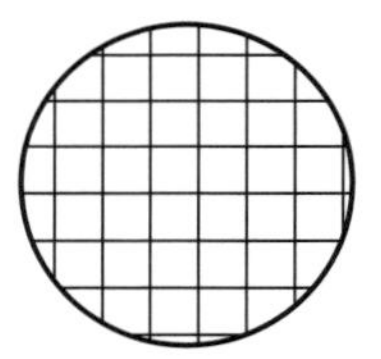
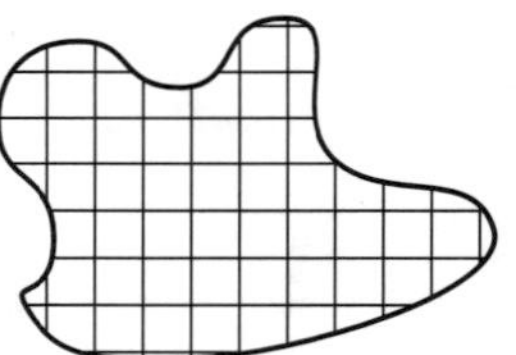

2) Have students practice their conjecture (Pick's Algorithm) on the fourth figure above. Emphasize the fact that the value of a square unit differs from the value of its side length (linear unit).

3) Seek out and harvest the gum drops. Have students find a black spot on the ground. Bigger is better. Encourage them to seek irregular figures as opposed to circular ones. Have students trace the gum drop onto paper. When they get back to class, they cut out these templates and trace them onto each of the three grids. This tracing procedure ensures that the figures on each grid will be congruent.

4) Have students calculate the area of each figure. While the procedure is the same for each one — count the interior units, divide the boundary units by 2, add these values and multiply by the value of one square unit — the students need to be made aware that the value of the square unit is equal to the area of the unit and not the length of the side. In other words, the area is $^1/_4$ in^2 for the $^1/_2$ inch grid, $^1/_{16}$ in^2 for the $^1/_4$ inch grid, and $^1/_{64}$ in^2 for the $^1/_8$ inch grid. If done correctly, the students should notice that the areas that they calculate are nearly equivalent.

The Gum Drop

Watch Your Step!

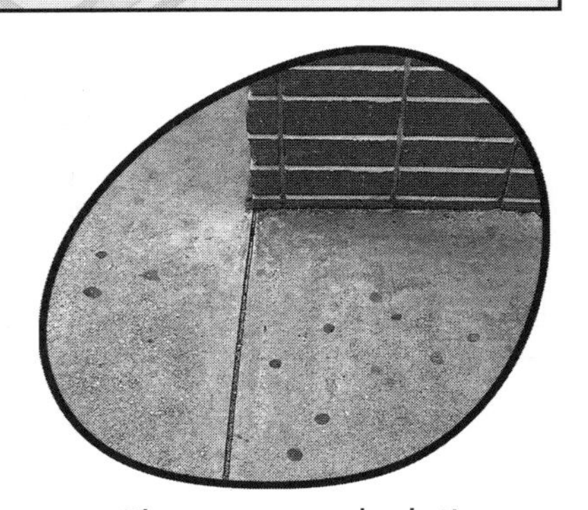

Find a *gum drop* — one of those black dots permanently fixed to the grounds of the campus — and trace it on a sheet of paper. Then copy that shape onto each of the grids below. Calculate the area of the gum drop. While the areas should obviously be the same, the different size grids will produce varying degrees of accuracy. Show your calculations.

$^1/_2$" GRID

Area of the Gum Drop = ________________ in^2

$^1/_4$" GRID

Area of the Gum Drop = ________________ in^2

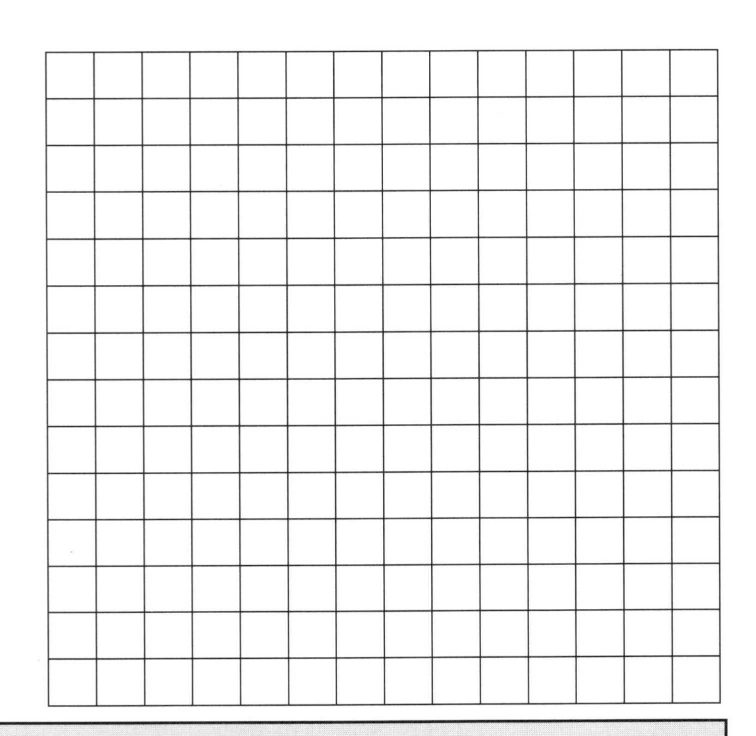

The Gum Drop

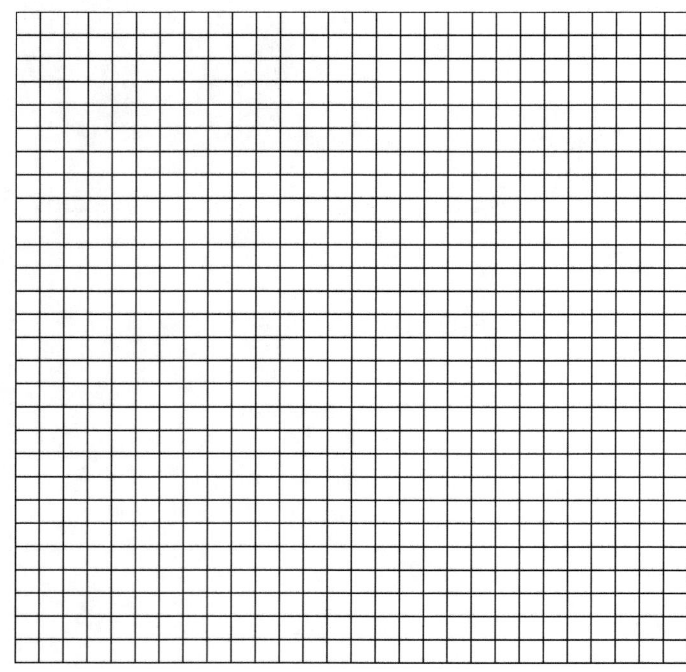

$^1/_8$" GRID

Area of the Gum Drop = ________________ in^2

Which grid do think offered the most accuracy and why?

Test Pick's Algorithm again on the following triangle. For each grid, calculate the area of the triangle by the traditional formula, then apply your new method.

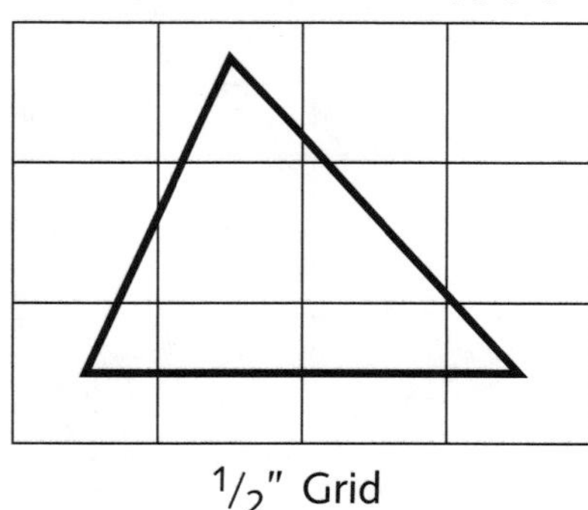

$^1/_2$" Grid

$^1/_4$" Grid

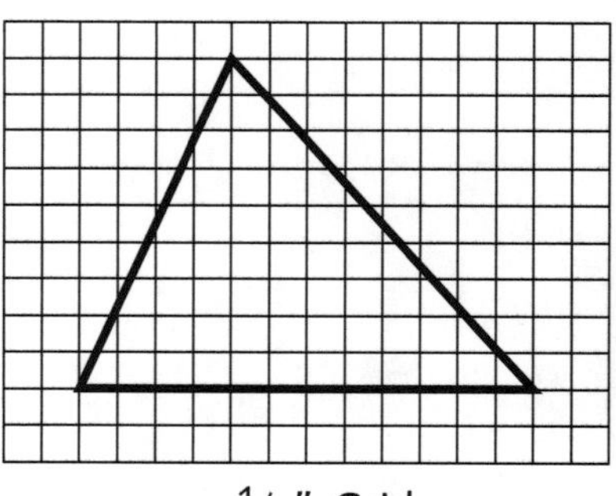

$^1/_8$" Grid

By the Formula:

By Pick's Algorithm:

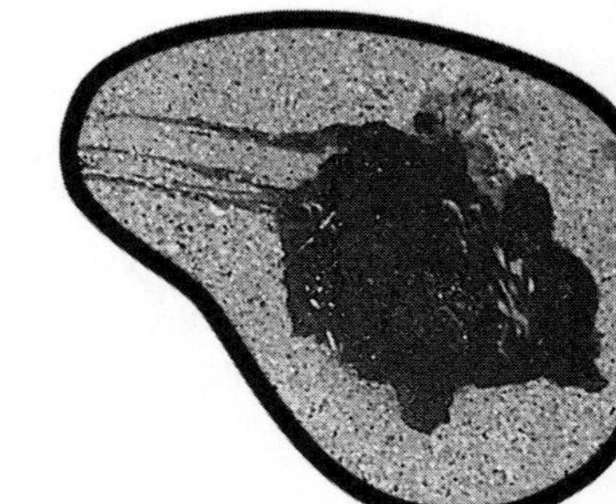

A Slice of Pi

An investigation on the ratio of circumference to diameter

Every high school student knows to substitute 3.14 for the symbol π during a calculation involving circles. Few, however, understand the concept of pi... that it is the ratio of the circumference to the diameter of any given circle. The following investigation is a very effective activity for teaching more than just the simple value of pi.

> ***Concepts***
> Pi, circumference of circles
>
> ***Time:*** 1hour
>
> ***Materials***
> Student handout, tape measure, six cylindrical objects
>
> ***Preparation***
> Determine which objects on campus the student will measure.

LESSON PLAN

Begin the lesson by asking students "What is pi?" Many of them will respond with "3.14," to which you then respond, "No, that is only an approximation, what is pi?" After much proding the students will be perplexed and that is when you claim that in this lesson, the students will be hunting for pi.

Each group measures and records the circumference and diameter ("How far around? How far across?") of the same six cylindrical objects. The lesson has the most impact when the objects are significantly different in size - a soup can, a waste basket, a planter, etc. These objects and their measurements are to be recorded in the given chart. The chart intentionally has the students record the circumference "over" the diameter.

The groups should also be given one object for which the diameter cannot be measured, as in a tree or pillar, and be asked to measure and record the circumference. The students should also be informed that a basketball hoop has a diameter of 18 inches and that this should also be recorded in the chart. Once the measurements are completed, the students calculate the ratio of the circumference to the diameter for each object, then write their findings on the board in a chart similar to the one below.

Object	Group 1	Group 2	Group 3	Group 4	Group 5
Soup Can					
Fence Pole					
Planter Box					

Your students will soon see that no matter what size the circle, the ratio of circumference to the diameter is suspiciously close to 3.14. Few of the values on the board will be that accurate due to the method of measurement, but the students understand that a consistent range of 2.7 to 3.3 has something to tell them. When you ask them to calculate the diameter of the final object (the tree, pillar, etc.), they should be able to do so without further instruction because they know more than the approximate value for pi; they understand the relationship of the circumference to the diameter. They will not instantly generate the formula, $C = \pi d$; that is the instructor's job. However, they will instinctively multiply the diameter of the hoop by 3.14 to get the circumference, and divide the circumference of the tree by 3.14 to get the diameter. Problems #5 & 6 are offered for practice or for teacher assessment.

Note: See the article "On a Good Day" to see how this lesson is woven with others to impart the idea of abstract generalizations and to act as a lead in for a lesson on arc length.

A Slice of Pi

An investigation on the ratio of circumference to diameter

The ancient thinkers discovered something rather curious about circles. This intriguing mystery of the natural world has to do with the relationship of "how far around" and "how far across." You are to investigate this pervasive pattern and use it to estimate the width of a tree and the circumference of a basketball hoop.

1. Write the name of the designated objects in the chart below. For each of these objects, measure the circumference and the diameter and record them in the chart. For the designated tree, measure and record the circumference of the tree. A standard basketball hoop is 18" across.

Object							B-ball Hoop	Tree
Circumference								
Diameter								
Ratio: $^C/_D$								

2. For each object, calculate and record the ratio of the circumference to the diameter.

3. Share your data with the class. What is the pervasive pattern?

____

____

____

WHAT IS PI?____________________

Object	Your Group's Ratios	Another Group's Ratios	Another Group's Ratios

4. Use your class data to estimate the diameter of the tree and the circumference of the hoop. Show the work.

5. Find the circumference of the following circles. Give both the exact and approximate values for your answers.
   a. 18
   b. 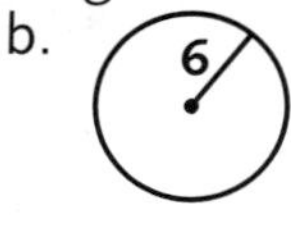

   c. A circle with a radius of 10

6. Find the diameter of the circle with the given circumference.
   a. 36π
   b. 50.26

PROJECT

POISON WEED

Adapted from CPM Educational Program

LESSON PLAN/SOLUTIONS

> ***Concepts***
> Area of sectors, circles, and triangles; area sub-problems, trigonometry, probability, and rates.
>
> ***Time:*** 1hour
>
> ***Materials***
> Student handout
>
> ***Preparation***
> Students should be familiar with computing the area of sectors.

1. This first of the four questions allows the students to begin with some easy sub-problems. There are several ways to calculate the area of the grass. In the case shown below, the entire yard was considered to be a 60' x 80' rectangle minus the barn-and-shed, which itself is another rectangle minus the concave portion.
   a) (60 / 3) • 6 = **120 bites/day**
   b) (60 • 80 - 30 • 30 + 10 • 20) • 40 = 4100 • 40 = **164,000 bites**
   c) 70 days • 120 bites/day = 8,400 bites; 8,400 / 164,000 ≈ **5%**

2. This second question introduces the students to area sub-problems involving circles, which will prepare them for the more difficult sector problems in the next section.
   a) $(^3/_4)\pi 10^2 = 75\pi \approx$ **235.6 sq. ft.**
   b) 120 bites / (235.6 sq. ft. • 40 bites per sq. ft.) ≈ **1%**
   c) $(^3/_4)\pi 20^2 + (^1/_4)\pi 10^2 = 325\pi \approx$ **1021 sq. ft.**
   120 / (1021 • 40) ≈ **0.3%**

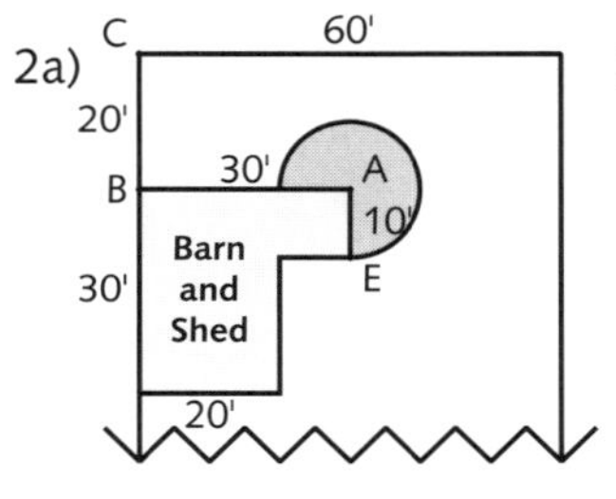

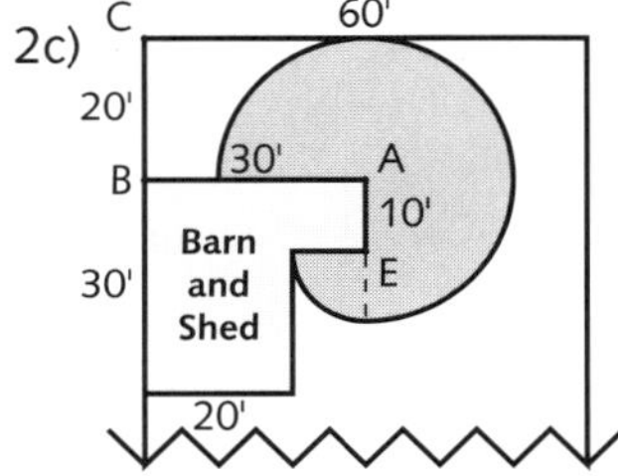

3. This question develops several skills that are required in the final question. The important step here is calculating the degree measure of the sector, which is the complement of one of the angles of the triangle.
   a) see right
   b) m∠DBC = $\cos^{-1}(20/30) \approx$ **48.2°**
   CD = $30^2 - 20^2 = 500 \approx$ **22.4 feet**
   c) 0.5(20)(22.4) = **224 sq. ft.**
   d) $(41.8 / 360)/(\pi 30^2) = 0.12(900\pi) \approx$ **328.3 sq. ft.**
   e) 224 + 328.3 = **552.3 sq. ft.**
   f) (7 • 120)/(552.3 • 40) = 840 / 22092 ≈ **4%**

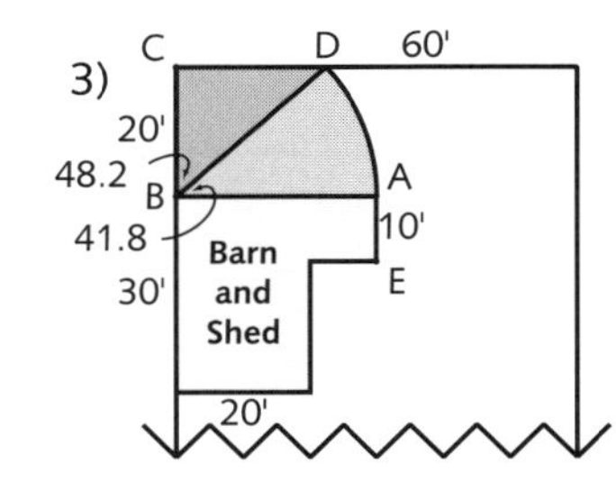

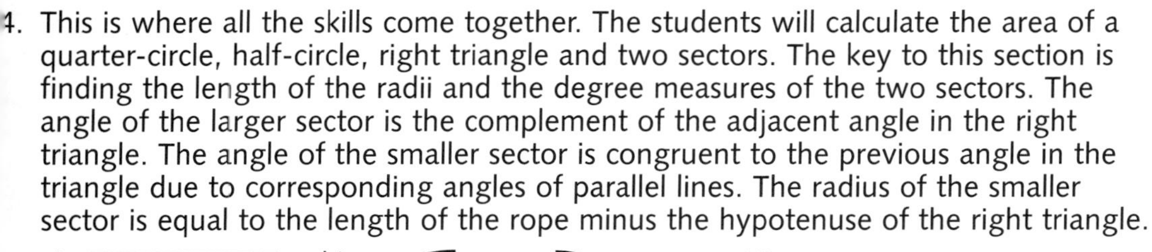

4. This is where all the skills come together. The students will calculate the area of a quarter-circle, half-circle, right triangle and two sectors. The key to this section is finding the length of the radii and the degree measures of the two sectors. The angle of the larger sector is the complement of the adjacent angle in the right triangle. The angle of the smaller sector is congruent to the previous angle in the triangle due to corresponding angles of parallel lines. The radius of the smaller sector is equal to the length of the rope minus the hypotenuse of the right triangle.

a)

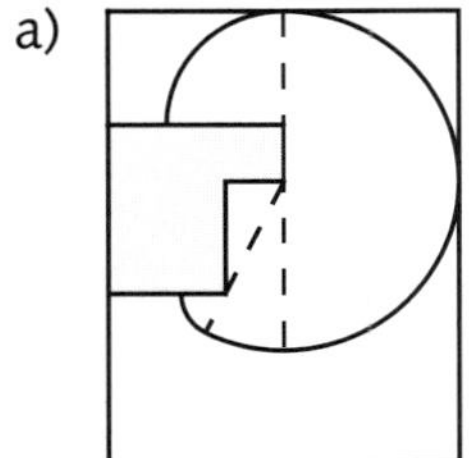

b)

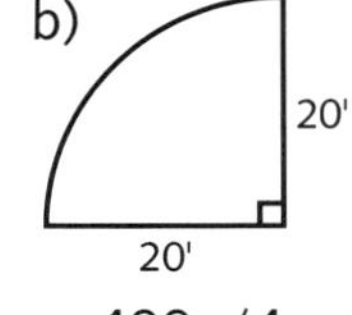

60' 30'

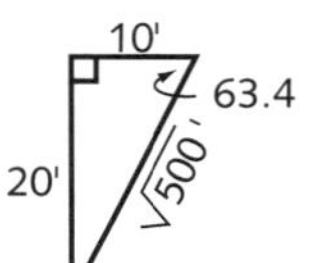

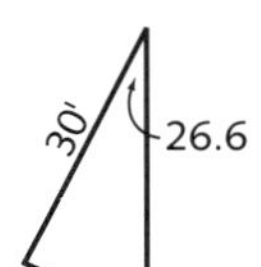

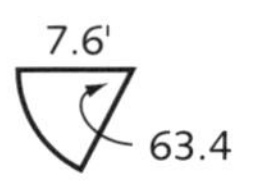

$400\pi/4 + 900\pi/2 + 0.5(200) + (26.6/360)(900\pi) + (63.4/360)(\pi 7.6^2)$
$100\pi + 450\pi + 100 + .07(2827.4) + 0.18(181.5)$
314 + 1413.7 + 100 + 209 + 32 ≈ **2068.7 sq. ft.**

c) (28 • 120)/(2068.7 • 40) = 3360/82748 ≈ **4%**

POISON WEED

Dimitri is getting his prize sheep, Zoe, ready for the county fair. He keeps Zoe in the pasture beside the barn and shed. What he does not know is that there is a single locoweed in this pasture which will make Zoe sick if she eats it; and she can eat it in one bite. Zoe takes about one bite of grass every three minutes for six hours a day.

1. Suppose that this diagram represents the shape of the field and building. Everything except where the building sits has grass growing on it. Each square foot of pasture provides enough food for about 40 bites.

a) How many bites of food does Zoe take during the day?

b) How many bites of food are available in the pasture?

c) What is the probability that Zoe will eat the locoweed if she grazes in the pasture for ten weeks and never eats the same patch twice?

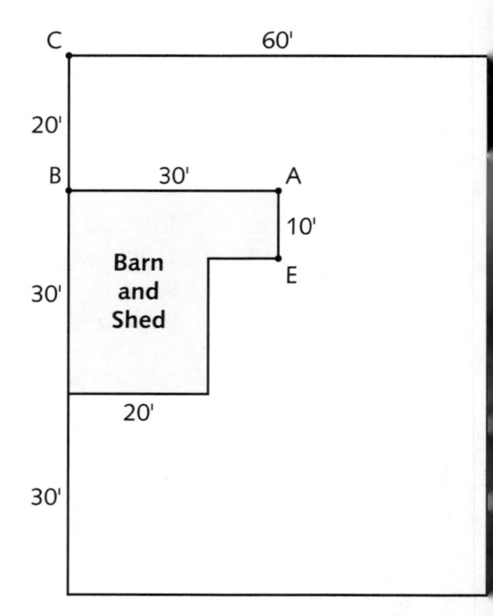

Diagram for Question 2a

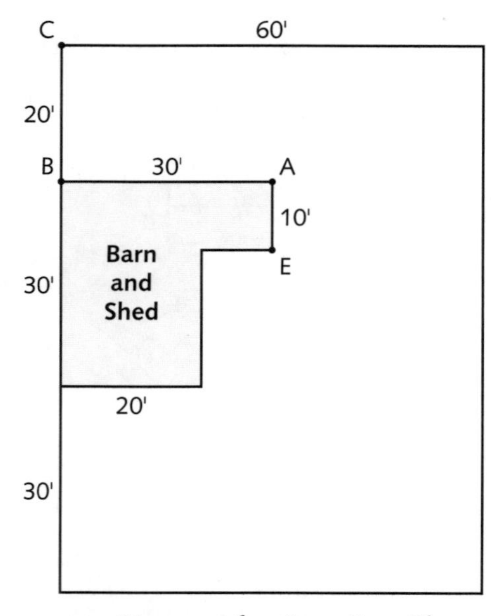

Diagram for Question 2b

2. Dimitri has heard about the possibility of the locoweed, so he decides to tie Zoe to a rope with one end at A in the diagram.

a) Over how much area can Zoe graze if the rope is 10 feet long?

b) If we know that the locoweed is within this area, what is the probability that she eats the weed in one day?

c) Suppose the rope is 20 feet long. Over how much area can Zoe now graze? What is the probability Zoe eats the weed in one day?

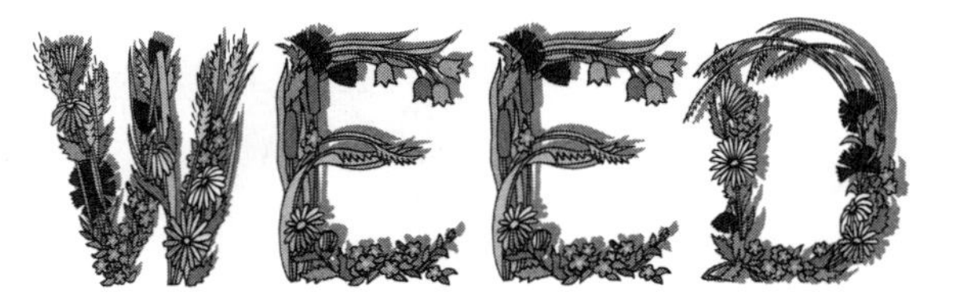

POISON WEED

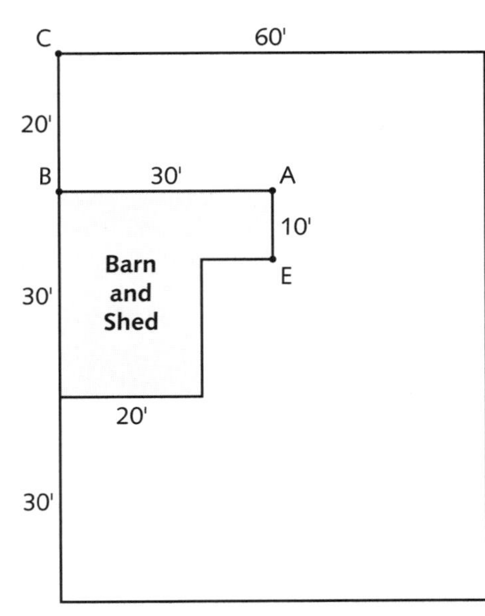

Diagram for Question 3

3. Dimitri's father moves the tied end of the rope to B and lengthens the rope to 30 feet. Over what area can Zoe now graze?

a) Sketch the situation carefully, noting that it is only 20 feet from B to the northwest corner of the field, C.

b) Let D be the point on the north fence 30 feet from B. Find $m\angle DBC$ and CD.

c) What is the area of ΔBCD?

d) What is the area of the grazing region not in ΔBCD?

e) What is Zoe's total grazing area for a 30' rope tied at B?

f) If we know that the weed lies in the region which Zoe can reach in the previous problem, what is the probability she eats the weed in a week, if she never eats the same patch twice?

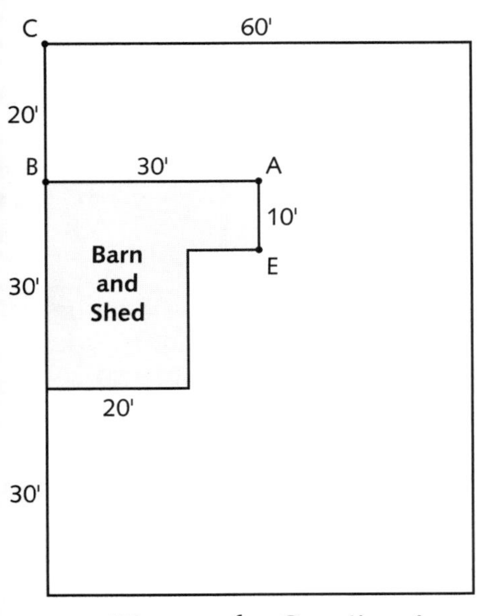

Diagram for Question 4

4. Suppose that Zoe is tethered to a 30 foot rope attached at E.

a) Identify all of the sub-problems you need to solve in order to find the area of Zoe's grazing region.

b) Find the total area.

c) Find the probability that Zoe will eat the weed in four weeks if she is tied to this rope and never eats the same patch twice?

PROJECT

LESSON PLAN

1. After introducing the concept of a house made of polygonal rooms with no hallways, have them sketch their creation. Make sure that the students are conforming to the specifications (the trapezoid is non-isosceles, no hallways exists, etc.). Also encourage students to draw all quadrilaterals either horizontally or vertically. Having a parallelogram oblique to the axes of the cartesian plane makes calculating the height very difficult. Another common trouble spot for students is being sure that the two radii of the sector are truly equal.

2. Establishing the coordinates of the pentagon is the most complex portion of this lesson. Figuring these coordinates does not tie in directly to the lesson's main concept, area of polygons, but it makes a powerful connection to trigonometry. This is best demonstrated if the teacher models this process with a given length of the side of the pentagon, and then the students generate coordinates for a different size pentagon.

   It is recommended that you demonstrate with a pentagon of side length three, then have students create their own pentagons of any length other than three. Begin by drawing a horizontal line three units long on the cartesian plane. Label the left end point A(0, 0), the right end point E(3, 0). Then share with the students the idea that we need to find point B such that AB = 3. Furthermore, $\angle ABC$ must be a specific angle. Indeed, in a regular polygon, it is 108 degrees [(n - 2)180/n]. Finding the coordinates of B becomes quite simple if we imagine a triangle (I) on the exterior of the pentagon with $m\angle A = 72$. The horizontal leg (x) and the vertical leg (y) can be found with the following equations:

   $x = 3\cos(72°) \approx .93$ $\qquad$ $y = 3\sin(72°) \approx 2.85$

   The same can be done for the second triangle (II), however, the angle of reference now is 54 degrees. This time, we use the following equations:

   $w = 3\sin(54°) \approx 2.43$ $\qquad$ $z = 3\cos(54°) \approx 1.76$

   Adding/subtracting these values to the previous coordinates as we work our way around the pentagon yields the following points: A(0,0) B(-0.9, 2.8), C(1.5, 4.5), D(3.9, 2.8), E(3, 0)

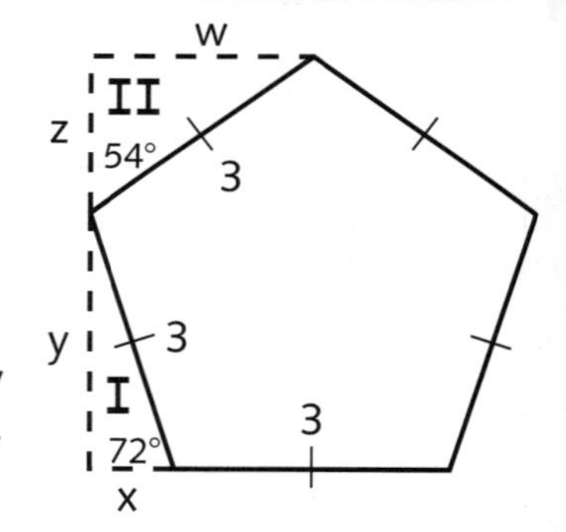

3. Once the calculations for the pentagon are complete, have the students finish their houses. They should be able to determine the coordinates of the vertices quite easily. Students should write the measures of all sides (in units, not feet) and angles on the diagram. Also, they should clearly state the dimensions required to calculate the area of the figure (e.g. lengths of both bases and the height of a trapezoid). Since all the measurements are written in units rather than feet, graphing should be a fairly straightforward task. They will need to decide what scale factor will give them reasonably sized rooms for people to live in.

4. The students will naturally calculate the area of the figures (according to their corresponding formulas) in square units. To convert to square feet, have them refer to their scale factor. For instance, if a student establishes a scale of 5′ to every quarter-inch, then every square unit represents 25 square feet. The student simply needs to multiply the answer obtained in square units by 25 in order to convert to square feet.

Concepts
Area of polygons, coordinate geometry, trigonometric functions, distance formula

Time: 2 hours

Materials
Graph paper, scientific calculator or trig table, student handout

Preparation
This is a culminating activity for area, trigonometry and graphing. Students should be able to find the area of sectors, regular polygons and quadrilaterals. They should also be able to use trigonometric functions and the Distance Formula.

In the town of Mathopolis, houses are built with rooms in the shape of a variety of polygons. These houses have no hallways. Each room opens into one or more of the others. The example given below shows a house made of a hexagon, a rhombus and a sector. You, being a Mathopolis architect, are commissioned to design a house that has exactly one room of each of the following shapes: regular pentagon, non-isosceles trapezoid, sector, parallelogram, rectangle, and a non-right triangle.

The submitted design must include the following:
- A sketch below of the overall design.
- Measurements for each room of all the walls and angles.
- A scale drawing of the house on quarter-inch graph paper. Show the scale.
- Calculations of the area of the floor plan of the house in square scale-units and in square feet.

Sketch of a sample polygon house

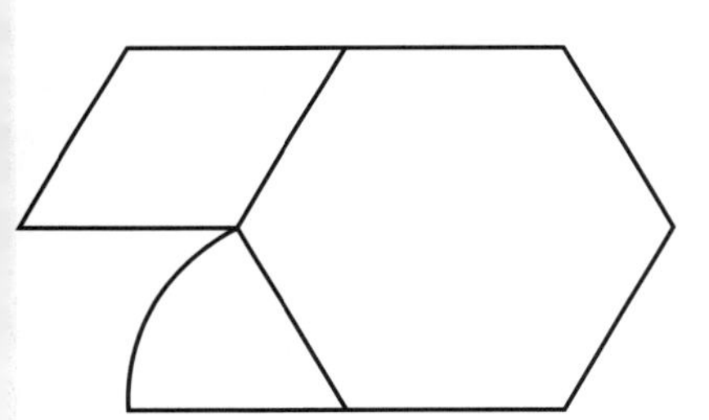

Scale: _____ ft = $^1/_4$ in

Sketch of your polygon house

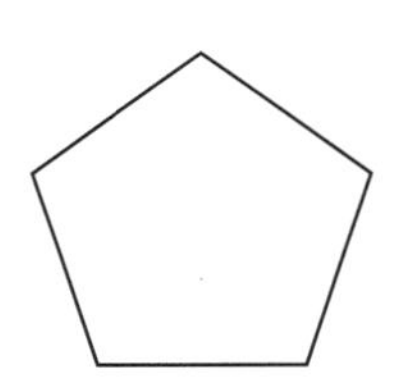

Total Area = ______ u^2 = ______ ft^2

Pentagon

Diagram & Measurements

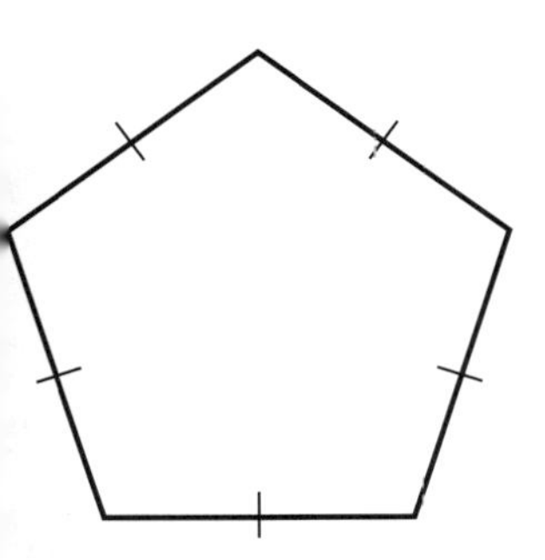

Coordinates

A(0 , 0) B(,)
C(,) D(,)
E(,)

Area

A = __________ u^2
= __________ ft^2

Triangle I

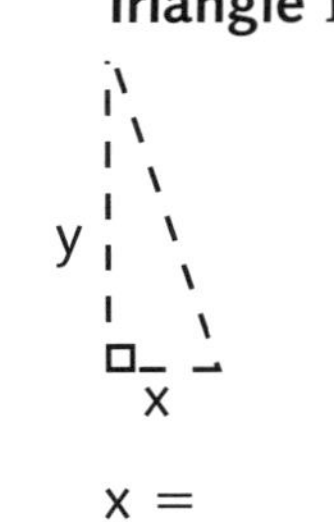

x =
y =

Triangle II

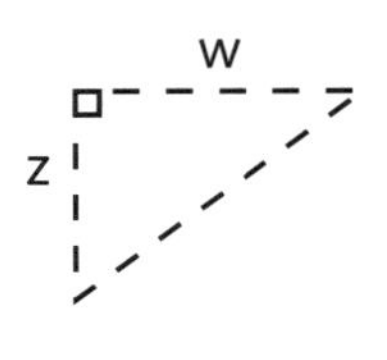

w =
z =

Non-Isosceles Trapezoid
Diagram & Measurements

Coordinates

(,) (,)
(,) (,)

Area

A = ________ u^2
= ________ ft^2

Parallelogram
Diagram & Measurements

Coordinates

(,) (,)
(,) (,)

Area

A = ________ u^2
= ________ ft^2

Sector
Diagram & Measurements

Coordinates

(,) (,)
(,)

Area

A = ________ u^2
= ________ ft^2

Rectangle
Diagram & Measurements

Coordinates

(,) (,)
(,) (,)

Area

A = ________ u^2
= ________ ft^2

Non-Right Triangle
Diagram & Measurements

Coordinates

(,) (,)
(,)

Area

A = ________ u^2
= ________ ft^2

Extra Credit: Carpet at Home Depot comes in 12-foot wide rolls, and sells for $4/yd^2. In order to carpet your polygon house, how long of a roll do you need to purchase, and how much will it cost?

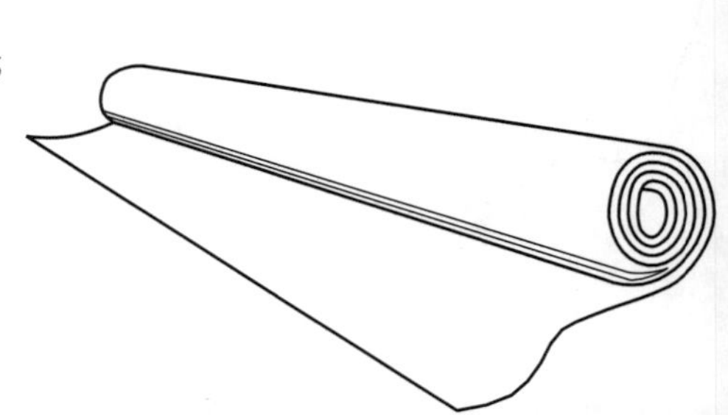

Finding the Surface Area and Volume of a

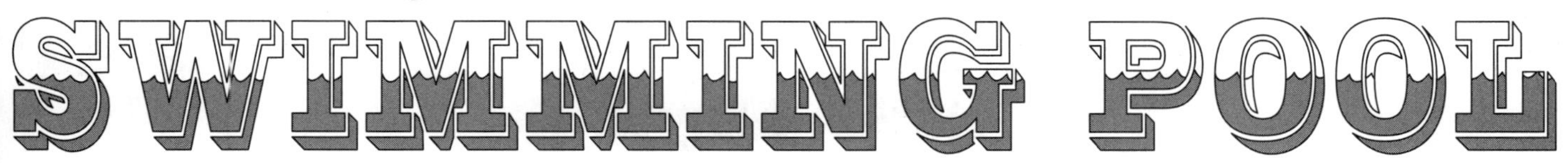

LESSON PLAN

This project may be implemented using the mock pool provided in the student handout or using your school's actual pool, if it has one. You may also choose to use the mock pool as practice, before tackling a real pool. There are basically three components of this project, each taking approximately one hour: Measuring, Drawing and Calculating. The pool featured in the student handout does not require any measurement. Instead the students will label the 3-D diagram with the data given. If the students do measure an actual pool, it will be very important to make clear that all measurements with the tape measure along the pool deck are HORIZONTAL distances between depth changes (markings). All the depth markings themselves represent VERTICAL distances.

Drawing can be quiet intimidating for students; so to help ease their anxieties, sample drawings are offered on the student handout. It is very important that students are particular about labeling the drawings with the proper lengths. This especially comes into play for the sloping portion of the pool. The length of the slope is neither the horizontal measurement nor the vertical depth. The Pythagorean Theorem must be used to find the length of the slope, the horizontal distance representing one leg and the change in depth representing the other. Be sure everyone has correctly drawn and labeled each face and prism before you move on to the calculations of surface area and volume.

If the drawings are done correctly, the calculations are fairly straight forward. Five of the seven faces are rectangles; so the only real work for the surface area is in breaking down the two sides of the pool. When calculating volume, it will help many students to know that the base of the trapezoidal prism is actually on the side of the pool, and the prism's height corresponds to the pool's width. Stress the need to clearly label all calculations.

Concepts
Measurement, surface area and volume of rectangular and trapezoidal prisms, Pythagorean Theorem.

Time: 2-3 hours

Materials
Student handout, tape measures and a swimming pool (optional).

Preparation
The students should already be familiar with the surface area and volume of prisms. If you choose to use the dimensions of your school's actual pool, rather than the diagram given on the student handout, you should get the measurements of the pool and calculate the solutions prior to the activity.

SOLUTIONS

Surface Area:

$F_1 = F_2 = 13 \bullet 28 + 0.5(12)(4 + 13) + 4 \bullet 35 = 606$
$F_3 = 40 \bullet 4 = 160$
$F_4 = 40 \bullet 35 = 1400$
$F_5 = 40 \bullet 15 = 600$
$F_6 = 40 \bullet 28 = 1120$
$F_7 = 40 \bullet 13 = 520$

$SA = 2 \bullet 606 + 40(4 + 35 + 15 + 28 + 13)$
$= 1212 + 3800 =$ **5,012 sq. ft.**

Volume:

$P_1 = 40 \bullet 28 \bullet 13 = 14{,}560$
$P_2 = 0.5(12)(4+13) \bullet 40 = 4{,}080$
$P_3 = 40 \bullet 35 \bullet 4 = 5{,}600$

$Vol = 14{,}560 + 4{,}080 + 5{,}600$
$= 24{,}240$ cu. ft.

Check:
$F_1 \bullet 40 = 606 \bullet 40 =$ **24,240 cu. ft.**

PROJECT

SWIMMING 

Solution Diagrams

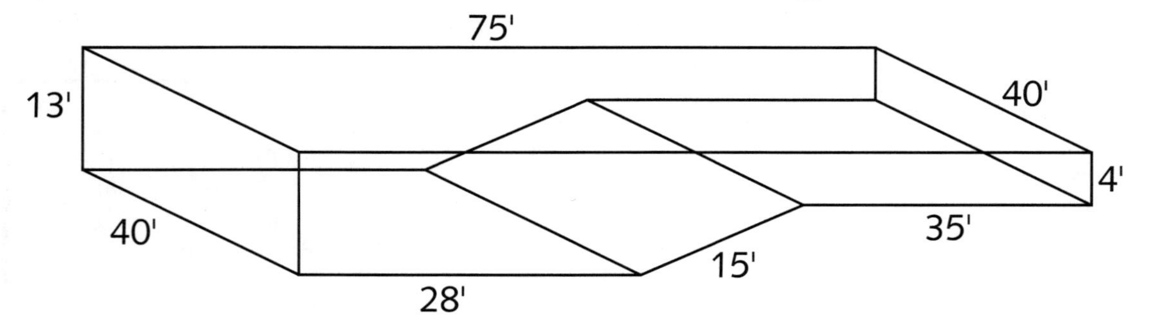

40'
4' F_3

75'
13' F_1 4'
35'
15'
28'

40'
35' F_4

75'
4' F_2 13'
35'
15'
28'

40'
15' F_5

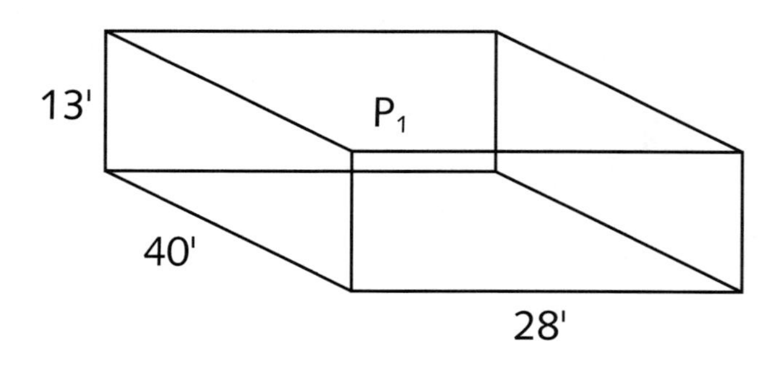

40'
28' F_6

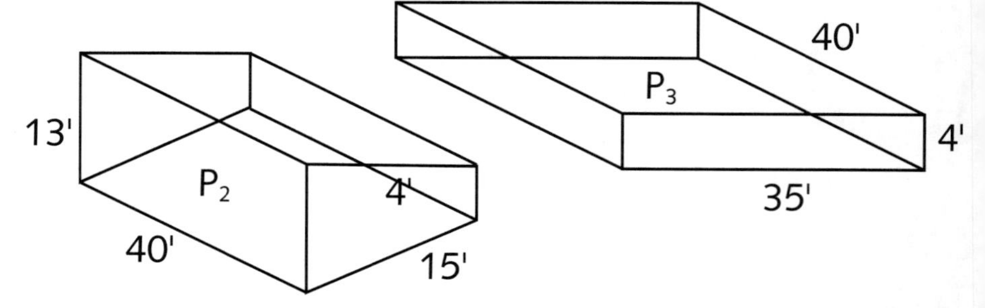

40'
13' F_7

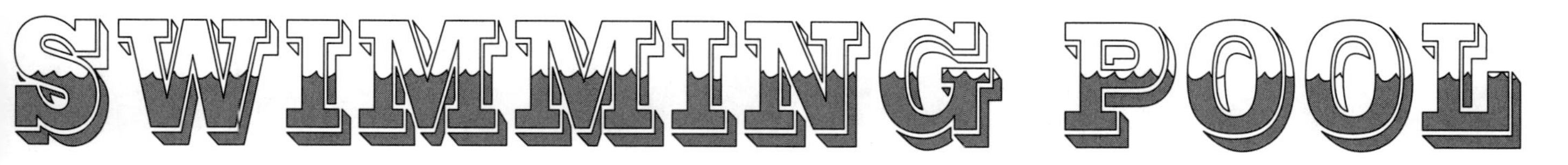

Your task is to calculate the surface area of the following pool and the volume of the water contained within it. You will do this by measuring, drawing and calculating as specified below. To make your task simpler, assume that each edge between faces is a line (sharp turn) rather than a gradual curve.

DIMENSIONS OF THE POOL

The pool is 75 feet long and 40 feet wide. It has two main depths: a shallow end of 4 feet and a deep end of 13 feet (with a slope connecting the two). The shallow end extends 35 feet from the right edge and the deep end extends 28 feet from the left edge.

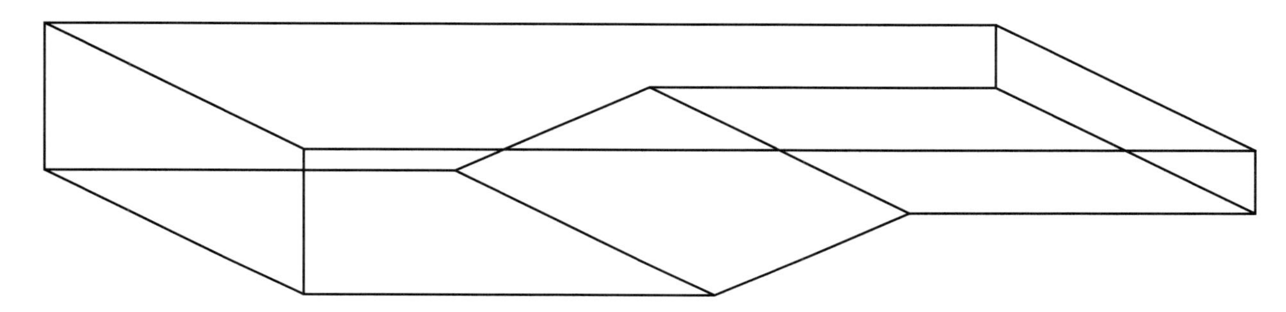

RECORDING THE LENGTHS OF THE POOL

1) From the data given, label the above diagram by writing the corresponding length next to each edge (line segment).

DRAWING THE POOL

2) Draw a view of each of the seven faces of the pool's surface: all four sides and the three surfaces along the bottom. Draw each view as if you're looking directly (perpendicular) at the face. Label them F_1, F_2, etc. Include the measurements of all the dimensions of each face.

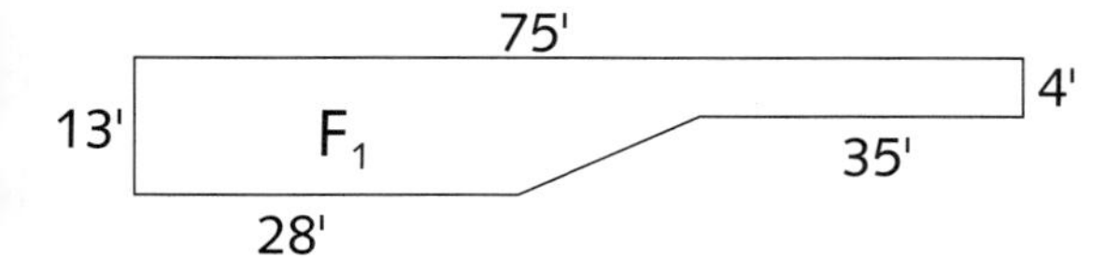

3) Visually dissect the pool's volume into prisms for which you can easily calculate the volume. Draw a 3-dimensional view of each of the prisms, labeling them P_1, P_2, etc. Include the measurements of all the dimensions of each prism. The first one is shown below.

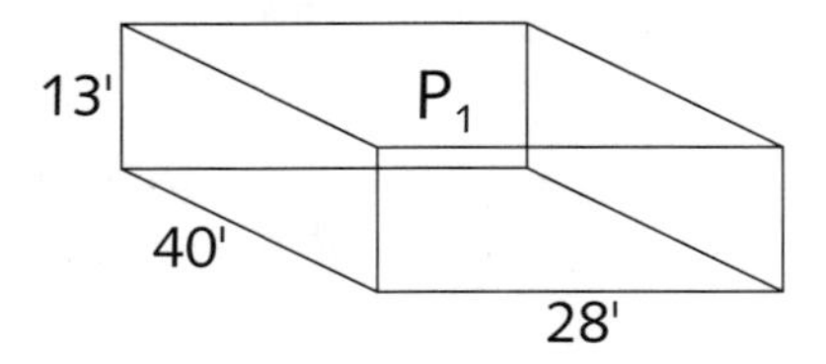

SURFACE AREA OF THE POOL

4) Show your calculations of the area of each of the faces. Be sure to accurately label each calculation with its corresponding notation. (F_1, F_2, etc.)

VOLUME OF THE POOL

5) Show your calculations of the volume of each of the prisms. Be sure to accurately label each calculation with its corresponding notation. (P_1, P_2, etc.)

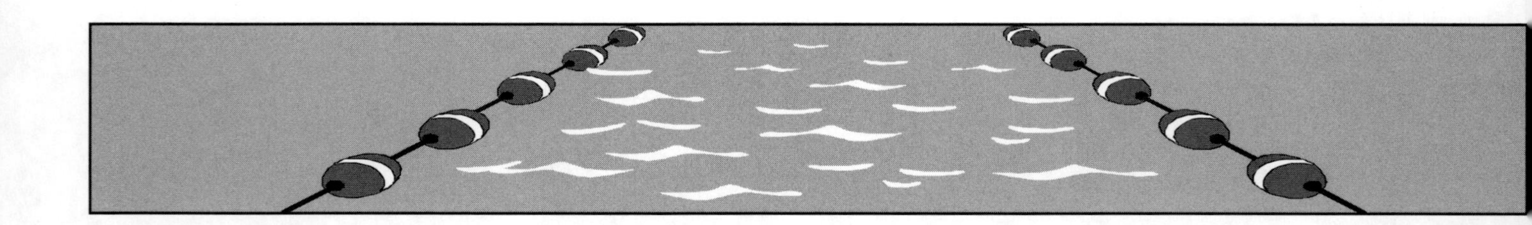

KING TUT: DECONSTRUCTING THE PYRAMIDS

1. Students (paired or tripled) are to construct the net for a square pyramid on a given sheet of construction paper. The net should be as big as possible, and yet still fit on one sheet of paper. The students are to generate their own strategies for creating a true square and for creating congruency among the triangles. The two most popular means to create the triangles are using a compass to construct equilateral triangles or using the perpendicular bisectors of the side to create isosceles triangles.

2. Students then cut out the net and measure the lateral and base edges, as well as the slant height and the diagonal. Emphasize that they are to record these measurements in their sketch.

3. Have students draw the solid of their pyramid. It should closely resemble their paper model. Having the students transfer the measurements from the net to the solid helps them visualize the various components of the pyramid.

4. Using the measurements of the lateral edge and half the base edge, calculate the slant height with the Pythagorean Theorem. Special right triangle ratios will generate the length of diagonal. Have students record these calculations on the handout, and compare them to the actual measurements that they found.

5. The height can be calculated with right triangles in two ways, by representing the hypotenuse with the slant height or with the lateral edge. Have the students predict the height by either or both methods and commit to this answer before they actually measure the height in the next step.

6. The directions for creating the dipstick are on the handout. Be sure the students place the dipstick such that the top edge is parallel with a side of the square. The fact that this height exists only in the solid and not in the net is the point of emphasis here.

Concepts
Identifying and calculating the height, slant height and lateral edge of a pyramid; surface area and volume; right triangle trigonometry; the relationship between a net and its solid.

Time: 1-2 hours

Materials
Scissors, ruler, protractor, adhesive tape or glue sticks, construction paper (one large sheets for every 2-3 students), student handout

Preparation
Students need only a general familiarity with pyramid analysis. The primary objective of this lesson is to develop their understanding of the difference between slant height and height. That understanding makes it easier to then visualize the right triangles necessary to calculate the variety of lengths and angles offered in the right pyramid. It is suggested that the instructor actually complete the activity before facilitating it with students.

7. The angle (from horizontal) that the dipstick is to be cut, should be the same as the angle that the pyramid's face makes with the ground. If you think of the problem in terms of slope, the ratio of the height to half of the base should be the same as the ratio of the distance d on the dipstick to half of the width of the dipstick ($h/b = d/0.5$). This part of the activity is a gem and can be expanded for an entire lesson itself.

8. Have the students calculate the angle between the lateral edge and the base of the triangle ($\tan^{-1}(\text{height}/0.5 \cdot \text{diagonal})$). Sliding a protractor into the seam of the pyramid makes for an easy measurement, but the dipstick will have to be moved out of the way.

9. The teacher should point out that, with the net, the students still have the surface area of the pyramid. However, they do not have volume until they fold the pyramid into a solid.

STUDENT HANDOUT

KING TUT: DECONSTRUCTING THE PYRAMIDS

You are to create the net of a square pyramid on the given sheet of construction paper. The net should be as big as possible and yet still fit on the paper. On this pyramid, you will then measure four attributes: the height, slant height, the angle of the lateral edge and angle of the slant height. Show all diagrams and calculations on the back side of this paper.

1. Devise a strategy to guarantee that all the triangular faces of the pyramid will be congruent.

2. Sketch the net of your pyramid. Include the measurements of all base and lateral edges, slant height and diagonal of the base.

3. Draw and label a three-dimensional model of your pyramid.

4. Using the measurements of the edges of your net, calculate the slant height and the diagonal of the base. Compare your two answers.

   Slant Height = ___________ Diagonal of the Base = ___________

5. Calculate the height of your pyramid and label it in your diagram. Measure the actual height according to the instructions in #6 and #7.

   Calculated Height = ___________ Measured Height = ___________

6. Create a "dip stick" in order to actually measure the height of your pyramid in the following manner. Cut a 1/2" wide rectangular strip of scrap paper. The length of this dip stick should well exceed the height of your pyramid. Fold a flap at the bottom of the dip stick and attach it to the center of the base. On the dip stick, draw a line where you calculated the height to be. Then fold up the opposite sides of the pyramid, and cut the dip stick to the height of the pyramid. How accurate was your calculation?

Calculated Height

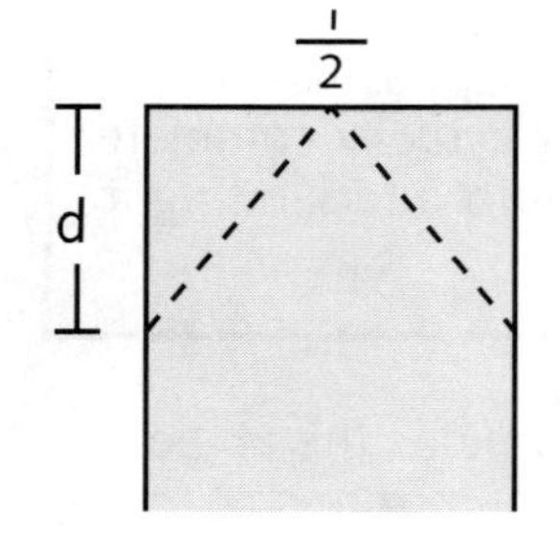

7. You are going to need to trim your dip stick so that the last two sides of the pyramid can meet the vertex. Before you actually cut, calculate the angle at which you need to make the cut. Then determine the vertical distance, d, down the dip stick that you will need to begin your cut in order to achieve that angle. Once you make your cut, fold up the side of the pyramid to test out accuracy. How did you do?

8. Calculate the angle of the lateral edge to the base. Insert a protractor into one of the seams of your pyramid and measure the angle to confirm your calculation. Compare your two answers.

   Calculated Angle = ___________ Measured Angle = ___________

9. Calculate the Surface Area and Volume of your pyramid.

   Surface Area = ___________ Volume = ___________

PROJECT

THE LUXOR

LESSON PLAN

This activity is designed for review and enrichment. It is assumed that the students have already been instructed on calculating the various dimensions of a pyramid, slant height in particular, in order to calculate the surface area and volume. The Luxor hotel has world-record attributes that help the student distinguish the various components of the pyramid. The lesson is self-directed, however, each question has a particular point of emphasis that the teacher should assess.

Concepts
Pyramids: slant height, lateral edge, surface area and volume; right triangle trigonometry.

Time: 1 hour

Materials
Student handout.

Preparation
Students need sound understanding of pyramids and their components.

1. The student should place the given height in the interior of the pyramid, perpendicular to the base, rather than along the surface of the pyramid.
2. Again, the student is to identify and label these dimensions. Each of these three requires the student to visualize a right triangle within the pyramid. Encourage the students to sketch these triangles.
3. The area of the base is needed for calculating both surface area and volume. Focusing on the size of the casino, helps the students distinguish between area of the base (B) and a base edge (b).
4. The students here are to use the given height, and not the calculated slant height which is often erroneously done on calculating volume. The idea of the atrium focuses student attention on the interior of the pyramid.
5. The amount of glass used on the hotel is intended to draw the students' attention to the lateral surface of the pyramid, so they may visualize the lateral area as four distinct triangles.
6. This question makes connections to previous material, specifically, trigonometry and corresponding angles.
7. The novelty of the inclinators helps distinguish the difference between the slant height and the lateral edge. The fact that their angles to the horizontal are different makes for a good discussion on which length should be longer: a smaller angle should yield a longer length and a longer horizontal distance.

SOLUTIONS

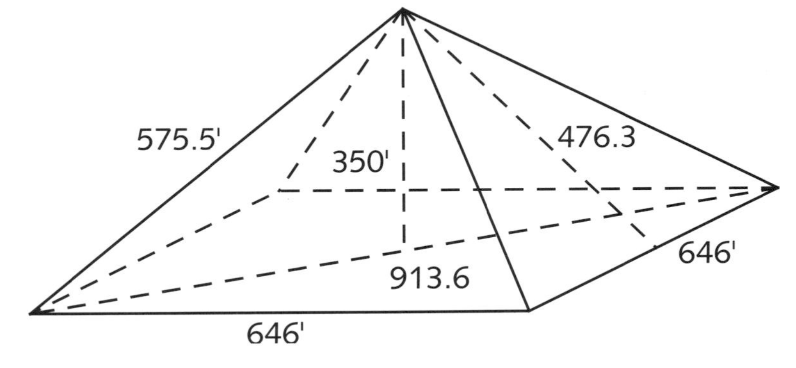

1. See diagram at the right
2. Slant Height = 476.3 feet
   Lateral Edge = 575.5 feet
   Diagonal of the Base = 913.6 feet
3. Base Area of the Casino = 120,000 square feet
   Percent of Total Base = $^{12,000}/_{417,316} \approx 29\%$
4. Volume = 48.6 million cubic feet.
   The actual atrium is only 29 million cubic feet. The various rooms in the hotel account for the difference.
5. Amount of Glass = 13 acres = 566,280 square feet.
   Lateral area = 615,380 sq. ft. The walls of the hotel are also comprised of steel and concrete, which account for the difference.
6. Angle of Glass and Floor = $\tan^{-1}(^{\text{height}}/_{\text{half of base edge}}) = \tan^{-1}(^{350}/_{323}) \approx \tan^{-1}(1.083) \approx 47.3°$

   Assuming all floors are parallel to one another and the glass wall represents a transversal, the angles made by the wall and each floor are corresponding angles.
7. Angle of Inclinators: $\sin^{-1}(^{\text{height}}/_{\text{lateral edge}}) = \sin^{-1}(^{350}/_{575.5}) \approx \sin^{-1}(.608) \approx 37.5°$

THE LUXOR

The Luxor Hotel in Las Vegas is modeled after the Great Pyramids of Egypt and boasts many world-record attributes. This activity investigates some of these awesome measurements. To begin, the building is 646 feet wide at the base and 350 feet high.

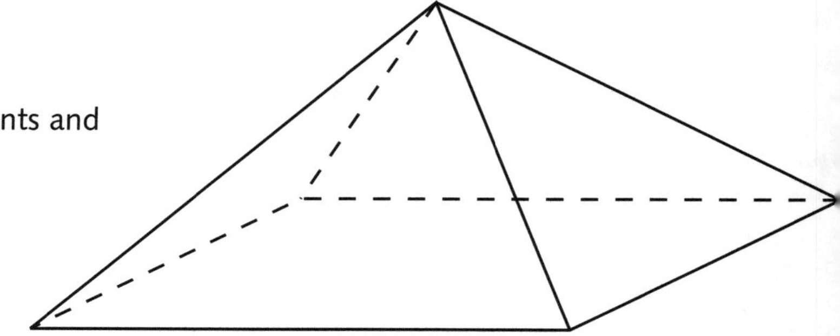

1. Place the dimensions in the diagram to the right.

2. Calculate the length of the following measurements and designate the answers on the diagram.

   Slant Height = ________

   Lateral Edge = ________

   Diagonal of the Base = ________

3. The casino at the Luxor is 120,000 square feet. What percentage of the pyramid's base is dedicated to the casino?

4. The Luxor boasts the largest atrium (a vaulted open space within a building) in the world, with a measurement of 29 million cubic feet. What is the volume of the entire pyramid? Your answer should differ significantly from the volume of the atrium. How do you account for this discrepancy?

5. The Luxor claims that its outer walls are covered by 13 acres of glass. There are 43,560 square feet in an acre. How many square feet of glass cover the Luxor? Calculate the lateral area of the pyramid. Your answers should be close, but not exact. How do you account for the discrepancy?

6. The rooms have an outer wall made entirely of glass. As you might assume, these glass walls are slanted. What angle do the glass walls form with the floor? What geometric principle guarantees that this will be true for all rooms along the same side of the pyramid?

7) Rather than elevators, the Luxor has inclinators (elevators that rise at an oblique angle) along the lateral edges of the pyramid. At what angle to the ground do the inclinators travel?

PROJECT

THE SHOPKEEPER'S

Guess how many marbles are in the jar and win a free gift!

Submitted by Greg Rhodes, Trabuco Hills HS, Mission Viejo, CA

We have all probably seen this type of contest in a local shop. As people walk by, they usually glance at the jar, scrutinize the marbles, and take a guess. Only rarely does someone use mathematics to improve their chances of guessing the correct amount in the jar. In this lesson, students will use a variety of measuring tools to mathematically estimate the number of marbles in a given jar. The overarching principle behind this lesson is that students can estimate the number of marbles by measuring the volume of one marble (of each type if there are different sizes), measuring the volume of the jar, and dividing to get the number of marbles that can fit.

Concepts
Volume (cylinders, and spheres), circumference and diameter, measurement, estimation.

Time: 1-2 hours

Materials
Large clear jar full of marbles (the more the better) graduated cylinders, calipers (if these are unavailable, compasses will suffice), measuring tape (sewing, not hardware style).

Preparation
Establish the five different measuring stations for each of the measuring methods. Each station should have one sample of each size marble in the jar. Talk to the science teachers at your school about borrowing any equipment you may need.

LESSON PLAN

The Hook

Have the jar of marbles prominently on display, when the students enter the room. Inform them of their task and have them make an initial guess. Once the students record this guess on the handout, explain the various measurements that they are to conduct. The classroom is organized into four different measuring stations explained below.

The Measuring Stations

Students measure attributes of the marbles and the jar and make some initial calculations. The first three are dedicated to measuring the marbles (in order to find the radius and thus the volume). Station #1 has a caliper (or compass), station #2 has a measuring tape; station #3 has a graduated cylinder. For large class sizes, it is advised to have several of the measuring devices at each station. Each station should also have one of each size marble that is in the jar. Using each method of measurement, the students should measure and record the indicated dimensions and make the requested calculations. Station #4 is focused on measuring the height, circumference and diameter of the jar in order to calculate its volume.

The four stations are discussed below. The students should be allowed to move freely among the different stations to measure and re-measure at their own pace. All measurements should be made in centimeters.

#1. Volume of Marble Using Diameter **Measuring Device:** Caliper

Here the students use a caliper to measure the diameter of the marble, calculate the radius, and then calculate the volume. If you cannot find any calipers, compasses work (albeit not perfectly). If you are using compasses, have the students "pinch" the marble with the ends of the compass and then measure that distance on a ruler or measuring tape. The students should then use the diameter to calculate both circumference and volume.

#2. Volume of Marble Through Circumference **Measuring Device:** Measuring Tape

Here the students measure the circumference by wrapping the tape around the marble. Suggest to the students to measure this a few times, because it is difficult to line the tape up exactly on the "equator" of the sphere. The students record the circumference, then use it to calculate the radius, and from that, the volume.

THE SHOPKEEPER'S 

Guess how many marbles are in the jar and win a free gift!

LESSON PLAN (continued)

#3. Volume of Marble Through Water Displacement **Measuring Device:** Graduated Cylinder
Here the students drop the marble in a graduated cylinder containing a small amount of water. The rise in the water level is the volume of the marble measured in cubic centimeters. Also discuss how to read a graduated cylinder (measure from the base of the meniscus, the curve of the water). The students are to use the volume to calculate the radius of the marble, and from that, the circumference.

#4. Volume of Jar Through Measurement **Measuring Device:** Measuring Tape
Here the students determine the volume of the jar through whatever measurements they choose. Give very little instruction on this to give students the creative freedom to problem-solve for the necessary components to calculate volume for your specific jar. If you have any empty jars that are identical to the one filled with marbles, place them here with plenty of measuring tape.

The Empty Space
One of the richest aspects of this lesson is the dilemma of the gaps between the marbles. Have a large graduated beaker of water handy. Pour just enough water into the jar, with the marbles in it, until the marbles are immersed. Have the students record the amount of water (cubic centimeters) that is now in the jar. This is the volume of empty space between all the marbles.

The Final Count
The students now determine what they perceive to be the typical marble. Most students determine the typical marble to be the average of the three volumes or the volume of the medium size marble. By subtracting the volume of empty space from the volume of the jar, and dividing this difference by the volume of the typical marble, the students can closely estimate the number of marbles in the jar. Once they do so, it's time to drain the water out of the jar, and count the marbles. Disseminate them on paper towels among the class and have everyone make a quick count. The students should record this count and calculate their margin of error.

TEACHER COMMENTS

- This lesson works best if you have a large number of marbles in the jar (at least one hundred). Several hundred marbles can be purchased cheaply from a toy store. This lesson actually began when I found my father's old marble collection from his childhood. He had saved over one thousand marbles! Thus, the shopkeeper's jar project was born.
- Different sized marbles work just as well as identical ones. In fact, it even challenges the students to take into account different sizes. This lesson is best done with three different sizes.
- An odd-shaped jar adds even more challenge to the problem. For example, I use a licorice tub (the kind you buy at a warehouse discount store) which is not a perfect cylinder; the top is a little wider than the bottom. Interestingly, the class usually produces two strategies for determining the average diameter of the jar. The first is to average the circumference of the jar at both the top and the bottom; the second is to measure the circumference of the jar at its mid-height. Both methods are correct.
- Encourage a "science lab" atmosphere of exploration and experimenting.

This is an example of the jar or marbles that we have used in our classrooms.

THE SHOPKEEPER'S 

Guess how many marbles are in the jar and win a free gift!

How many marbles do you GUESS are in the jar? ______________

MARBLES: Estimate the volume of each marble by using a ...

Caliper	Small	Medium	Large
Diameter/Radius	____ / ____ **cm**	____ / ____ **cm**	____ / ____ **cm**
Circumference	__________ cm	__________ cm	__________ cm
Volume	__________ cc	__________ cc	__________ cc

Tape Measure	Small	Medium	Large
Diameter/Radius	____ / ____ cm	____ / ____ cm	____ / ____ cm
Circumference	__________ **cm**	__________ **cm**	__________ **cm**
Volume	__________ cc	__________ cc	__________ cc

Graduated Cylinder	Small	Medium	Large
Diameter/Radius	____ / ____ cm	____ / ____ cm	____ / ____ cm
Circumference	__________ cm	__________ cm	__________ cm
Volume	__________ **cc**	__________ **cc**	__________ **cc**

The volume of a typical marble in the jar is: ________cc

JAR: Circumference ______ cm Radius ______ cm Height _____ cm Volume ______ cc

SPACE: Volume of space in the jar: __________ cc

How many marbles do you ESTIMATE are in the jar? __________

How many marbles are actually in the jar? __________

What was your margin of Error? ________%

Super Size It! Perimeter, Area, & Volume

LESSON PLAN

The task here is for the student to "Super Size" a popular brand name product. In other words, the student is to create a package that is larger, yet proportional to the product's original package. For example, a student may create a snickers bar that is 3.2 times the size of the original, or a cereal box that is 1.5 times as big as the original box.

Concepts
Area and Volume of Prisms, Fundamental Theorem of Similarity

Time
1 hour plus time for students to construct the box

Materials
Rulers and Student Handout. Students must also provide materials to create the super-sized box.

Preparation
Students should have a familiarity with ratios of perimeters, areas and volumes.
i.e. $(P_1/P_2)^2 = (A_1/A_2)$

1. Have students bring in a product package to class. It is recommended that most students use a rectangular prism, although, if they feel adventurous, they may explore cylinders and other solids. Have students measure the dimensions of the package and record these dimensions along with the name of the product on the student handout. Students should then decide on the ratio by which they wish to enlarge the package. Encourage them to use a ratio that is not a whole number. The students should figure what the new dimensions of the packages will be according to the ratios that they chose.
2. Have the students calculate the surface area and volume of both the original and new packages. Be sure they show their solutions.
3. Have the students then calculate the ratio of the new surface area to that of the original, and the new volume to that of the original. This is the crux of the lesson: The Fundamental Theorem of Similarity. The students are to understand that the ratio of the areas is equal to the square of the ratio of the perimeters, and that the ratio of the volumes is equal to the cube of the ratio of the perimeters. They should then show how this principle regarding the ratios can be used as a short cut to find the new surface area and volume. For example, assume that the original package was a 1 x 2 x 3 box and that we want to enlarge it to twice the dimensions: 2 x 4 x 6. The ratio of the new to original surface area would be 88/22 = 4 and the ratio of their volumes is 48/6 = 8. Thus, the new area could have been calculated by multiplying the original area by the square of the ratio of the dimensions. The volume could have been calculated similarly, but with the cube of the ratio of the dimensions.
4. The students are then asked to determine which ratio to use in order to enlarge the price. While the cost of a real product also involves packaging, advertising and overhead, we should base the pricing on the quantity of product. In our example above, by doubling the dimensions, we should increase the price eight-fold.
5. The final step, of course, is to create the package. This phase of the activity is intended to help students actually SEE these ratios. The students should create the package to the exact dimensions that they calculated. The real challenge for them will be to create the logo's and insignia proportional to the original design. Allow them to accomplish this by any means they choose, then discuss it when they submit their completed products. It is interesting to discover how many students voluntarily devise a methodical strategy that requires proportional reasoning.

Additional activities: (a) Give the students the two nets of a package that was enlarged by a scale factor of two. Offer each group of students four copies of the original net, and have them cut them into pieces in such a way that they fit exactly within the confines of the enlarged net. (b) Have samples of the two completed packages and fill the original with some type of small candy. Pour the candy into the large package. Ask the students to predict how many times this can be done until the large package is full, and then show them the true result (8 or 2^3 times total).

Super Size It! Perimeter, Area, & Volume

You work for Geo-Foods. The company wants to super size its most popular product. You are to determine the new dimensions for the product so that they are proportional to the old one. Your are also to choose an appropriate price for your product. Choose any product that has a rectangular prism for a package and CREATE the new package. Record all measurements and calculations below, and be sure to decorate the new package. Extra Credit for a ratio of similitude that is not a whole number.

Name of the product: ________________ **Ratio of similitude:** ________

Dimensions:
Original: _____ x _____ x _____ New: _____ x _____ x _____

Surface Area:
Original: ________ New: ________ Ratio: ________

Volume:
Original: ________ New: ________ Ratio: ________

Compare the three ratios above. ____

____

Pricing:
Original: ________ New: ________ Ratio: ________

Can There Be GIANTS?

LESSON PLAN

Hook the students with the two questions at the top of the student handout. The record for the tallest person is 8′11″. Why no bigger? While we can lift much more than an ant in absolute terms, why can we not carry something ten times our size like the ant can?

Focus the attention of the students on the first chart titled "Dimensions of Creatures," in which we have a hypothetical person who is 6′ tall, able to lift 200 pounds, and weighing 200 pounds. One day this person wakes up in the morning twice as big (tall, wide and deep). How much will this person be able to lift and how much will this person weigh?

To get these answers we turn to the blocks and the chart titled "Dimensions of Block Piles." Start by having students place a single block on the desk, and define the edge to be one unit long. Discuss the surface area and volume of this one block. These values already exists in the chart. Now have students build the 2 x 2 x 2 block pile, as shown in the figure on the handout. Have students discuss the surface area and volume of the pile, and their ratios to the original cube. Fill in the second column of "Dimensions of Block" chart and the "Ratio to Original." The ratios of 2, 4 & 8 can then be used to determine the ratios and dimensions of the Giant column in the top charts. (Note: The ratio of strength correlates to the ratio of skin — or surface area.) Complete the two block charts until the students discover the relationship in column n: n, n^2, n^3. Have students verify this conjecture in the next two charts. The charts should support the principle that the ratio for the surface area is equal to the square of the ratio of the dimensions, and the ratio of the volumes should be equal to the cube of the ratio of the dimensions. This principle can then be used to complete the charts.

Concepts
Ratios of perimeter, surface area and volume, writing algebraic expressions

Time: 1-2 hours

Materials
Student handout, uniform blocks or multi-link cubes

Preparation
The teacher needs to acquire 64 uniform blocks or multi-link cubes for each group of students.

COLORFUL POINTS OF DISCUSSION

A really bad B-movie titled "Attack of the 50-Foot Woman" depicted a woman who, by radiation poisoning, was enlarged to fifty feet tall and then ransacked a city. We rounded her height to 48 feet to make the ratios easier to calculate. In another movie, "Austin Powers: The Spy Who Shagged Me," Dr. Evil claims that Mini-Me is "one-eighth of me." Assuming that Mini-Me is half the height of Dr. Evil, this is actually one of the few times that Hollywood got the math right. Also, the tallest man on record died by having his internal organs crushed under his own body weight. In other words the ratio of his weight, overcame the ratio of his strength.

SOLUTIONS TO THE CHARTS

Dimensions of the Creatures

1"	3'	6'	12'	48'
.04	50	200	800	12,800
.005	25	200	1600	102,400

Ratios to Human

$\frac{1}{72}$	$\frac{1}{2}$	1	2	6
.0002	$\frac{1}{4}$	1	4	64
.000003	$\frac{1}{8}$	1	8	512

Dimensions of Block Pile

1	2	3	4	10	n
$6u^2$	24	54	96	600	$6n^2$
$1u^3$	8	27	64	1000	n^3

Ratio to Original

1	2	3	4	10	n
1	4	9	16	100	n^2
1	8	27	64	1000	n^3

Ratio to Previous

2	$\frac{3}{2}$	$\frac{4}{3}$	$\frac{5}{2}$	$\frac{n_e}{n_o}$
4	$\frac{9}{4}$	$\frac{16}{9}$	$\frac{25}{4}$	$(\frac{n_e}{n_o})^2$
8	$\frac{27}{8}$	$\frac{64}{27}$	$\frac{125}{8}$	$(\frac{n_e}{n_o})^3$

Ratio Between Any 2

$\frac{4}{2}$	5	$\frac{10}{3}$	$\frac{n_e}{n_o}$
4	25	$\frac{100}{9}$	$(\frac{n_e}{n_o})^2$
8	125	$\frac{1000}{27}$	$(\frac{n_e}{n_o})^3$

Can There Be GIANTS?

Can there be giants? Why can ants carry so much more than we can, relative to their own body weight? The answer to these questions can be found in the following investigation.

	Dimensions of the Creatures					Ratios to Human				
	Insect	Mini-Me	Human	Giant	Monster	Insect	Mini-Me	Human	Giant	Monster
Height (perim)	1"	3'	6'	12'	48'			1		
trength (area)			200lbs					1		
Weight volume)			200lbs					1		

	Dimensions of the Block Pile						Ratio to Originial					
Height	1	2	3	4	10	n	1	2	3	4	10	n
Surface Area	$6u^2$						1					
Volume	$1u^3$						1					

	Ratio to Previous					Ratio Between Any 2			
Height	2				$\frac{n_e}{n_o}$	$\frac{4}{2}$	$\frac{10}{2}$	$\frac{10}{3}$	$\frac{n_e}{n_o}$
Surface Area	4					4			
Volume	8					8			

Can There Be GIANTS?

1. Draw a figure similar to the one below, with a ratio of similitude of 1.5. Find the perimeter and area of both figures. What are the ratios of the perimeters and areas of the two figures? How do these ratios compare with the ratio of their dimensions?

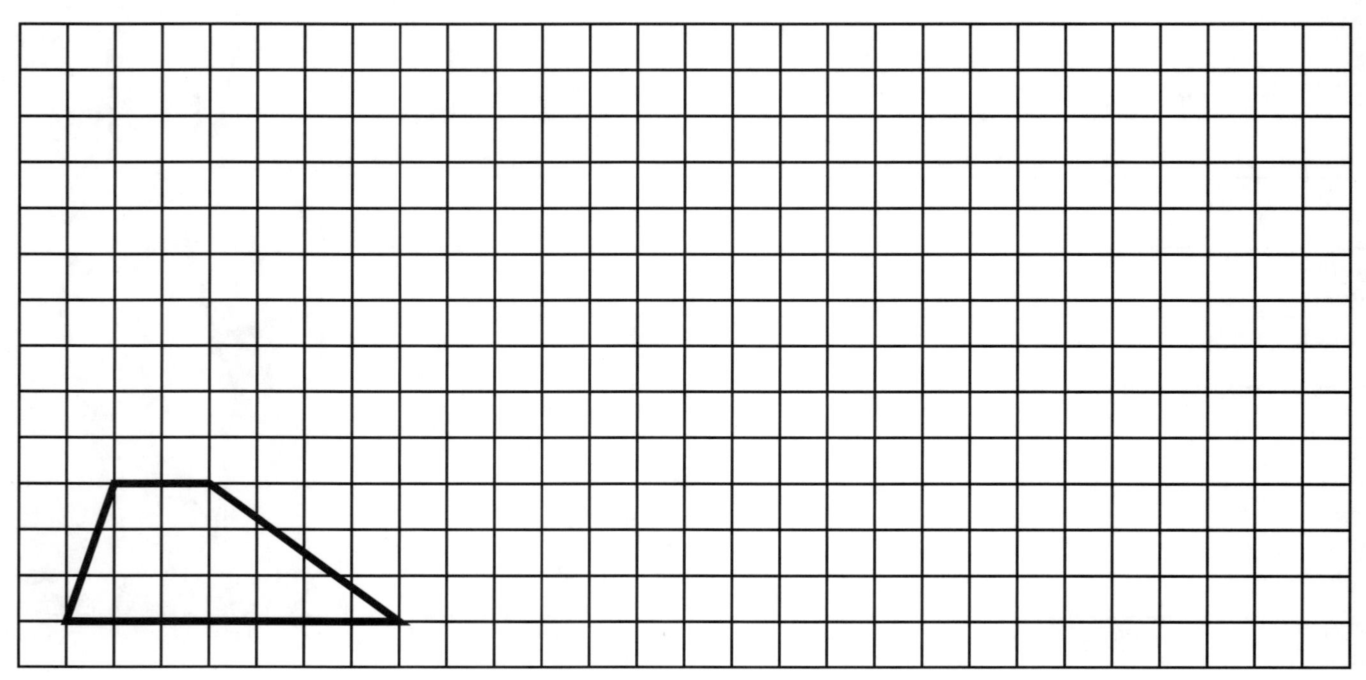

2. The dimensions of the solid on the right are doubled.

What are the new dimensions?	Height _____	Length _____	Width _____
What are the surface areas of the two solids?	Original _____	Enlarged _____	Ratio E/O _____
What are the volumes of the two solids?	Original _____	Enlarged _____	Ratio E/O _____

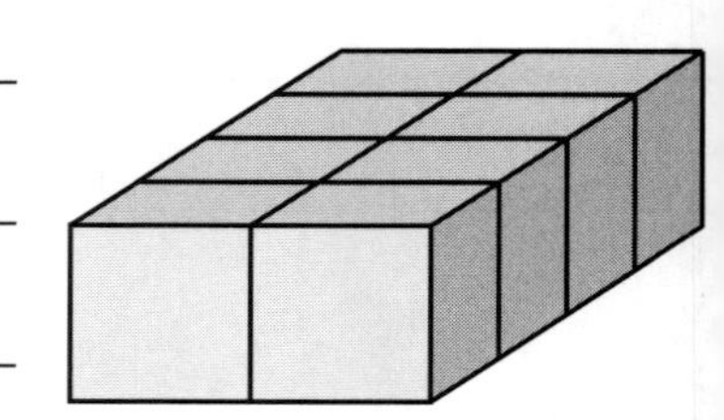

3. How many times larger are the surface area and volume of a solid for which the dimensions are . . .
   a) doubled?
   b) tripled?

4. The ratio of the heights of two similar solids is 5:2. What are the ratios of their surface areas and volumes?

5. The height of a trophy is 20 inches. The replica of the trophy is 50 inches. How much more gold plating is required for the new trophy? How much more will the new trophy weigh?

6. The surface area of a model bike is 75 in^2; the surface area of the real bike is 300 in^2. If the model bike is 15 inches tall, and weighs 3 lbs, what is the height and weight of the real bike?

Whatever Floats Your Boat

Submitted by Randy Brouwer, Trabuco Hills High School, Mission Viejo, CA

LESSON PLAN

BUOYANCY: The buoyancy worksheet prepares the students for the actual investigation with the penny canoe. The objective is to teach them how to calculate the weight from a given draft and vice-versa. The vehicle that allows them to do that is volume. For example, given ANY penny canoe made from a half-gallon milk carton, the volume of water displaced can be determined by the following formula: $V = d^2h$, where d is the draft (vertical distance under water) and h is the length of the canoe (the keel). It is derived from the formula of a prism: $V = Bh$. The area of the base for a triangular prism is: $B = \frac{1}{2}bd$, where d is the draft and b is the length of the water line at the bow/stern of the canoe. Since the triangle is an isosceles right triangle, $\frac{1}{2}b = d$. Therefore, $\frac{1}{2}bd = d^2$. Thus, we arrive at $V = d^2h$. (Have the students calculate the volume conceptually several times, before showing this formula.) Since a standard half-gallon milk carton is 19 cm long, we get $V = 19d^2$. Given a maximum draft of 6.7 cm for a standard milk carton, $V = 6.7^2(19) = 853$ cc. Since a cubic centimeter of water weighs one gram, the canoe will sink under the weight of 853 grams. For the purpose of this portion of the activity, assume the weight of the canoe to be negligible. Since we have a quadratic equation for our volume, the students should see a parabolic graph when they plot the data.

PENNY CANOE: Have students take turns at one or more of your water stations. The first thing the students will do is determine the weight of the empty canoe (approx. 26.75 g). In order to do this, the students will have to buttress the canoe on either side with a finger. Then they should weigh the designated number of pennies. Since our process of weighing is very crude, students will get a variety of answers. However, their final estimation for the weight of a single penny should be fairly accurate at 2.5 grams per penny. This places our estimation of the maximum number of pennies that the canoe will float at about 330 pennies. The nickel problem is a homework assignment to further reinforce the skills obtained during the lesson and is not intended to be an actual activity. A nickel weighs about 5 grams.

MAIDEN VOYAGE: The students' ultimate objective is to create a cardboard boat that will support the weight of one or more members of their group for a trip across the full length of a swimming pool. The students are permitted to cover the outside of the boat with duct tape, but must account for the extra weight, and may line the inside with clear plastic, if they wish. These are the only materials allowed in the construction of the boat. The students may use any design they wish (most will choose a flat bottom), but they must show their drawings and estimate the draft according to the weight of the passengers. The boat will be towed by another student along the side of the pool. Someone must be responsible for recording the actual draft of the vessel once it is in the water with all passengers aboard. The most effective boats will be those that have a relatively large base.

Concepts
Volume of prisms, area of triangles, unit conversions, solving equations, parabolic graphs.

Time: 2-4 hours depending on which segments of the project the instructor chooses to use.

Materials
Student handout; a class set of scissors and rulers; one or more water stations consisting of approximately 10 plastic baggies filled with 50 pennies each and a few large containers of water; student provided materials: half-gallon milk carton, cardboard and duct tape.

Preparation
Create one of the penny canoes from a milk carton as described in the lesson. Have the penny-bags and a water-filled container to demonstrate the lab and walk the students through the instructions. Have students, paired or in groups, create their own canoes from the milk carton they supply, as explained on the Penny Canoe handout. (You can view photographs of this lesson at www.mathprojects.com)

Whatever Floats Your Boat

MAIDEN VOYAGE

KEY CONCEPTS

Displacement: An object floating (or submerged) in water will displace a mass (weight) of water equal to the mass of the object.

Buoyancy: A boat's buoyancy is determined by the sum of the weights of a boat and its cargo divided by the volume of the boat. Measured in the metric units below, if this ratio is less than one (i.e. the weight of the boat and it's cargo, measured in grams, is less than or equal to the volume of the boat, measured in cubic centimeters), then the boat will float. On the other hand, if the ratio is greater than one (the weight is greater than the volume), then the boat sinks.

Unit Conversion: One gram of water = One cubic centimeter of water = One milliliter of water

Diagram of Boat:

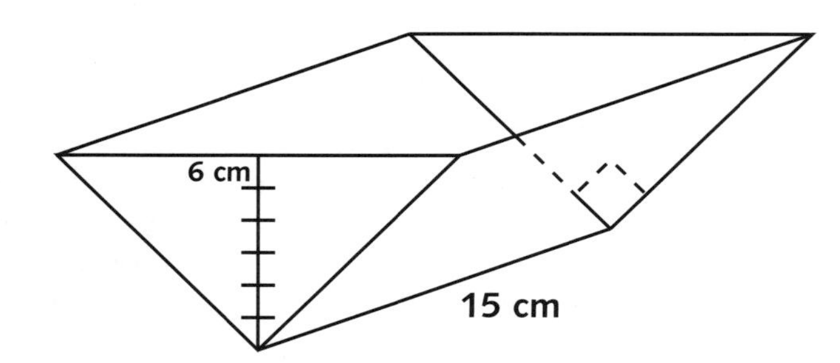

Though the lengths may vary, this is the model boat that we will be using throughout this lesson.

1. Complete the chart below by determining the mass of the cargo for the given draft of the boat.

Draft (cm)	1	2	3	4	5
Cargo (g)					

2. Complete the chart below by determining the draft of the boat for the given cargo.

Draft (cm)				
Cargo (g)	125	240	400	1000

3. What is the boat's maximum cargo weight?

4. Graph the data from the charts above. Before you do, predict what kind of relationship describes the data (linear, parabolic, hyperbolic, no correlation). After you graph, describe your findings.

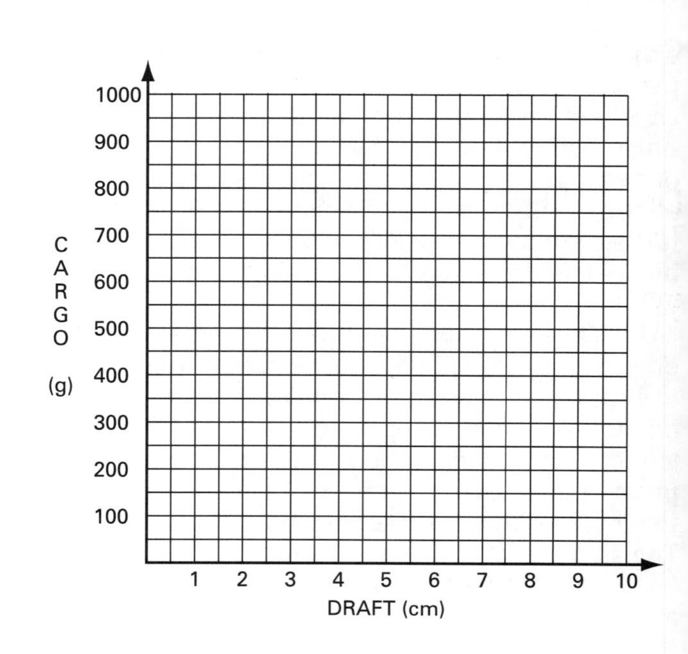

Whatever Floats Your Boat

BUILD A PENNY CANOE

Cut a half-gallon milk carton in half vertically along the diagonal of the base. If you then fold and staple the top portion flat, you will form a triangular prism. Along one of the bases, draw the draft -- a line from the keel to the top of the canoe, and mark it off in millimeters, labeling each centimeter. See the diagram on the buoyancy page.

DETERMINE THE WEIGHT OF A PENNY

Use your milk carton canoe, to determine the weight of a penny. Place the canoe in a container of water with each of the designated number of pennies in it:

a) Record the draft.
b) From the draft, calculate the volume (cm^3) and the weight (g).
c) Then determine the weight of one penny.

# of Pennies	Draft (cm)	Volume (cm^3)	Total Weight (g)	Weight per Coin (g)
100				
150				
200				

Average weight per penny: ______________ grams

Predict the maximum number of pennies that the boat can float.

Can you fill the boat with pennies, and still have it float? Why or why not?

DETERMINE THE WEIGHT OF A NICKEL

The experiment has been repeated using nickels, instead of pennies, and the results have been recorded in the chart below. Use this data to predict the weight of one nickel and the maximum number of nickels that the boat can float.

# of Nickels	Draft (cm)	Volume (cm^3)	Total Weight (g)	Weight per Coin (g)
25	2.8			
50	3.8			
100	5.2			

Average weight per nickel: ______________ grams

Whatever Floats Your Boat

THE MISSION

With the information and understanding that you now have about buoyancy, design a boat that will hold one or more of your group members while floating in a pool of water. Permissible materials are cardboard and duct tape. A plastic lining for the inside is optional. The boat may be of any shape. It does not have to resemble the milk carton boat that you built previously.

THE BOAT DESIGN

3-D Drawing | Top View | Front View | Side View

THE WEIGHT

. . . OF THE BOAT

Determine the weight of your boat by using a scrap piece of cardboard. Calculate the surface area of the scrap and weigh the scrap to determine the weight of the cardboard. Then use that information, along with the surface area of your boat, to determine the approximate weight of your boat.

a) Surface Area of the Scrap:

b) Weight of the Scrap:

c) Weight of the Cardboard: (in grams per square inch)

d) Surface Area of the Boat:

e) Weight of the Boat:

. . . OF THE CARGO

f) Total Weight of All Passengers Listed in Kilograms:

. . . OF THE TOTAL

g) Total Weight of the Boat and Cargo:

THE DRAFT

h) Estimated Draft:

THE CHRISTENING

i) Actual Draft:

Our Boat . . . Floated/Sank. Why?

What was common among the effective boat designs?

LESSON PLAN

The lesson is divided into three phases: 1) practicing the skills and algorithms used in the project, 2) taking the measurements at the shot put pit, and 3) calculating the radius, the degree measure of the arc and the arc length.

PHASE ONE: Practicing the Principles: On the first page of the handout, have students complete Part A: Dissecting a Circle (steps #1-8). The completed solution diagram is shown to the right. Since the students will be generating their own chords within the circle, their actual measurements will vary. The important points are that (AP)(PB) = (CP)(PD), the radius is $\frac{1}{2}$ (CD), and the degree measure of the arc can be found by using trig functions. In step #8, the use of the string in measuring the actual length of the arc is important because this is what the students will actually do in the shot put pit.

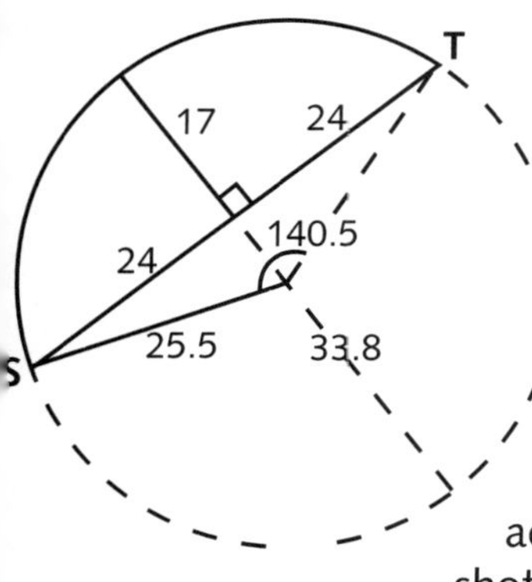

In Part B: Completing a Circle (step #9), the students have an opportunity to test their new understanding of these principles. The solution diagram is given here. Assuming all the students measure in millimeters, their measurements should be very similar.

PHASE TWO: Measuring the Pit: The next phase of the project is to have students actually measure the various parts of the arc in the shot put pit. The first two measurements, the chord and distance from the midpoint of the chord to the arc, are necessary for carrying-out other calculations later. The actual measuring of the radius and the arc are for verification purposes at the end of the project. It makes a very strong impression in the minds of students when their calculated answers are close to the measured answers.

PHASE THREE: Calculating the Arc: Once the measurements are acquired, the students should return to the classroom to make their calculations. The same process that they practiced on the first page of the worksheet should be implemented here, so limited guidelines are supplied to the students. In fact, the only two components that they are asked to record are the radius and the degree measure of the central angle/arc. From these two pieces of information, they can then calculate the length of the arc.

Concepts
Circles, lengths of chords, perpendicular bisector of a chord, arc length, trigonometric functions

Time: 2-3 hours

Materials
One ball of string (70 feet long), one compass and one tape measure for each group of 3-4 students; student handout.

Preparation
Verify that the shot put pit at your school has the appropriate distance markings (arcs are typically spaced ten feet apart). If a shot put pit is not available, then any arcs will do, like a curved set of stairs. Avoid conducting this project on a windy day, or your students' strings may end up a knotted mess.

SHOT PUT ARC

Your task is to calculate the length of one of the arcs in the shot put pit. Before you do so, here are two exercises to study the principles that you will need to accomplish this task. You will be applying your knowledge of chords, trigonometry, and arc length. For these two exercises, make all measurements of length in millimeters.

PART A: Dissecting a Circle

1. Draw chord AB anywhere in the circle such that it is NOT a diameter.

2. Draw a chord that is the perpendicular bisector of AB. Label the intersection of these chords point P and label the new chord CD so that CP is shorter than PD. Measure the following distances (in millimeters):

   AP = __________ PB = __________ CP = __________

3. Use the formula (AP)(PB) = (CP)(PD) to calculate the length of PD. Then measure PD to verify your answer:

   Calculated Distance of PD = __________

   Measured Distance of PD = __________

4. Find the center of the circle and label it M. Draw in radii AM and BM. Calculate the radius of the circle, then measure it to verify your answer.

   Calculated Radius = __________

   Measured Radius = __________

5. Verify your measurements for AP, PM, and AM by using the Pythagorean theorem.

6. Calculate m∠AMP and m∠AMB.

7. Use m∠AMB and the radius to find the length of arc $\overset{\frown}{AB}$. $\overset{\frown}{AB}$ = __________

8. Measure the length of arc $\overset{\frown}{AB}$ by laying a small piece of string atop the arc. Mark on the string, the length of the arc. Straighten the string and then measure the arc's length along the string. How accurate were you?

   $\overset{\frown}{AB}$ = __________

PART B: Completing a Circle

9. Use the algorithm above to find the length of arc $\overset{\frown}{ST}$, and to complete the circle.

T

S

SHOT PUT ARC

PART C: Measuring the Pit

10. With the string, form a chord from one end of your arc to the other. Measure the length of this chord.
11. Measure the distance from the midpoint of the chord to the arc, along the perpendicular bisector.
12. Measure the radius of your arc. Be sure to measure to the true center of the circle.
13. Now lay your string along the arc itself. Mark the length of the arc on the string itself. Straighten the string and then measure the length of the arc along the string.

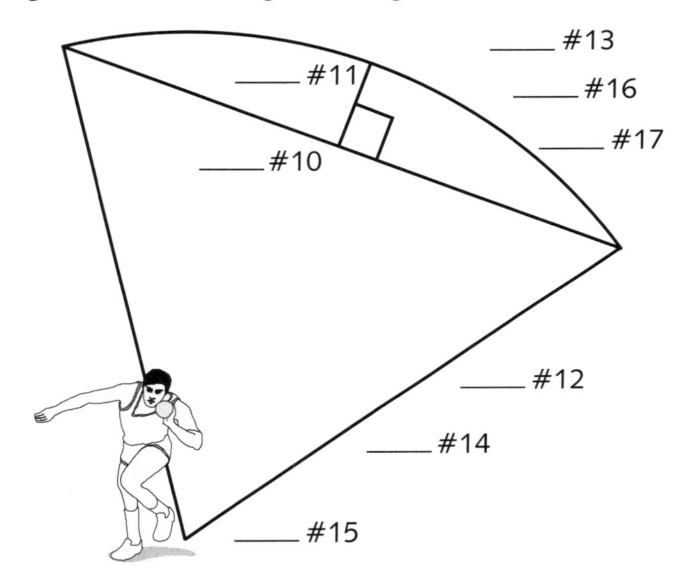

PART D: Calculating the Arc

14) Calculate the radius of the arc. It should be close to the the value that you measured, but not necessarily exact.

15) Calculate the degree measure of the arc.

16) Calculate the length of the arc.

17) Calculate your margin of error between your calculated length and the length that you measured in the pit.

STRIKE A CHORD

Recognizing Pervasive Patterns

"Within a circle, two intersecting chords cut each other into pairs of segments. The product of the segments of one chord equals the product of the segments of the other." This concept seems to perplex students year after year. The trouble may be that all the different scenarios — two chords, two secants, a secant with a tangent — are "dumped" on them all in one lesson. The general practice of presenting students with a barrage of numbers and theorems makes it difficult for them to develop a concrete understanding of the principles. This activity allows them this opportunity by having them measure actual examples of the various theorems.

Concepts
Circles, lengths of chords, perpendicular bisector of a chord, arc length, trigonometric functions

Time: 1-3 hours

Materials
One compass and ruler per student.

Preparation
This activity is an investigative, introductory lesson, so no prior knowledge by the students is required.

LESSON PLAN

The activity is divided into two lessons, one related to the measures of angles, the other related to the lengths of segments. For each of the four components, part (a) has the students draw and measure the scenario in order to discover the pattern, while part (b) allows them to verify it. Pause after each question to check that the entire class has successfully drawn it.

Do not give students any initial information about the chords, secants and tangents except for their definitions. To help focus the students, tell them that they will be searching for a pattern regarding these segments and lines. For example, in the second lesson, the students should measure the four segments "from point to circle, point to circle" (PA, PB, PC, PD) in millimeters. Have a diagram on the board that displays their common measurements. Prompt them to find a relationship: the product of the segments will be equal.

To test this conjecture, have each student draw a circle on the back of the student handout and place two intersecting chords anywhere within the circle (no diameters). The students should then measure the segments of these chords and validate the conjecture. Even with small errors in measurement, the pattern will make a strong impression in the students' minds and help them understand how to calculate the lengths of the chords. Stress that the size of the circles and lengths of the chords chosen by the various groups are all different; and yet the pattern persisted. Students can then practice some simple examples with given values for each segment. The process (with the same chart) can be repeated for problems involving two secants or a tangent and a secant.

CONJECTURES

- For two chords intersecting on a circle, the measure of the inscribed angle is **half of the measure of the intercepted arc.**
- For two chords intersecting in the interior of a circle, the measure of the angle of intersection is **half the sum of the measures of the two intercepted arcs.**
- For two secants or a secant and a tangent intersecting in the exterior of a circle, the measure of the angle of intersection is **half of the difference of the measures of the two intercepted arcs.**
- For two chords intersecting in the interior of a circle, **the product of the two segments of each chord are equal.**
- For two secants or a secant and a tangent intersecting in the exterior of a circle, **the product of the two lengths measured from the point of intersection to the circle, on each line, will be equal.**
- A radius perpendicular to a chord, **bisects the chord.**
  (This is proven by drawing the radii from center T to each end of the chord, A and B, and then proving these two triangles are congruent by the Hypotenuse-Leg Theorem.)

STRIKE A CHORD
Recognizing Pervasive Patterns

1. a) Measure ∠Q and arc $\widehat{AB}$. Designate three points C, C' & C" on circle Q such that the three points are NOT on arc $\widehat{AB}$. Measure ∠AC*B. What do you notice?

   b) Draw circle T. Designate two points A & B anywhere on circle T. Designate one more point C on circle T such that the point is NOT on arc $\widehat{AB}$. Measure the two angles shown. Do these measurements support your conjecture?

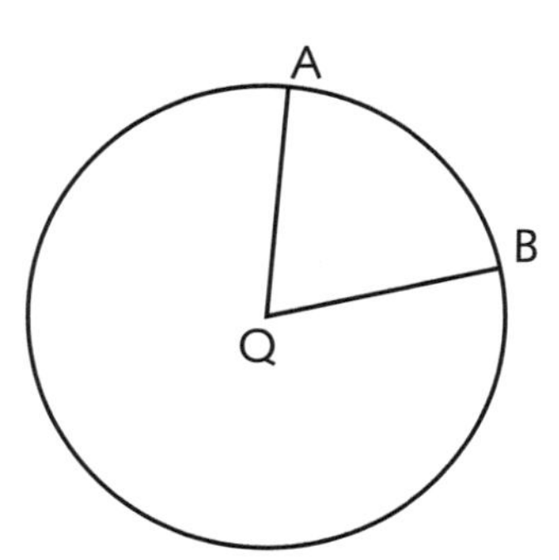

For two chords intersecting on a circle, the measure of the inscribed angle is ____________________

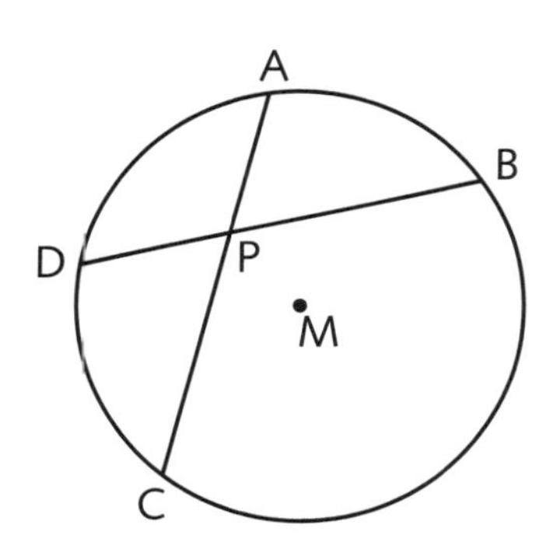

2. a) For circle M, measure arcs $\widehat{AB}$ and $\widehat{DC}$ and ∠APB. What do you notice?

   b) Draw circle S. Designate four points A, B, C & D on circle S. Draw chords $\overline{AC}$ & $\overline{BD}$ intersecting at point P. Use the diagram to verify your conjecture.

For two chords intersecting in the interior of a circle, the measure of the angle of intersection is

____

3. a) In circle L, measure arcs $\widehat{AB}$ and $\widehat{DC}$ and ∠APB. What do you notice?

   b) Draw circle R. Designate four points A, B, C & D on circle R. Draw secants $\overline{AD}$ & $\overline{BC}$ intersecting at point P. Use the diagram to verify your conjecture.

4. a) In circle O, measure arcs $\widehat{AB}$ and $\widehat{BC}$ and ∠APB. What do you notice?

   b) Draw circle W. Designate three points A, B, & C on circle W. Draw secants $\overline{AP}$ & tangent $\overline{BP}$ intersecting at point P. Use the diagram to verify your conjecture.

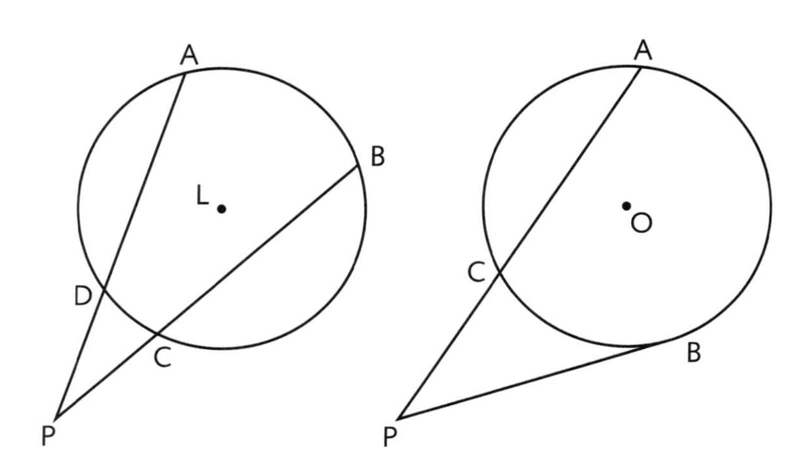

For two secants or a secant and a tangent intersecting in the exterior of a circle, the measure of the angle of intersection is ____

STUDENT HANDOUT

STRIKE A CHORD

Recognizing Pervasive Patterns

1. a) For circle M, measure segments $\overline{AP}$, $\overline{BP}$, $\overline{CP}$ & $\overline{DP}$. What do you notice?

   b) Draw circle S. Designate four points A, B, C & D on circle S. Draw chords $\overline{AC}$ & $\overline{BD}$ intersecting at point P. Use the diagram to verify your conjecture.

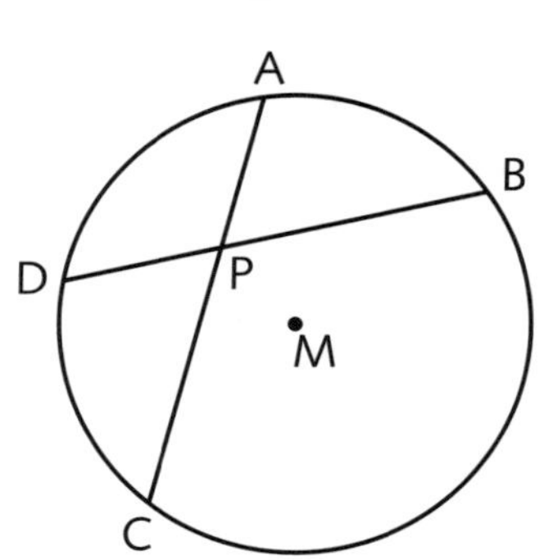

For two chords intersecting in the interior of a circle, ________________________________

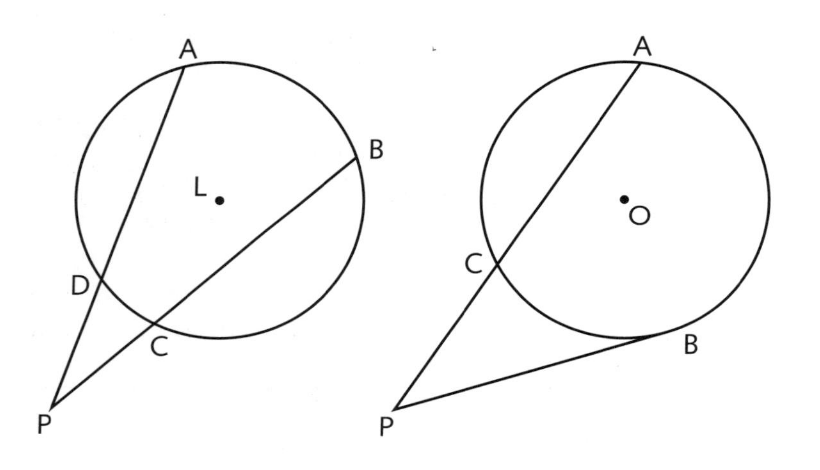

2. a) In circle L, measure segments $\overline{PA}$, $\overline{PB}$, $\overline{PC}$ & $\overline{PD}$. What do you notice?

   b) Draw circle R. Designate four points A, B, C & D on circle R. Draw secants $\overline{AD}$ & $\overline{BC}$ intersecting at point P. Use the diagram to verify your conjecture.

3. a) In circle O, measure segments $\overline{AP}$, $\overline{BP}$ & $\overline{CP}$. What do you notice?

   b) Draw circle W. Designate three points A, B, & C on circle W. Draw secants $\overline{AP}$ & tangent $\overline{BP}$ intersecting at point P. Use the diagram to verify your conjecture.

For two secants or a secant and a tangent intersecting in the exterior of a circle, ________________

____

4. a) In circle Q, draw any chord. Then draw a radius from the center of the circle perpendicular to the chord. What do you notice about the two segments of the chord?

   b) Prove the conjecture above; given circle T with $MT \perp AB$, prove that $AM \cong BM$.

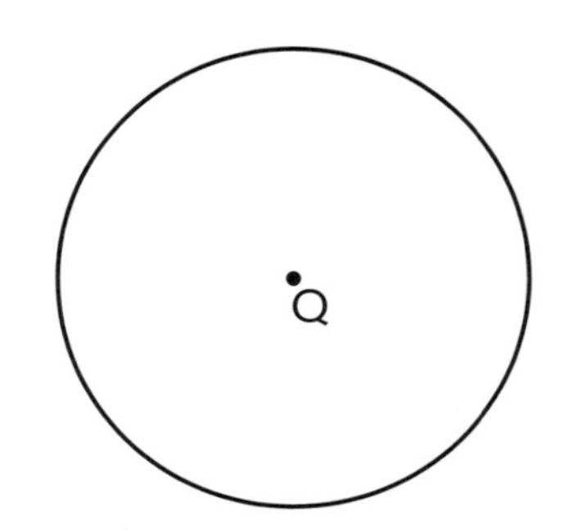

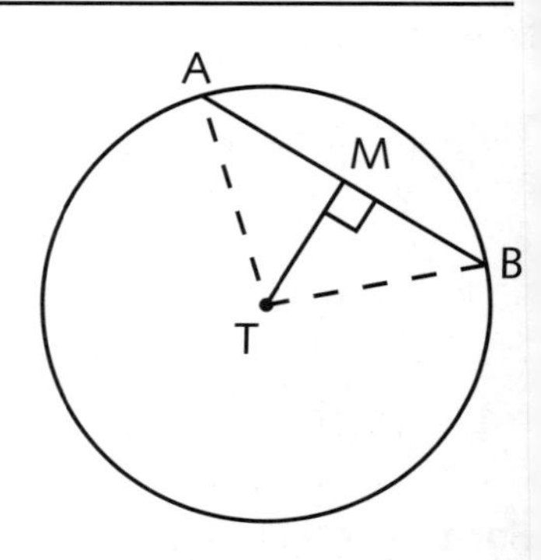

A radius perpendicular to a chord ________________________________

PROJECT

Reflection Golf

THE TECHNIQUE

This activity demonstrates how a reflection through a line can be used to determine the path of a bank shot in games like miniature golf and billiards. For a single bank, the ball is reflected over the wall that the ball is to strike, and from that reflected image, draw a line to the hole. Where the line crosses the wall of the golf hole is where the bank should occur. (see Figure 1) For two banks, the first reflection is reflected over the second wall — or its extension. (See Figure 2)

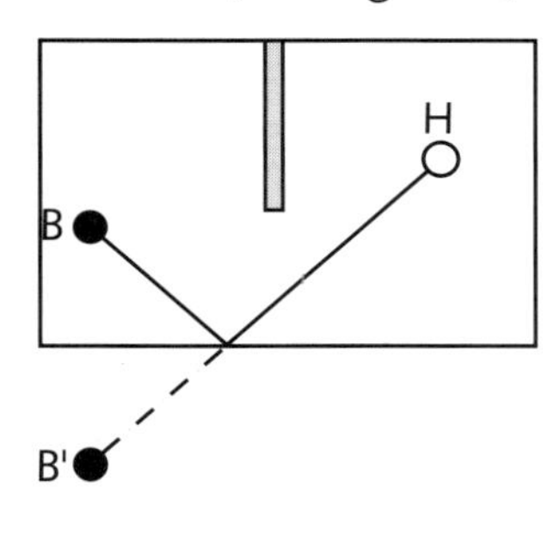

Figure 1 - One Bank

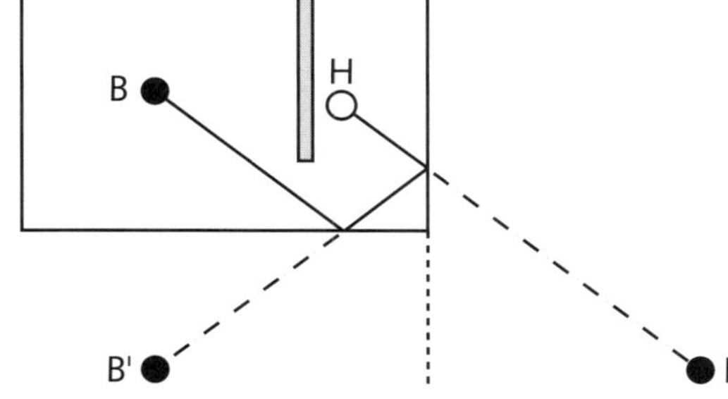

Figure 2 - Two Banks

Concepts
Transformations, reflections, and composite reflections.

Time: 1 hour

Materials
Student handout, rulers, colored pencils.

Preparation
Students should be familiar with basic reflections of a point through a line and with composite reflections.

THE LESSON

Have students practice with models like those shown above. Explain the activity as stated on the student handout with three major points of emphasis. The first is that the students are to show the reflections through each wall of the miniature golf hole. The second point is that the students need to remember to reflect the reflection for multiple banks. The final point is that the hole must require the designated number of banks. In other words, the student must create obstacles that mandate the determined path of the ball to the hole.

PROJECT

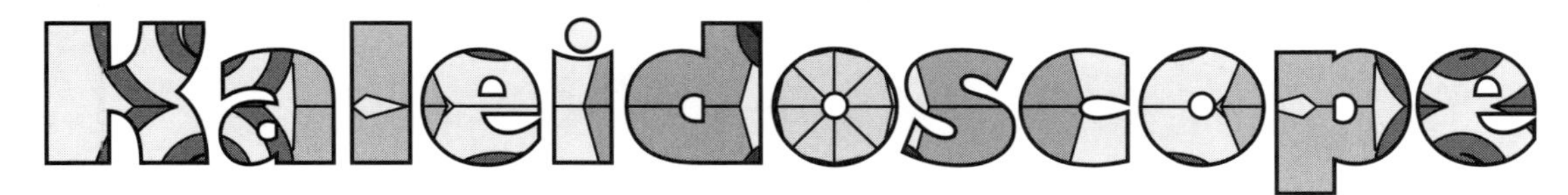

An actual kaleidoscope creates its unique design by reflecting a fundamental image (e.g. a group of colorful pebbles) through a series of mirrors. The kaleidoscope design in this activity is created in the same way. However, the fundamental image is an artistic rendering of the student's name, and reflecting lines are substituted for the mirrors.

The fundamental image (the student's name) is reflected over the nearest line, and then the reflected image is subsequently reflected over each successive line until all eight cells of the paper are filled. The next step is to label each type of transformation as a reflection or rotation (the cells alternate between the two options). Stress to the students that the goal is to create a complex pattern in which the original name is no longer recognizable. A unique design should be the final result (see figure on the left). This is best accomplished by making sure that the name reaches from one side of the cell to the other.

Concepts
Transformations, reflections, rotations, composite reflections.

Time: 1 hour

Materials
Student Handout, rulers, paper and colored pencils.

Preparation
Students should be familiar with basic transformations, reflections and rotations in particular.

Reflection Golf

You are to design one or more golf holes for a miniature golf course. Show each drawing on a separate sheet of paper. The following stipulations apply:

1. Each hole will require at least the designated number of banks from the tee to the hole.
2. You are to show the path of the ball with a solid line, the projection of the path outside the course will be designated by a dotted line.
3. Show each reflection of the ball, designated by B, B', B" etc.
4. Use dashed lines to show wall extensions.
5. Show your obstacles - these will necessitate the banks.
6. Have fun by coloring your course.

For example:

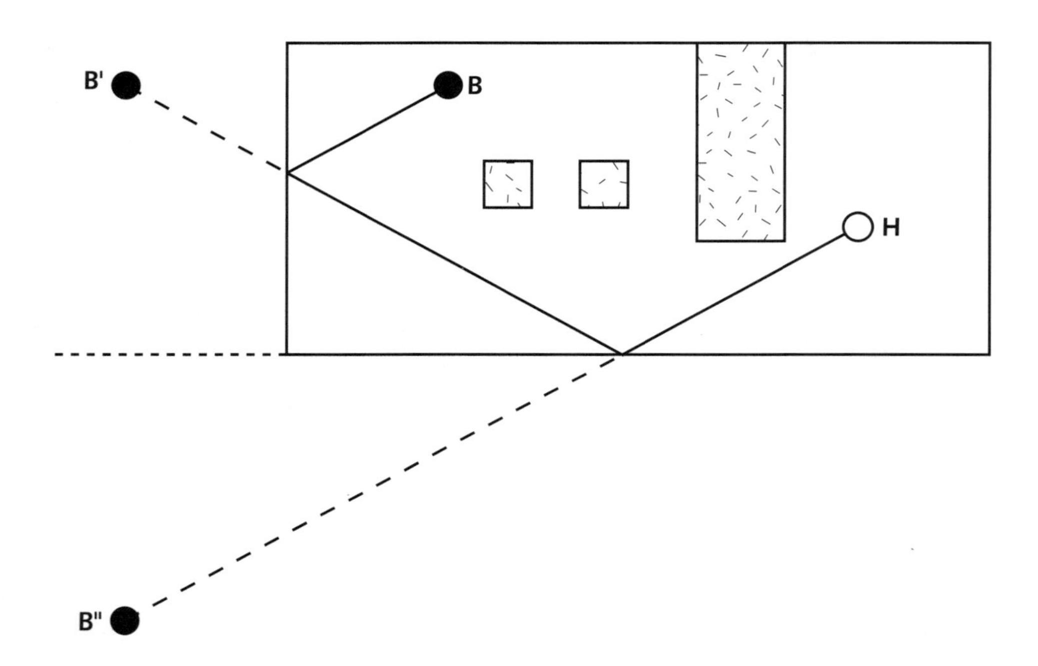

NOTES ON GRADING

To receive at least a C on the project you must successfully complete one hole, Hole #1, which requires one bank. A grade of B will be given for a second hole, Hole #2, which requires two banks (in addition to the first hole). For an A, you must also complete a third hold, Hole #3, which requires three banks (in addition to the first two). Extra credit will be given for designing a fourth hole, Hole #4, requiring more than three banks.

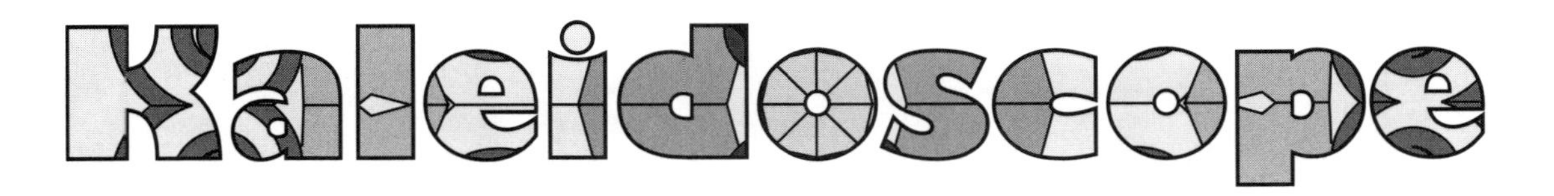

Design a kaleidoscope image using your first name and its transformations.

1. Begin with a square sheet of paper (trim an 8.5" x 11" to a square). Create a fold along the paper horizontally, vertically, and along each diagonal, so that eight triangular regions are formed.

2. In one of the regions, place an artistic rendition of your first name. Use some kind of block or puffy letters so that your name fills the entire region, and the borders of the letters actually touch the borders of the region.

3. Then reflect your letters through one of the folds to fill the next region. (see diagram below). Continue to do this all the way around the page until all regions are full. Keep the color scheme of each letter consistent throughout the design. (e.g. "C" below is orange through the design, while "H" is always green.)

4. On the backside of the original region, write the word "original." On the backside of the first reflected region, write the word "reflection." On the back of each of the other regions, write whether that region is a reflection, translation, rotation, or glide reflection in regards to the original.

For example:

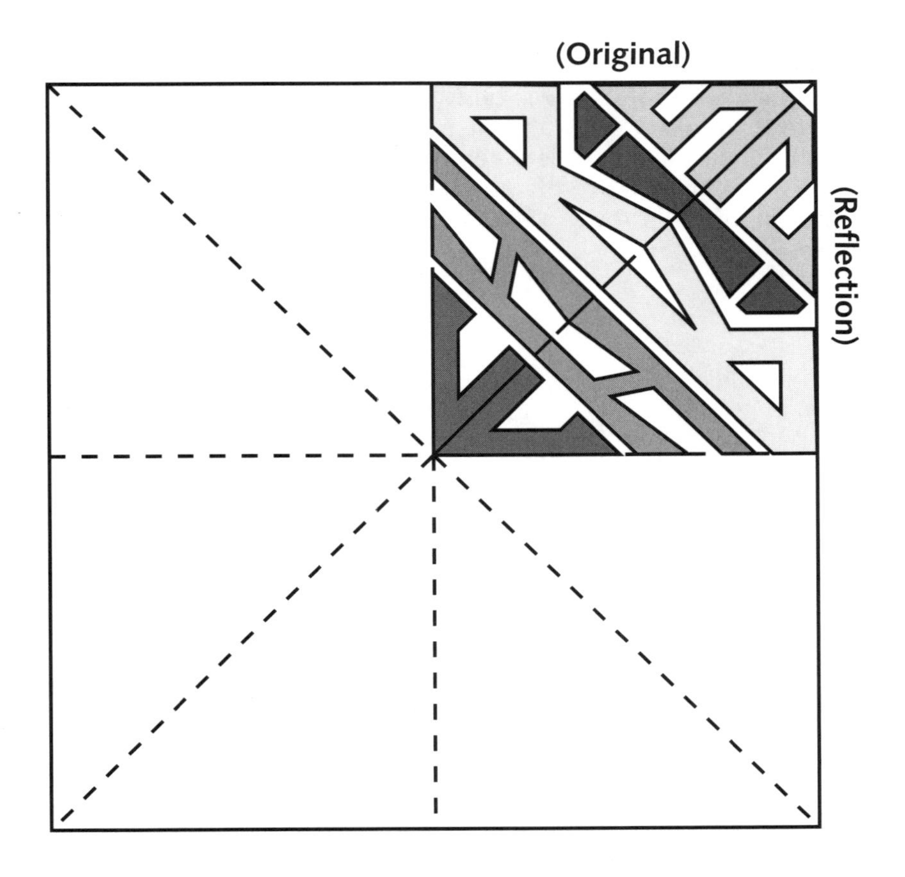

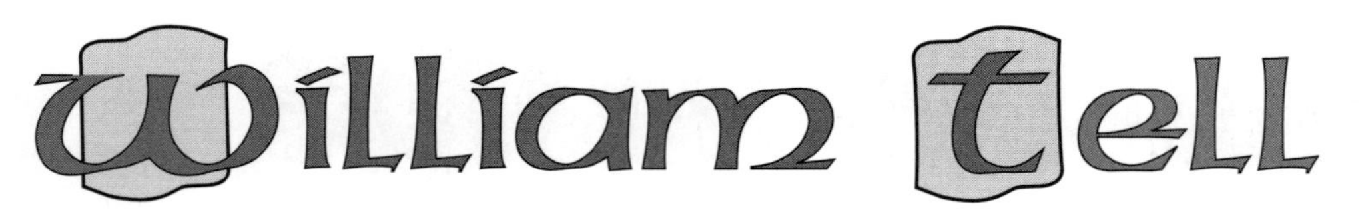

William Tell

THE GAME

You are at the annual Renaissance Festival and because of your excellent geometric mind you have decided to try your luck at the William Tell Game. Remember the famous archer who had to shoot an apple off his own son's head? In this version, there are three mannequins, three balloons representing apples, and an arrow that you use to shoot at the balloons. On the student handout, you will find a description of the mechanical function of the game.

The object of the game is to place two lines of reflection so that on your first turn the arrow will pivot along the arc through the first balloon (thus popping it) but not through the second. On your next turn you must place the lines of reflection so the arrow will pivot along the arc, through the second balloon, but not the third. On your third and final turn, you must show the lines of reflection so that the arrow will rotate exactly to the third balloon, but no farther. After each turn, the arrow returns to its original position. Just like any other carnival game, you get to upgrade your stuffed animal for each successful attempt.

Concepts
Rotation by reflection over two intersecting lines and rotation of a figure around a given point.

Time: 1 hour

Materials
Protractor, straightedge, student handout

Preparation
Familiarize students with rotations and reflections.

THE LESSON PLAN

Explain the game to the students and show them the example diagram below. Stress that the magnitude of the rotation will be twice the degree measure between the lines. For example, if we want the arrow to come to rest 120 degrees from its origin, then the angle between the lines should measure 60 degrees. Also stress the importance of showing the intermediate reflection. It may help the students to reflect both the tip and the tail of the arrow, so they keep its orientation correct.

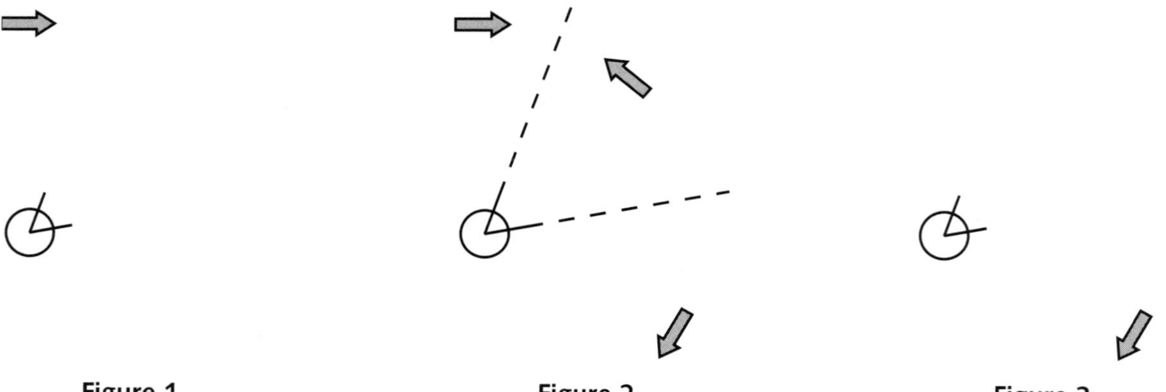

Figure 1 Figure 2 Figure 3

Have the students work in groups on the first problem (or balloon) on the worksheet and discuss as a class. The main issue for the students will be where to place the pointers (lines of reflection). Again, reiterate that the key is the angle between these lines. Share out several samples on the board so it is very evident that the angle is the important component, not the location of the lines.

Work out the question for the second balloon and share out the results. For the third balloon, be sure the students are aware that the arrow tip must come to a stop exactly ON the balloon, not through it. If the ballon hits the wall behind this ballon, they lose all the prizes that they have acquired so far.

You are at the annual Renaissance Festival and because of your excellent geometric mind you have decided to try your luck at the William Tell Game. Remember the famous archer who had to shoot an apple off his own son's head? In this version, there are three mannequins, three balloons representing apples, and an arrow that you use to shoot at the balloons.

William Tell is a mechanical game that works like this: You stand at a large metallic cylinder (much like an oil drum). Projecting from the right of the drum is a thin steel rod with an arrow at the end of it. Atop the cylinder is a large button and two pointers that look like antennae. When you hit the button, the arrow pivots clockwise around the cylinder. Along this arc are the three mannequins, each with a balloon atop its head.

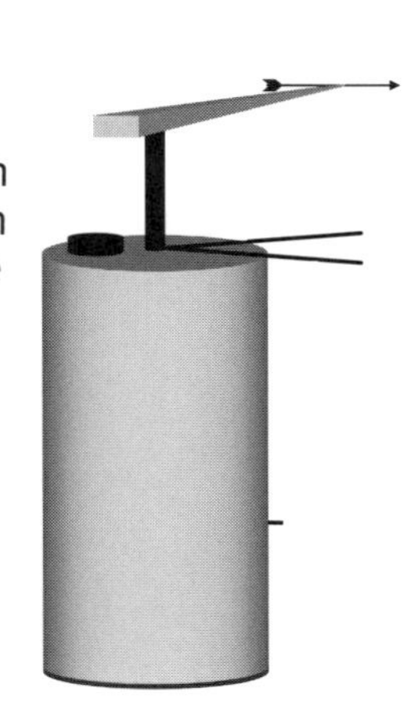

The distance that the arrow rotates all depends on the position of the pointers. The machine calculates the final resting position of the arrow by calculating the composite reflection (rotation) of the arrow over the two lines represented by the pointers. The arrow itself moves only along the arc, the reflections are done strictly within the machine's computer.

Your task as the player is to set the two pointers in such a position that the arrow will pop the designated balloons.

1. Determine the position of the pointers such that the arrow pops the first balloon, but not the second.

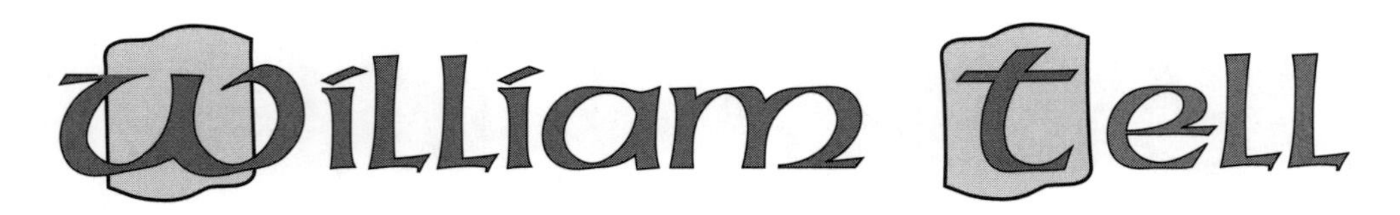

2. Determine the position of the pointers such that the arrow pops the second balloon, but not the third.

3. Determine the position of the pointers such that the arrow pops the third balloon, but does not strike the backboard behind it.

The $elling of AMERICA

Adapted from Supplemental Materials, The University of Chicago Mathematics Project
Randy Hoffman, Trabuco Hills High School, Mission Viejo, California

THE FOUR-COLOR CONJECTURE

Given any 2-dimensional map, wherein no adjacent regions may share the same color, the minimum number of colors required to color the entire map will never exceed four.

LESSON PLAN

Introduce the conjecture by laying out the ground rules for map coloring, and then having students experiment with simple models. The rules are simple. No two regions (states or countries) may share the same color — otherwise they would look like a single region. Regions that share only a corner (vertex) may share the same color. The obvious maximum number of colors that may be used is equal to the total number of regions on the map. However, the challenge is to use as few colors as possible. With these rules, have students color the following regions with the fewest number of colors possible. (Note: if colored pencils/pens are not available, have the students use numbers to represent colors. For instance all regions with a number "1" in them would be the same color).

Concepts
Four-Color Conjecture

Time: 1 hour lesson, several days for student homework

Materials
Student handout (map and chart). Have many extra copies of the map available.

Preparation
None, unless you intend to have students reproduce their charts within a spreadsheet.

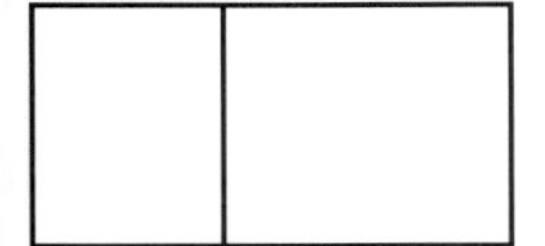
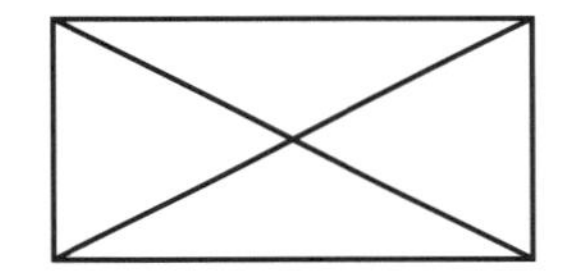
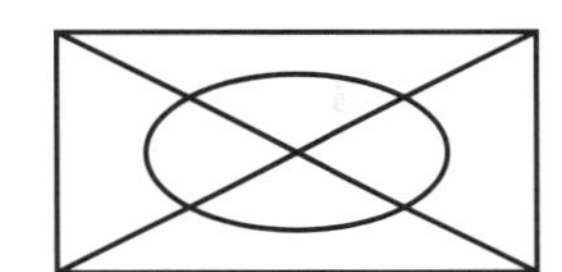
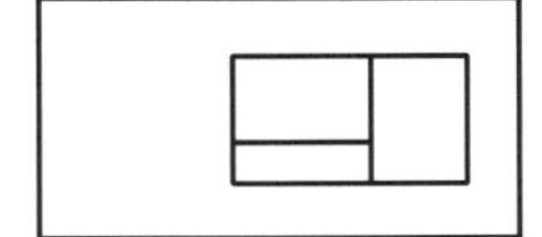
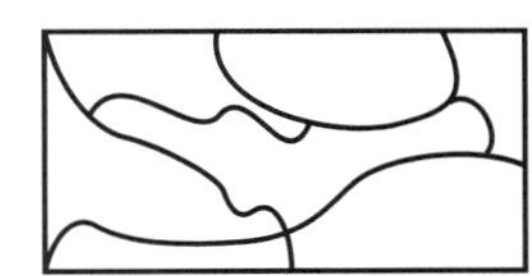

Have students place their responses on the board, and invite others to check the accuracy of the work. (Is this truly the fewest number of colors needed for this map?) Then solicit students for a conjecture on the fewest number of colors required for the maps on the board (four). Ask them to draw maps that require exactly four colors, and have some of those solutions placed on the board. Finally, challenge the class to create a map that requires more than four colors. You will probably get several confident answers to this one, all of which will be able to be colored with four or fewer colors. This is a concept that they will enjoy and can grasp very quickly. Once they do, inform them that they must now use this knowledge to sell off the United States of America!

THE CHALLENGE: Color and Sell the United States of America

Someone (the teacher) has offered to buy the United States of America. The country has been opened up to bidding, and will be purchased from the lowest bidder. The objective is to sell the United States for the cheapest price, but the map of the country must first be colored. The selling price per square mile of each state is determined by its color: Red $1 (per square mile), Yellow $2, Green $3, Blue $4. So for instance, Oklahoma is 69,956 square miles. If it is colored red it will sell for $69,956; if it is colored blue, it will sell for four times that amount, or $279,824. Each student is to color the map accurately according to the rules discussed and list the sale price of each state and the total sale price for the USA in the chart provided.

Give students some time in class to get started on the assignment, and then give them several days at home to complete it. During the class time, observe their strategies for optimizing the price. They should quickly recognize that large states and isolated states (Hawaii & Alaska) should most likely be colored red, while smaller states should be colored blue. However, some comprises will need to be made because of the juxtaposition of the states. Let them have fun with it, and be sure to show off the lower bids when the assignment is due. For those teachers who like to take their classes to the computer lab, the completing of the chart makes for a great spreadsheet activity.

The $elling of AMERICA

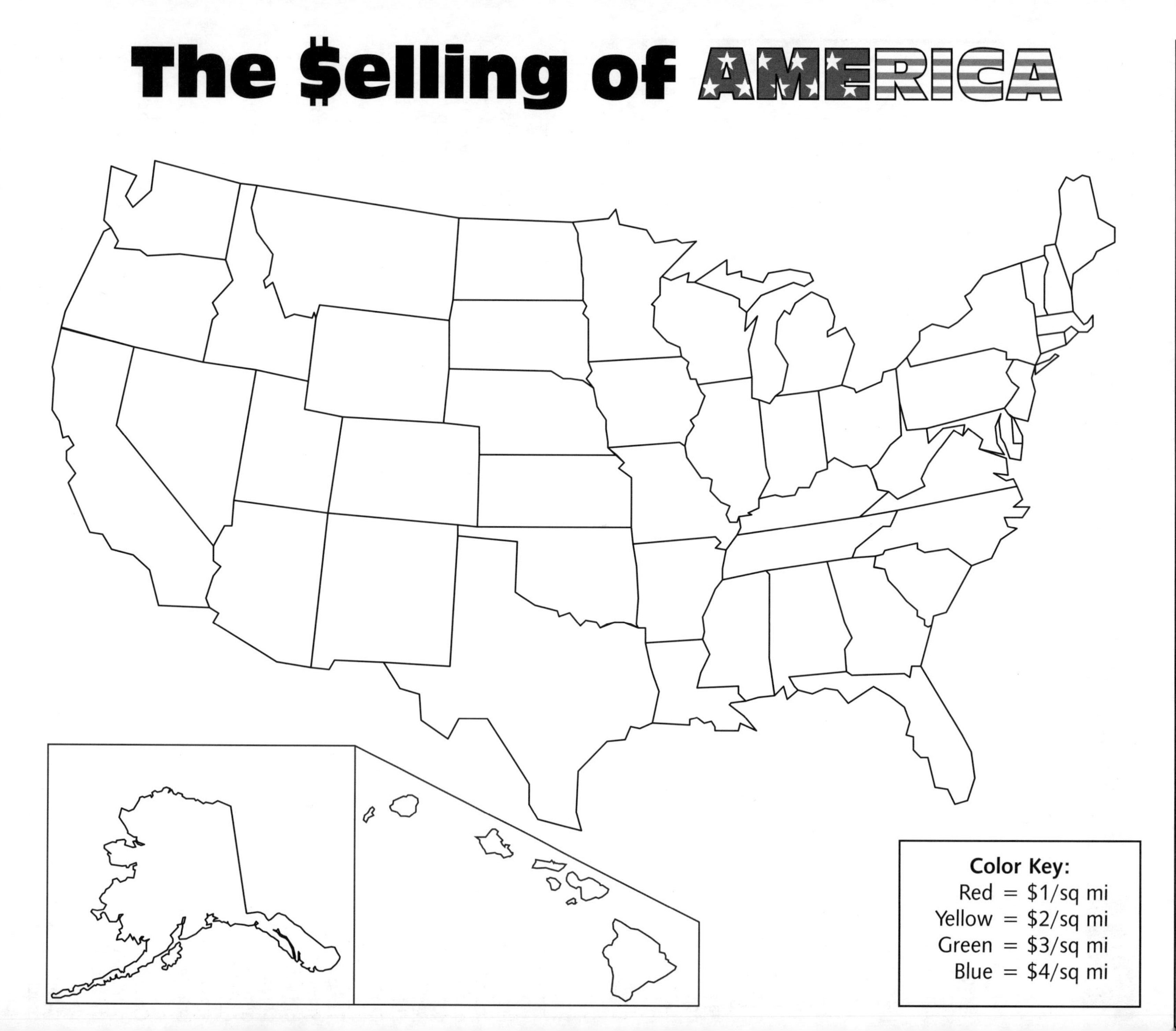

The $elling of AMERICA

STATE	AREA	COST	TOTAL	STATE	AREA	COST	TOTAL
Alabama	51705			Missouri	69697		
Alaska	591004			Montana	147049		
Arizona	114000			Nebraska	77355		
Arkansas	53187			Nevada	110560		
California	158706			New Hampshire	9279		
Colorado	104091			New Jersey	7787		
Conneticut	5019			New Mexico	121593		
Deleware	2044			New York	52735		
Dist. of Columbia	69			North Carolina	52669		
Florida	58664			North Dakota	70703		
Georgia	58910			Ohio	44786		
Hawaii	6471			Oklahoma	69956		
Idaho	83566			Oregon	97073		
Illinois	57871			Pennsylvania	46043		
Indiana	36413			Rhode Island	1213		
Iowa	56275			South Carolina	31112		
Kansas	82280			South Dakota	77116		
Kentucky	40409			Tennessee	42144		
Louisiana	47751			Texas	266807		
Maine	33265			Utah	84899		
Maryland	10460			Vermont	9614		
Massachusetts	8284			Virginia	40766		
Michigan	97102			Washington	68138		
Minnesota	86614			West Virginia	24231		
Mississippi	47690			Wisconsin	66215		
				Wyoming	97809		
	SUB-TOTAL				SUB-TOTAL		
					GRAND TOTAL		

THE VECTOR MAP

In this lesson, students will use vectors to draw a scaled map of a variety of objects, such as trees, lampposts, buildings, or pillars. Students will determine the magnitude and direction for each object, calculate the components of the vector, and write the coordinate pair that represents the object's location. The final product for this lesson is a map of all the objects, their coordinates, and their vectors (see bottom of page).

Concepts
Vectors, components, trigonometric functions, coordinate graphing.

Time: 1 hour

Materials
Protractors and straws

Preparation
Preview the venue and establish the origin, the landmarks that determine the axis, and the various objects.

ESTABLISHING THE COORDINATE PLANE

Prior to the lesson, choose an area of campus of which the students will make a map. Requirements for the area are:

a) In the center of the area should be some landmark (pole, trash can etc.) which can be used as the origin of an imaginary set of axes.
b) There should be four objects to the imaginary north, south, east and west compass points that establish the x- and y-axis. Portable objects (books and trash cans) may be used if no permanent landmarks are available.
c) You will need 4-6 objects to represent the terminal points of the vectors on the map. They can be almost anything, but try to spread them out 360° around the origin (at least one in every quadrant). The more objects you choose, the more practice the students will have in calculating their components. If there are not enough trees, lampposts, etc. you can provide your own with chairs, trash cans, or even books. The path from the origin to the objects must be unobstructed. The students will be pacing off the distance between these objects.

MEASURING THE VECTORS

For each object, the students need to measure two attributes of the vector: the magnitude and the angle.

Magnitude: The students are to pace the distance from the origin to the object. As long as the same person within each group paces-off each object with consistent strides, the scale of the map will be accurate.

The Angle: While one student in each group is pacing-off the magnitude, another student should be measuring the angle formed by the object, the origin, and the x-axis. This is done by the student holding a protractor flat (horizontally) with the straightedge aligned with the x-axis. The student then aligns a straw through the vertex of the protractor so that the object is visible through the straw. Another student then reads the angle measure determined by the straw on the protractor. This is the angle of the vector measured from the x-axis.

This simple procedure should be repeated for each object. The students will need about 30 minutes to measure the angles and distances of all objects. If there is any remaining time, return to class and allow the students to begin their calculations.

SAMPLE VECTOR

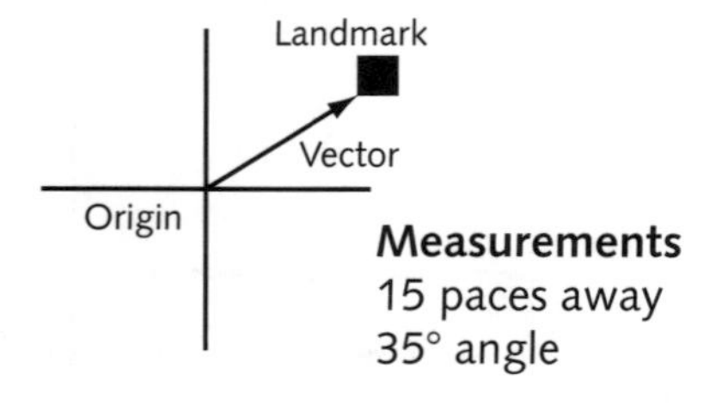

Measurements
15 paces away
35° angle

Calculating Components

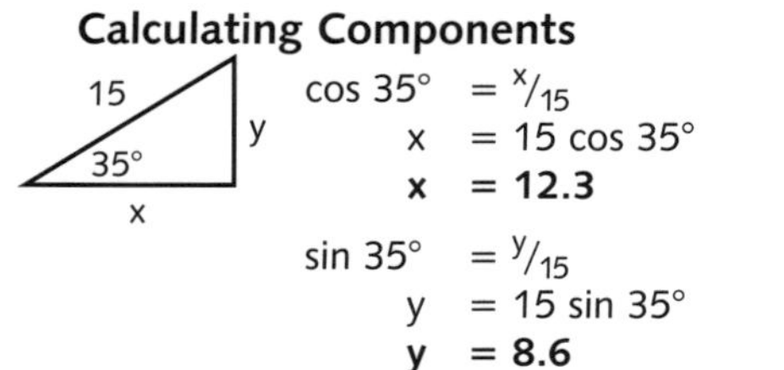

$\cos 35° = x/15$
$x = 15 \cos 35°$
$\mathbf{x = 12.3}$

$\sin 35° = y/15$
$y = 15 \sin 35°$
$\mathbf{y = 8.6}$

Vector

15 @ 35°
North of East

Coordinates & Graph

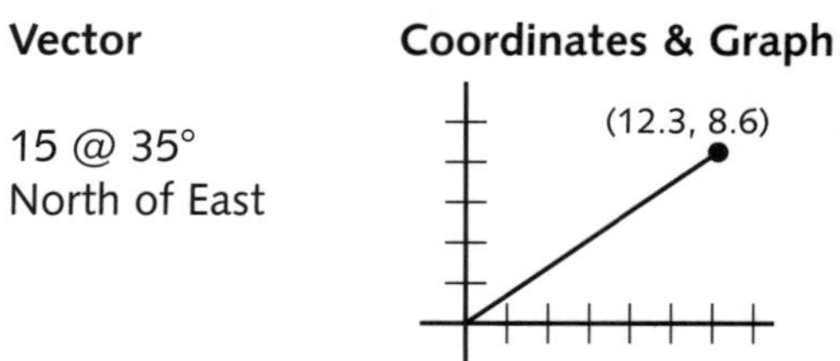

THE VECTOR MAP

Draw a scaled map of a given area of campus using vectors and coordinates to identify all the designated objects. First pace off the distance from the origin to the object. Record the number of paces as an estimate of the vector's magnitude. Next, with the compass and straw measure the angle from the x-axis. This will give you the direction of the vector. An example piece of data is: 17 paces @ 40° North of East (first quadrant). Repeat this process for each object, having the same student pace off all the objects. Then, use the trig functions to calculate the components of your vectors. Plot the coordinates of each object on the graph below.

Object #1: Magnitude __________ @ _____° ______ of ______ (,)

Object #2: Magnitude __________ @ _____° ______ of ______ (,)

Object #3: Magnitude __________ @ _____° ______ of ______ (,)

Object #4: Magnitude __________ @ _____° ______ of ______ (,)

Object #5: Magnitude __________ @ _____° ______ of ______ (,)

Object #6: Magnitude __________ @ _____° ______ of ______ (,)

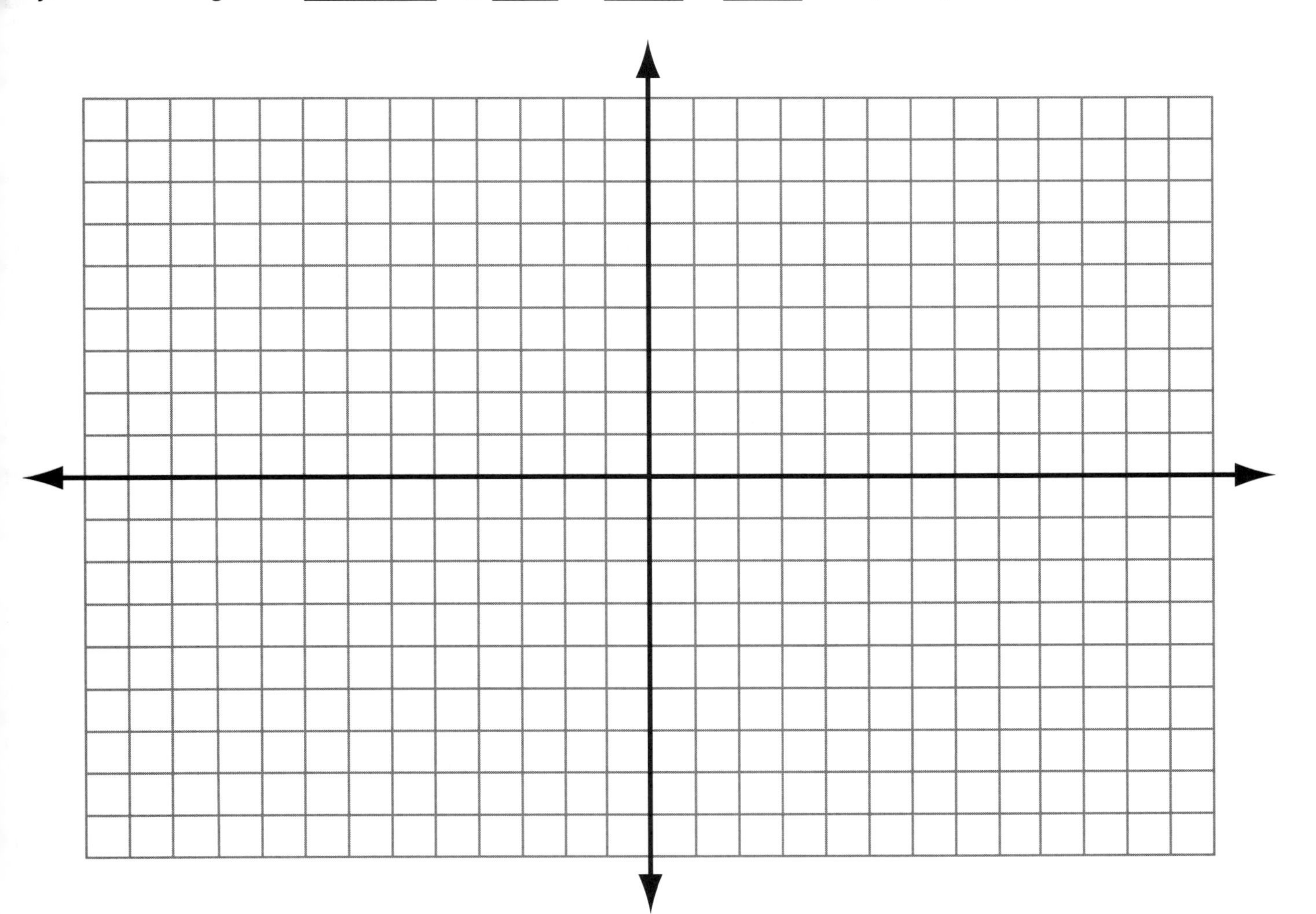

Princess Dido: The Geometry of an Ox Skin

THE LEGEND OF PRINCESS DIDO

According to the epic *Aeneid*, Dido (pronounced "Dee Dough") was a Phoenician princess from the city of Tyre (now part of Lebanon). Her treacherous brother, the king, murdered her husband, so she fled the city and sailed with some of her loyal subjects to Carthage, a city on the northern coast of Africa. She wished to purchase some land from the local ruler in order to begin a new life. However, he didn't like the idea of selling land to a foreigner. In an attempt to be gracious and yet still spoil Princess Dido's request, the ruler said, "You may purchase as much land as you can enclose with the skin of an ox." Undaunted, Princess Dido and her subjects set about the task by slicing the ox skin into thin strips and then tying them together to form a long band of ox hide, and foiling the ruler's malicious plan. (Hildebrandt & Tromba, *Mathematics and Optimal Form*, 1985)

According to this legend, how large of an area could Princess Dido have feasibly enclosed with the ox skin?

The isoperimetric principle (for a given perimeter, the circle offers the greatest area) is the central mathematical idea that drives the ancient legend of Princess Dido as well as this lesson. However, since much of this lesson is a review of the major concepts in geometry it best serves as a culminating activity of the course.

Concepts
Area, perimeter, isoperimetric principle, fundamental theorem of similarity, measurement, unit conversion.

Time: 3-5 hours

Materials
One king size bed sheet, with lines drawn lengthwise 1 inch apart, class set of scissors & rulers, student handouts

Preparation
Prepare the bed sheet and its pillow case by drawing parallel lines one inch apart. Also choose a area, such as a football field, on which to cut the sheet on the final day of the project.

LESSON PLAN

THE PRACTICE: Randy Rat (2-3 Hours)

Share the legend of Princess Dido with the class while a king size bed sheet is laid out on the floor. After the story, tell the students that they will be cutting this bed sheet into one inch strips and finding how big of an area Princess Dido could have conceivably encompassed. Ask them to take a guess as to how large of an area the bed sheet could surround. Most students estimate it to be about the size of two classrooms.

The students will first investigate the elements of the Ox Skin challenge on Randy Rat. The imaginary rat skin is superimposed onto conventional graph paper, and is small enough to experiment with on a desk.

Part 1: The area of the rat skin is **20 in^2**. Hopefully, all the students will breakdown the problem into sub-problems (the four rectangles representing the nose, head, torso and tail). From here, the students will implement one of two strategies: a) convert the dimensions into inches and then calculate the area or b) most commonly, they will find the area of the rectangles in square units (**320 total**), then convert to square inches. Many of those who must convert the area from square units will make the common error of dividing by 4 linear units instead 16 square units. Randy Rat's head is a square inch in which exactly 16 square units can be seen.

Part 2: Next, challenge the students to determine a cutting strategy that produces the longest half-inch wide rat skin strip. (There are actually many ways to cut, but only ONE final length: **40 inches**). The rules for cutting are simple: The strips must be a half-inch wide (two units). Yes, two quarter-inch wide units may be set side by side. The intermediate strips must be rectangular (squared ends). In other words, when the students are done cutting they should form one long rectangle that is a half-inch wide. (This is the key which you do not want to reveal until after the students have investigated it for themselves: The strip is actually a rectangle with an area of 20 in^2 — conservation of area — and a width of one-half. The only possible length is 40). Have the students complete this in three distinct steps: PREDICT the length of the strip, CUT the strip laying the pieces end-to-end, and then MEASURE the strip to confirm the prediction. It is suggested that each group be given one extra copy of Randy Rat to cut up, and that all calculations be shown on the backside of their original copy.

Princess Dido: The Geometry of an Ox Skin

LESSON PLAN (continued)

Part 3 : Once the students have discovered the answer of 40 inches, have them now choose three shapes (one triangle, one non-square quadrilateral, and any other shape such as a polygon circle or irregular figure). Each of the three should have a perimeter of 40. The students are to find the shape that yields the largest area (Princess Dido really wants to stick it to the king). The students will more than likely investigate isosceles triangles, rectangles that are close to squares, regular polygons (pentagons and octagons because the number of sides are factors of 40) and a circle. Some will try a right triangle, because of the ease of calculating the area. However, they will be challenged to find a Pythagorean triple that has a sum of 40 (8, 15, 17) and will eventually try an alternate strategy. Browse the room and choose several students to place one of their figures on the board. Have the triangles to the left with the number of sides increasing as you view the board until the circle is found to the far right. Then point out to the students how within each category of figures, the regular figure yields the greatest area. Also, the area increases as the number of sides increases until we see that **"for a given perimeter the circle yields the largest area" (isoperimetric principle)**. The students then can test their accuracy by forming a circle with the strips (standing on their edge) to form a circle, and measuring the diameter to confirm that it is indeed what they calculated (just under 13 inches).

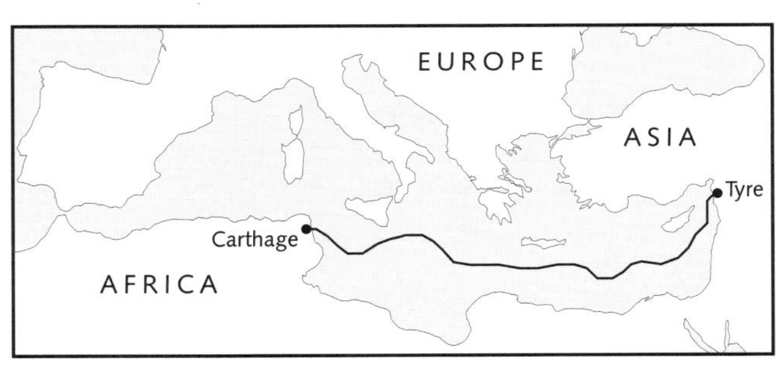

Princess Dido's journey through the Mediterranean Sea

Part 4: To later understand how Princess Dido could have encompassed an entire city with the ox skin, have the students revisit the idea that the ratio of the areas of two similar figures is the square of the ratio of their perimeters.

THE PREDICTION: Ollie Ox - The Bed Sheet (< 1 hour)
Once the students are familiar with the process that lies ahead with the ox skin (bed sheet), they will make the calculations on how big they believe the ox skin shape (circle) will be. A standard King Size bed sheet is about 8 x 9 feet with a pillow case of about 2 x 3.5 feet. Calculate only one side of the pillow, but allow the students to cut both to make up for any overlap. (Yes, we fudge here a little.) Depending on the exact dimensions of the bed sheet, the students will calculate a circumference of over 11,000 inches. Demand that they put this in terms that are commonly understood. The students should get an area of about 70,000 square feet (1.5 acres) and a circumference of 1,000 feet, and a diameter of about 300 feet (the length of a football field!!).

THE PRACTICUM - Princess Dido (1-2 Hours)
The climax for the students will be the actual cutting of the bed sheet to test their calculations. Ahead of time, determine which large area on your campus you wish to enclose with the strand (e.g. gym, administration, or the football field). The actual cutting is the longest phase. It is suggested that you have two students on opposite sides of the sheet cut the sheet in two, then add two more students to cut each half in half, and so on. Soon many will be cutting at the same time. Assign a few people to be staplers, being sure that they staple as shown in the diagram given them. Also have runners, who deliver the strips from the cutters to the staplers. Have the digital camera ready when they are all done!! This will take a little more than one hour, so if possible, seek permission to combine two classes into a block for this day only.

Use the last half hour to debrief the mathematics involved and explore what Princess Dido could have accomplished with strips that are thinner than those used by the students. Hildebrandt & Tromba claim that the semi-circle wall around Carthage (which is a port city along the coast, way to go Dido) encloses approximately 60 acres. That would require cutting the bed sheet into strips that are between a quarter and an eighth of an inch wide...or about the width of a rawhide shoelace. Could the legend be true?

Princess Dido: The Geometry of an Ox Skin

PART ONE: The Puzzle
How large of an area do you believe that Princess Dido could have feasibly enclosed with the skin of an ox?

PART TWO: The Practice
Use Randy Rat to familiarize yourself with the process involved in approximating the possible area. Answer Questions 1-8 on the handout.

PART THREE: The Prediction
Using the actual dimensions of the ox skin (bed sheet), calculate the perimeter, area and width of the shape made by cutting the ox skin into one-inch wide strips.

Dimensions of the bed sheet: _______ x _______ Dimensions of the pillow case: _______ x _______

Perimeter: __________ Width: __________ Area: __________

PART FOUR: The Practicum
Cut an ox skin sized sheet into 1 inch strips. Staple the ends as shown. Once completed, measure the width and perimeter of the enclosed area, and calculate the solution to Princess Dido's puzzle.

Perimeter: __________ Width: __________ Area: __________

Compare your results with your predictions and explain possible reasons for any discrepancies.

EXTRA CHALLENGE: What would the perimeter, area and width be for 1/2, 1/4 and 1/8 inch strips?

SUPER CHALLENGE: The *Aeneid* claims that Dido connected the two ends of the strand to a straight beach along the Mediterranean. If so, how much land could Dido have encompassed with the ox skin?

SUPER DUPER CHALLENGE: Hildebrandt & Tromba claim that Dido enclosed over 60 acres. If so, how thin would the strips needed to have been? (1 acre = 43,560 sq. ft.)

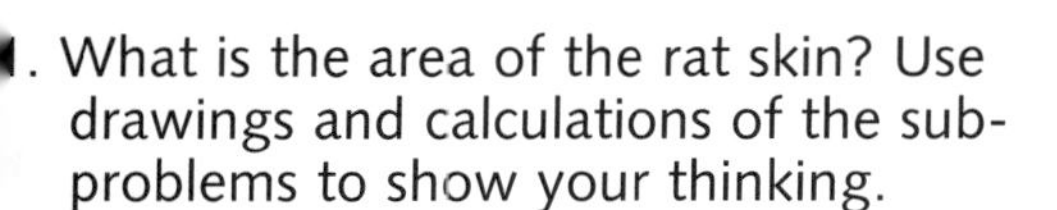

Randy Rat

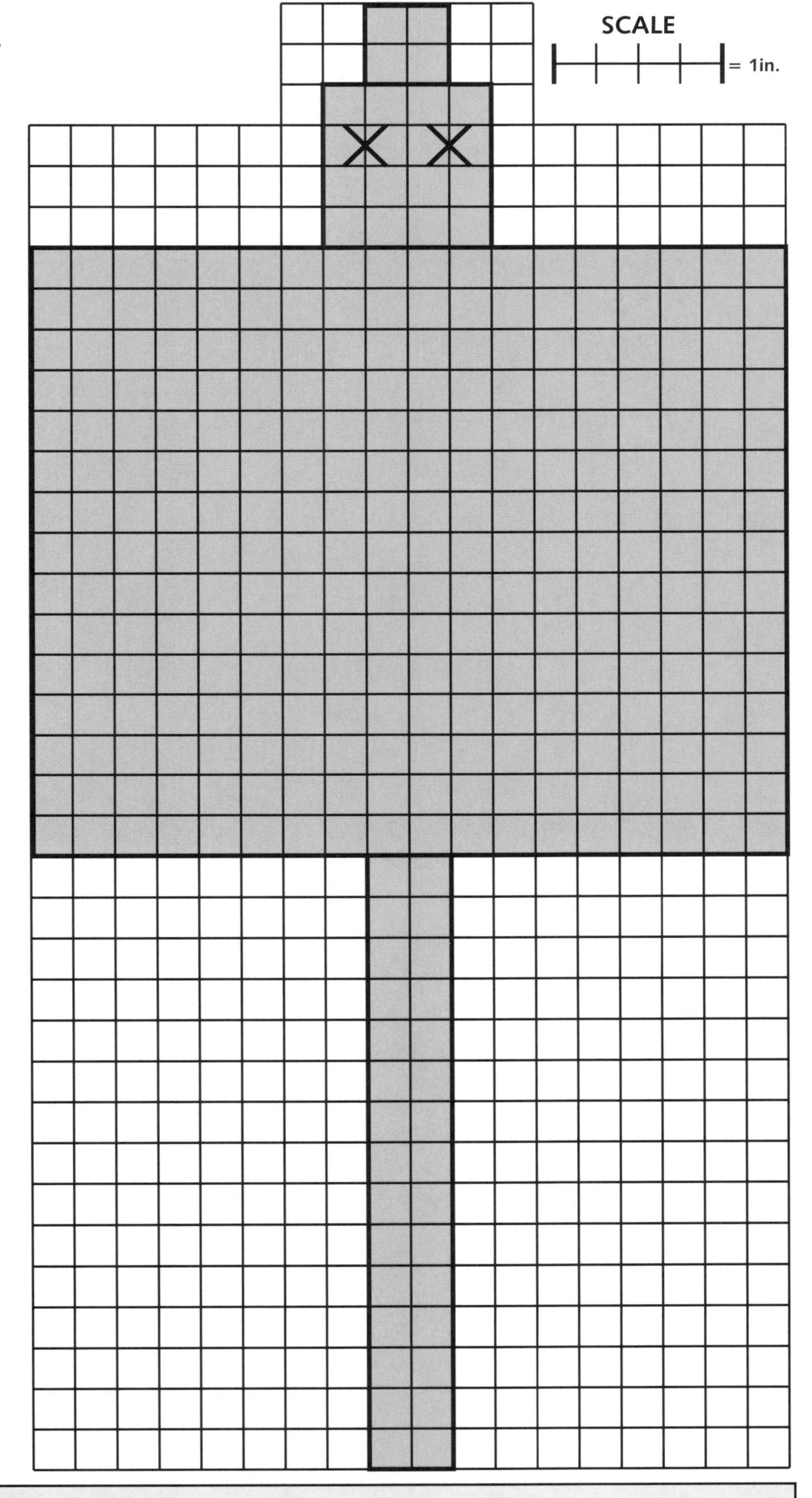

1. What is the area of the rat skin? Use drawings and calculations of the sub-problems to show your thinking.

   Area = ______ in^2

2. If you were to cut the rat skin into 1/2 inch strips, what would be the length of the rat skin band, barring any overlap? (Hint: visualize the band as a long rectangle, 1/2 inch wide.)

   Length = ______ in

3. Choose three different shapes and use the perimeter from #2 to determine the area of each. At least one should have more than 4 sides.

   a. Which of your three shapes yields the largest area?

   ____

   b. What shape do you think will give the largest area for any given perimeter? If you have not done so already, calculate the perimeter and area for the optimal shape.

   A ___________ with a

   perimeter of __________ and an

   area of ____________

4. Assume that you cut the rat skin into $^1/_4$ inch strips. Calculate the perimeter and area of one of your shapes.

   a. What is the ratio of the perimeters between the two similar shapes?

   P_1/P_2 = _____ A_1/A_2 = _____

   b. What is the ratio of the areas between the two similar shapes?

   c. How do these numbers compare?

Articles

MPJ

NOW AVAILABLE
Q.M.I.
VERSION 4.5
QUALITY
MATH
INSTRUCTION
UPGRADE YOUR
TEACHING FROM:
· COVERING THE MATERIAL 1.0
· REGURGITATION 2.2
· BEHAVIOR MOD 3.1
· EDUCATIONAL TRENDS 3.5

THE $4\frac{1}{2}$ PRINCIPLES OF QUALITY MATH INSTRUCTION

What makes a good teacher? This question was posed to me by a colleague, Dr. Howard Judson, a high school economics teacher with a doctorate in political philosophy. He asserts that everyone on campus knows who the good teachers are and by examining their common traits, the qualities of a good teacher can be deduced. A study based on this premise was actually conducted by two educators in New York, and they summarized their findings in an article in *Phi Delta Kappan* (Cohen & Seaman, March 1997).

Dr. Judson has formulated his own theory, based on their findings, as to what these traits are: intelligence, enthusiasm and a good rapport with the students. This is a very strong hypothesis, especially since Dr. Judson possess all three qualities in abundance and is a reputable instructor in his own right. However, these three points do not make up the entire picture. After all, if anybody who is smart, enthusiastic and works well with kids were thrust into a classroom, that person would not necessarily be an effective teacher.

Dr. Judson further offers that good teachers are traditional in their methodologies. I would have to disagree here, for as I look back at the good teachers in my own experience, many of them stuck out because their styles were different. In fact, methodology is not part of the answer at all.

That lesson can be found in the Third International Mathematics and Science Study (TIMSS). In March of 1998, *The Math Projects Journal* was granted an interview with Dr. William Schmidt, the American Coordinator of TIMSS. He claims that among the top-performing countries in mathematics (no, the United States is not one of them) there is no common methodology, but there are common principles of instruction that all the top-performing countries share: teaching to conceptual understanding and teaching with mathematical substance. In his writings and public presentations he stresses two additional components: standards and accountability.

Q.M.I. VERSION 4.5

Here, then is an answer for Dr. Judson, regarding what makes a good math teacher. It is titled "The Four and a Half Principles of Quality Math Instruction," or in the vernacular of the information age, "Q.M.I. version 4.5." The first four principles come from the research shown in the international comparisons of the TIMSS report:

1) **STANDARDS**: Focus on a limited number of topics. Know in advance what you want your students to know; don't just cover the textbook.
2) **CONCEPTS**: Teach students to understand the topic, not just to memorize an algorithm.
3) **SUBSTANCE**: Intellectually challenge students; raise your level of questioning.
4) **ACCOUNTABILITY**: Hold students to knowledge and performance expectations that go beyond grades and unit credit.

The fifth principle comes from professional experience and opinion rather than research, and therefore, its emphasis is demoted to a half-principle.

5) **RAPPORT**: No philosophy, technique, methodology, instructional material or textbook can replace the student-teacher relationship. You must reach 'em before you teach 'em.

This list is by no means comprehensive. However, it would be difficult to refute that a good teacher has a clear idea of what students are to learn and aspires to have them understand the material beyond a superficial level. Furthermore, the most impactful teachers are usually those that intellectually challenge students and hold them to high expectations of achievement. After all, the best teachers are often the most demanding ones.

Continued on next page

Q.M.I 4.5 (CONTINUED)

Dr. Judson will find most of his criteria in this list of four and a half principles, too. A teacher must be both intelligent and enthusiastic in order to accomplish all of the above. And what of rapport? It goes beyond popularity and reaches towards respect. Look back on the great teachers that you have had, and you will probably see that they all had the respect of their students and gave respect in return. In most cases, the teacher and students truly enjoyed each other.

At the time that Dr. Judson posed this question to me, I was reading a book titled *Six Easy Pieces* by Dr. Richard Feynman. It caught my eye because the subtitle of the book was, "Physics Taught by its Most Brilliant Teacher." The preface of *Six Easy Pieces* is full of insights into the teaching philosophies and methods of one of the finest teachers of contemporary academia. Here are some quotes by Dr. Feynman regarding teaching:

Dr. Feynman on Standards
"First figure out why you want the students to learn the subject and what you want them to know, and the method will result more or less by common sense."

Dr. Feynman on Concepts
"I wanted to take care of the fellow who cannot be expected to learn most of the material in the lecture at all. I wanted there to be at least a central core or backbone of material which he could get...the central and most direct features."

"I couldn't reduce it [a particular scientific principle] to the freshman level. That means we really don't understand it."

Dr. Feynman on Rapport
"The best teaching can be done only when there is a direct individual relationship between a student and a good teacher — a situation in which the student discusses the ideas, thinks about the things, and talks about the things. It's impossible to learn very much by simply sitting in a lecture, or even by simply doing problems that are assigned."

So where is "the use of math projects" in the list? Math projects are not on the list, because in-and-of themselves they are not critical to quality math instruction. Projects are effective tools of instruction only when they embody these four and a half basic principles of teaching discussed herein — in particular, teaching to conceptual understanding and with mathematical substance. To gain further verification of the potential effectiveness of math projects, though, I once again call upon Dr. Feynman.

Dr. Feynman on Projects
"I think one way we could help the students more would be by putting more hard work into developing a set of problems which would elucidate some of the ideas in the lectures. Problems give a good opportunity to fill out the material of the lectures and make more realistic, more complete, and more settled in the mind the ideas that have been exposed."

Thank you, Dr. Feynman, for the encouragement to keep creating and implementing quality math projects and problems, and to persist in developing good student-teacher relationships. Thank you, Dr. Schmidt, for revealing to us the value in teaching to conceptual understanding and with mathematical substance, and for pressing us to see the need for standards and accountability. And thank you, Dr. Judson, for the impetus to explore the very nature of teaching, and for reminding us that in this "people business," the people — teachers and students — are the key. All your contributions have given our profession the gift of The Four and a Half Principles of Quality Math Instruction.

A Call for Substance:
An Interview with Dr. William Schmidt

William H. Schmidt is the National Research Coordinator and Executive Director of the U.S. National Center which oversees participation of the United States in the TIMSS studies. As a professor at Michigan State University, he is also widely published in both journals and books on mathematics education. In March 1998, The Math Projects Journal *had an opportunity to sit down with Dr. Schmidt to discuss the TIMSS 1995 report and what it has to say to American educators.*

MPJ: Can you give an example of a model lesson from one of the top achieving countries, either Germany or Japan, which are the focus of the videos?

Dr. Schmidt: Let's cut to the chase. If you look around the world, there just isn't a single way to teach that is dominate among the top achieving countries. Some of them are very didactic, lecture-oriented classes. Some of them are like the kind that you see in the Japanese tapes. If teachers know their mathematics well, they can be just as engaging through a lecture format, as they can teaching as the Japanese do. It is very clear to me that there isn't one way to do this. Instead, the more analysis that I do, the more I believe that there are some principles involved here that just might go across countries.

MPJ: What is that common thread?

Dr. Schmidt: I think the common thread that makes for the top-achieving countries is pure, honest-to-goodness mathematical substance. If the teachers really know and understand the mathematics, then they bring that to the students, through whatever means they know best. Also, a large part of this idea is to develop this stuff conceptually and not just algorithmically. I think many people misunderstand the Japanese videos. It is not so much the methodology, as it is the mathematics. You watch those lessons and the instructor really understands the mathematics, engaging those students in more ways than we do in this country.

MPJ: So, if a teacher were to do a dog-n-pony show lecture with drill-n-kill practice, and do it well, would it work?

Dr. Schmidt: The dog-n-pony show lecture, yes. The drill-n-kill, no. That's what I said about there being some principles. I think the common element is a clear understanding of the subject matter and then going through it much more conceptually than algorithmically.

MPJ: Can you give us a model of how to teach math conceptually rather than algorithmically?

Dr. Schmidt: A U.S. lesson typically starts out with the algorithm. For instance, there is the example in the videos of a guy teaching geometry. He says to the kids, "Here are two supplementary angles, one is thirty, how much is the other?" A student says, "a hundred and fifty." And the teacher says "Good, now why is it that?" And her response is, "Because they are supplementary." Instead, conceptually, you could show them that if they measure a straight line, it's always one hundred and eighty degrees. Then they realize that if they put a line anywhere its going to cut it into two parts. That's conceptual; you start with understanding why, so if you forget the stupid name, supplementary, and you see a line with an angle you'll know what the other one is. That's the difference.

MPJ: How is a strong conceptual understanding of the mathematics developed among teachers?

Dr. Schmidt: It comes from two sources. In some countries, they must major in these fields. The other thing we don't think about is that they are products of their own systems. For instance, Japanese teachers don't necessarily take more mathematics at the university level than we do. But look at what they already know before going to the university. They are already ahead.

Continued on next page

A Call for Substance (continued)

MPJ: In regards to the things that our readership is looking at, active learning, projects, manipulatives, do you have any models from these other countries, or that you think could be done here?

Dr. Schmidt: You don't find very much of that anywhere else. They seem to be uniquely American inventions. Especially the cooperative learning. We asked teachers how much they use groups, and it's pretty much nonexistent. We are too much into the methodology in this country, and we miss the substance. We start talking about small groups and manipulatives and it just becomes process. Therefore, the substance behind it gets lost in the shuffle. And for a lot of these ill-prepared teachers, they grab onto this because that's what they understand.

MPJ: We hear that the US teachers assign more homework and spend more class time dealing with homework than the top achieving countries.

Dr. Schmidt: Yes. The dominate activities in the U.S. lessons were reviewing homework and doing seatwork. One thing that was startling is that the typical American lesson had only 10 minutes or less of instruction.

MPJ: What role does homework play in some other countries?

Dr. Schmidt: It varies a lot. Japan doesn't give a lot of homework, but the kids study for the next lesson. There's a difference, of course. Studying is what you do at the university, and homework is what you do in grade school. But Japan is unique. Worldwide, homework and seatwork are still the dominant activities. I think if you do that and you do it well, and develop the topics conceptually, it can work.

MPJ: Is this a curriculum issue instead?

Dr. Schmidt: It is the core issue, but just putting that in place by itself wouldn't work. You have to help teachers teach in ways that engage kids.

MPJ: So, that is something that teachers could start doing today. We could focus on engaging students and developing topics conceptually?

Dr. Schmidt: That is my point. We must start paying much more attention to the subject matter and teach it more conceptually and less algorithmically. And that is why we are in a catch-22. The Japanese teachers grew up in their system seeing math developed conceptually, no matter what they learned at the university level.

For our teachers it is a lot more difficult; they have to break out of a mold that they've been put into. But I think that is something that teachers can do - Get off the algorithmic side. Don't just give an equation and when a kid asks why, say "'Because that's the equation." Try to get them to understand what lies underneath some of this stuff.

MPJ: It seems that you are suggesting a lesson should move from concept to algorithm to application.

Dr. Schmidt: Not necessarily. A lot of the lessons that we've seen, like in France and such, start out with an application as a motivator. An example is a science one about transformers. They started out by looking at a map of the city and looking how electricity would flow. This got them hooked on the issue, then they hit them with some good hard science about the transformer.

That's very often how it happens: hook them with some kind of application, then take them into it conceptually, let them flounder - that's where I think what the Japanese do is a good idea - let them talk about some of their ideas, then give them an algorithm, a formula and a few examples. Whereas we typically start with the formula with a few sentences about it, and then have them do worksheets.

Continued on next page

A Call for Substance (continued)

MPJ: The report states that American textbooks cover too many topics, yet they typically have only fifteen chapters.

Dr. Schmidt: That is mistaking the notion of what a topic is. The definition of topic has to do with the substance of the mathematics, and when we define it that way, the measurement across all these topics is not how many chapters are in each book.

MPJ: Can you give us an example of four or five topics?

Dr. Schmidt: Congruence and similarity, three dimensional geometry, linear equations, and fractions. We actually tested 44 topics and determined how many of these topics were in any given textbook. Our 700 page books address about 35 topics. The Japanese, on the other hand, spend half of the eighth grade year on congruence and similarity alone, and their gain in that year is higher than in any other country.

The dilemma I have in telling you what to do is that the teacher shouldn't decide which five to ten topics should be studied in a year. It only works if somebody coherently lays this thing out as to what needs to be done.

MPJ: Do you have any last things to add.

Dr. Schmidt: People still think that there are general things a teacher should do, like cooperative learning. That's what people push. We push all the things that have nothing to do with subject matter. I'd like to challenge the notion that there is a single way to do things. If you listen to the ideological left, they say that there is only one way to teach. And the data just do not support that. Among the top achieving countries you cannot find one dominate way of teaching.

On the other hand, the ideological right are calling for "the basics." Yet, the latest analysis shows that the United States, through 8th grade, does average or above average in all the standard arithmetic skills. This is not the place were we are hurting the most. That is all we teach. That is what's wrong, we never go beyond the basics.

If I wanted to become rich and be an advisor to schools to jack their scores up, I know how to do it. We have certain areas of math that we have the international comparisons on. I can tell you the seven items that we are the weakest on, and if schools just did something in those areas, we'd go up in the international rankings. None of those areas is anything that we would consider the basics.

MPJ: What are those areas of weakness?

Dr. Schmidt: Measurement, error analysis, geometric shapes, perimeter, area and volume, congruence, similarity, vectors, geometric transformations, and three dimensional geometry. These are not the basics.

MPJ: Tomorrow, our readers will not be able to change the textbooks or create national standards. What can a teacher do in the classroom today that will model the type of change that you and the TIMSS report call for?

Dr. Schmidt: That's a tough question, because most of what I have argued is that, based on the data, these really are systemic issues. However, the data also shows that how we teach is as important as what we teach. Teachers should challenge students with more mathematical substance and develop the ideas more conceptually rather than algorithmically.

A Tale of Three Countries:
A Synopsis of the TIMSS Video Classroom Study

In conjunction with the TIMSS study, a video survey was conducted of eighth-grade math lessons in Germany, Japan, and the U.S. This survey was the first attempt to collect videotaped records of classroom instruction from nationally representative samples of teachers. Below are descriptions of three of the recorded lessons.

GERMAN LESSON: Volume and Density

PART ONE: Sharing Homework. A student uses an overhead projector to show and explain her results. Afterwards, the teacher leads the class in a discussion of the results. When mistakes are pointed out by other students, the presenter makes corrections on the transparency.

PART TWO: Revisiting Previous Materials. The teacher asks for the three formulas from the previous lesson and the students provide them. The teacher says that they will develop a fourth method today.

PART THREE: Posing a Problem. The teacher presents a problem, "An iron sheet with a length of 0.5 m and a width of 20 cm weighs 3.9 kg. Calculate the height (thickness) of the sheet (7.8 g/cm^3)." The teacher leads an introductory discussion of the problem, reminding students to think about "our three step [method]: given, wanted, and calculation path."

SAMPLE PROBLEM FROM THE GERMAN LESSON

A receptacle has a rectangular shape and the base area is 40 cm times 30 cm. It is filled with oil (specific weight = 0.93) and has a weight of 55.800 kg. Calculate the clear height of the receptacle.

Equations Used:
Surface Area $= 2(a \cdot b + a \cdot c + b \cdot c)$
Volume $= a \cdot b \cdot c$
Mass $= V \cdot r$

PART FOUR: Working the Problem Together. A volunteer at the board tries to solve the problem while taking suggestions and corrections from the other students. One student says "I would convert that into centimeters." The volunteer then responds "I wouldn't." The teacher says, "Would you give him a reason" and the volunteer explains his reasoning. During this, several other students take their turns at the board. The activity concludes when the students agree on the answer.

PART FIVE: Summarizing the Result. When the teacher asks for a summary of the lesson, students volunteer that they have learned to calculate the length, width, and height of a rectangular solid.

PART SIX: Assigning Seatwork. The teacher lays out three types of problems that differ in their level of difficulty and asks the students to choose those they would like to do. During part of this work time, the teacher meets with four students at the board who are having specific difficulties with the earlier problem.

JAPANESE LESSON: Areas of Triangles

Students explore the areas of triangles and apply this concept to solve a problem involving irregularly shaped quadrilaterals.

PART ONE: Linking Yesterday's Lesson to Today's Topic. The teacher begins the lesson by asking, "Do you remember what we did last period?" A student replies that they studied how to obtain the area of a triangle constructed between parallel lines. Then the teacher walks to the TV monitor in front of the classroom, which is connected to a computer, and shows a series of triangles between two parallel lines (Figure A). He reminds the students that the areas of these triangles are the same because the base and the height are always the same.

Continued on next page

A Tale of Three Countries (continued)

PART TWO: Posing the Problem. The teacher draws a figure on the board representing two pieces of land. The boundary is a line bent in the middle (Figure B). The owners want to know how to make the boundary straight without changing the areas of the two pieces of land. Then the class engages in a brief question and answer session to clarify the problem.

PART THREE: Working on the Problem. The students work individually on the problem, while the teacher circulates around the room. The teacher mainly gives hints to the students instead of showing them what to do. For example, the teacher asks one student, "Is there a method that uses the area of triangles?" and says to another student, "Are there parallel lines anywhere." After three minutes, the teacher suggests that students may want to work together. He adds: "And for now I have placed some hint cards up at the board so people who want to can refer to them." For about ten minutes the students discuss the problem with each other, the teacher, or the assistant teacher.

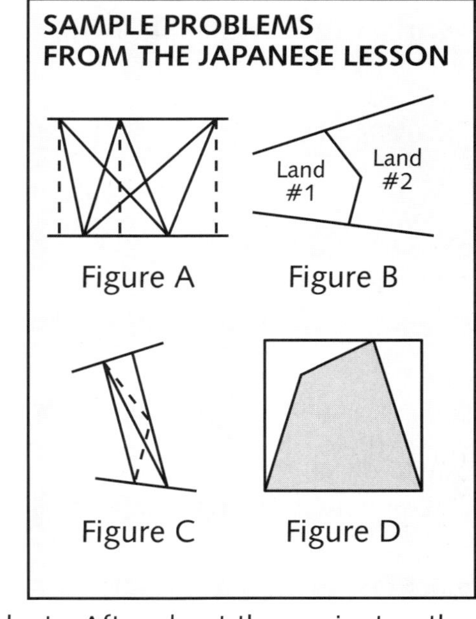

PART FOUR: Students Presenting Solutions. The teacher asks two students to draw their solutions on the chalkboard (Figure C). While they explain their solutions, the rest of the students and the teacher ask questions and request clarifications.

PART FIVE: Reviewing Students' Methods and Posing Another Problem. The teacher reviews and clarifies the students' methods. Then the teacher presents a follow-up problem, which is to change a quadrilateral into a triangle without changing the area (Figure D). The students work on the problem at their desks, and the teacher walks around assisting individual students. After about three minutes, the teacher again tells them to discuss their methods with one another.

PART SIX: Summarizing the Results. The teacher encourages them to find as many solutions as possible. He draws ten quadrilaterals on the chalkboard and asks students to show their solutions. After 23 minutes, he briefly reviews the solutions and asks which students found each solution. The teacher ends the lesson by suggesting that, for homework, the students try to change other polygons, such as pentagons, into triangles with equal areas.

U.S. LESSON: Measures of Angles

Students practice calculating the sizes of various angles. The teacher concludes by presenting the formula for finding the sum of the interior angles of any polygon.

PART ONE: Warm-Up Problems. The teacher draws four diagrams on the chalkboard and asks students to find the missing ten angles (Figure E). The teacher helps find four of the angles, and then he asks students to find the rest. After about 40 seconds, the teacher works through the remaining problems, by eliciting responses from students using questions such as "If this angle is a right angle and this is thirty degrees, what does F have to be?... And what's left for angle E?..."

PART TWO: Checking Homework. The teacher reviews the previous homework, a worksheet entitled "Types of Angles." It includes definitions of terms (such as "supplementary"), sample problems with solutions, and about 40 problems. The teacher checks answers in a question and answer format: "The complement of an angle of eighty-four, Lindsay, would be...? Albert, what's number four...?"

Continued on next page

A Tale of Three Countries (continued)

PART THREE: Assigning Seatwork. The teacher hands out a worksheet that contains two sample problems with solutions and 15 problems. The teacher works through the first several problems with the students, asking the same style questions as before. While the students work individually, the teacher circulates around the room, assisting individual students.

PART FOUR: Providing Help on Problems. The teacher receives many questions on two of the problems and decides to work these problems with the class. On one of the problems, for example, he says "Two angles are supplementary. Therefore they must add up to one hundred eighty degrees. But they are equal. Each one of them has to be...?" After both problems have been answered and discussed, the students return to their worksheets.

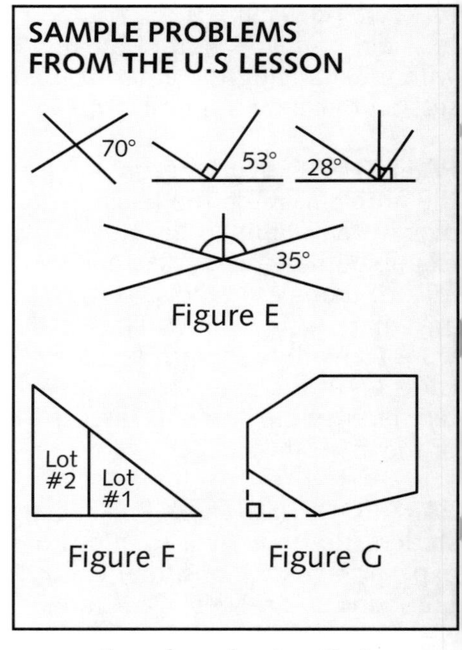

PART FIVE: Checking More Homework and Introducing a New Formula. After the students finish the worksheet, the teacher asks them to get out a previous worksheet, containing two problems. The first is a map of two streets intersecting in a 45 degree angle with a triangular shaped piece of land between the streets (Figure F). A boundary line divides the piece into two parking lots. The task includes finding the measure of the angle formed by the property line and the diagonal street, and suggesting a more equal way to divide the lots. The second problem involves finding the sum of the interior angles for any polygon (Figure G.) The teacher asks students for the answers they found using their protractors. The teacher then presents the formula (180(n - 2)) and asks students to try it out.

PART SIX: Previewing the Upcoming Schedule. The teacher concludes the lesson by announcing the topic for the next day and informing students of the dates for the next quiz and the next exam.

The lessons included in this article are excerpts from the public domain web site of the National Center of Educational Statistics at: http://nces.ed.gov/timss/video/index.html. Copies of the videos are also available through the Center.

TIMSS Statistics

How high is the mathematical content of the lessons?

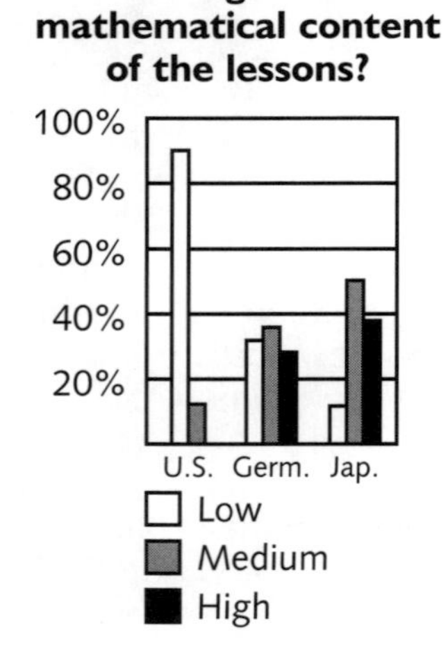

How often are math concepts stated vs. developed in the lessons?

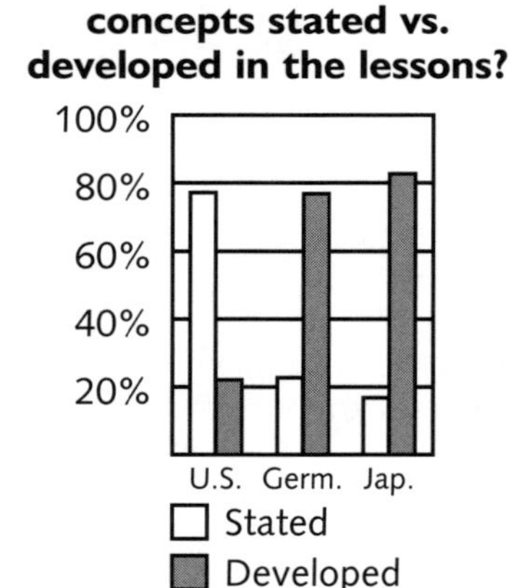

How many teachers view the goal of their lessons as skills vs. thinking?

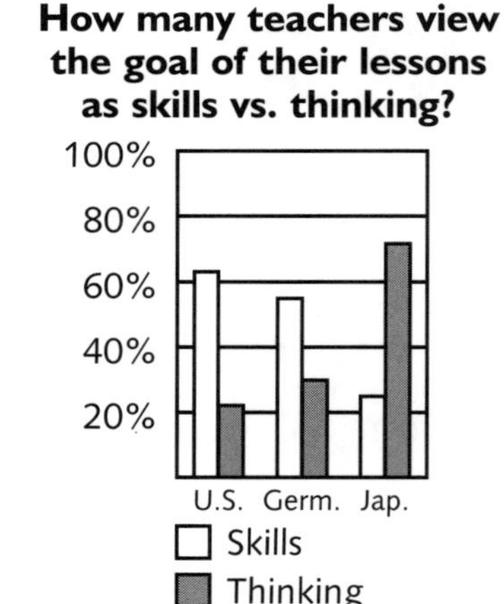

What is the purpose of seatwork in the lessons?

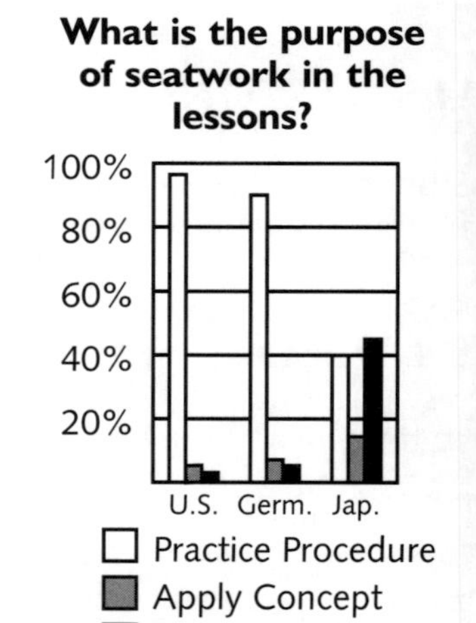

Data is taken from Lessons in Perspective: How Culture Shapes Math Instruction in Japan, Germany and the United States, *sponsored by The California State University Institute for Educational Reform.*

A TIMSS Primer:
A Crash Course in the Findings of the Third International Mathematics and Science Study

THE STUDY
The Third International Mathematics and Science Study (TIMSS) was the most comprehensive study of its kind. Forty-two countries tested students at the beginning and end of 4th, 8th and 12th grade in 40 topics. The study also investigated teaching practices as well as curriculum. It was the first international study regarding education that analyzed what goes on in classrooms around the world, as well as what comes out of them.

THE RESULTS
- In the 4th grade, our students are above average in the international comparison in mathematics. Yet they slip to only average in 8th grade, and then plummet to the bottom of the rankings in 12th grade.
- In 4th and 8th grade, other countries make a large gain in at least one topic area, while the U.S. is the only country that did not make any significant gain in any topic area.
- Even our most advanced students (Pre-Calculus and AP Calculus) emerged at the bottom of the rankings. The best of our best are not world-class.

THE MYTH BUSTERS
- U.S. students spend more time in math classes and are assigned more homework than in other countries.
- Student diversity and poor discipline are issues in other countries as well.
- The study sample did not consist of select groups in other countries. TIMSS monitored this very carefully.
- The scope and sequence of mathematics instruction is not universal. Countries vary in regards to the grade level in which they introduce a topic, the method in which they teach it, and the amount of time given to it.
- Back-to-basics is not the answer. Through the eighth grade, our students perform above average on arithmetic. However, the other countries begin teaching higher level math concepts in the eighth grade, while the U.S. continues to re-teach fractions, decimals, and percentages.

THE REAL REASONS
- In the United states, we predominately teach students "how to do" something, while in other countries they teach students to understand mathematical concepts.
- We are the only country to track students before high school. In all other countries there is a national standard of what every eighth grader is to know.
- We lack a nationally defined curriculum, or even a national consensus.
- Our curricula are a mile-wide and an inch-deep. We teach 30-40 topics a year with little or no depth, while most countries teach only 5-15. This is the number one factor in our lack of significant gain in any of the topic areas between grade levels.
- We do not intellectually challenge our students. Transcripts from video-taped lessons were given to college math professors who were asked to rate the level of questioning in each lesson. These were their results:

Level	Germany	Japan	U.S.
Low	40%	13%	87%
Middle	37%	57%	13%
High	23%	30%	0%

THE SOLUTION
The United States needs to:
- Establish coherent national standards.
- Limit the number of topics taught.
- Support the training of teachers.
- Increase the quality of the instruction that our students receive.

ARTICLE

a Mountain of VISION creates a Landslide of ACHIEVEMENT

A Case Study in Raising Standardized Test Scores

Every Algebra and Geometry teacher at Trabuco Hills High School was in a meeting with the principal to discuss our enthusiasm for a new instructional program that we had implemented that year. We shared how we established standards, rewrote curriculum, and created innovative testing methods, and by doing so, we had improved our students' competency. We spoke enthusiastically of our students' increased ability to reason and problem-solve. At the end, our principal was very encouraged, but told us, "We need something to show for it. You can say how you think the students are improving, but we are consistently at the bottom of the district pile when it comes to the standardized tests. We need the numbers to reflect what you are claiming." He was right, and we knew it.

The Golden State Exam (GSE) was to be our next standardized test. These exams are California state tests given in a variety of subjects. They are achievement tests rather than aptitude tests. That is, the questions are more rigorous than what you would find on a basic competency test. Therefore, though any student may take the exams, schools usually reserve them for their top students.

Someone on the team suggested that we give the exam to all of our Algebra and Geometry students. The initial reaction was that our scores were already hurting, so including the lower-end students would only lower our averages. Eventually we decided that if we truly believed in the program and the students as much as we claimed, then we should take the chance. We did.

The exams were given in May of 1998. The numbers which our principal was looking for came that October. The State awards one of three distinctions: High Honors, Honors, and Recognition. Each is based on a rubric, and once the tests are graded, the percentage of students achieving each level is reported to the schools. The percentages represent how many students achieved that level "or higher." In other words, the percentage for recognition represents all three levels combined, which is why some of the percentages total more than 100%.

A comparison of the performances of the students statewide versus those at Trabuco Hills is displayed below.

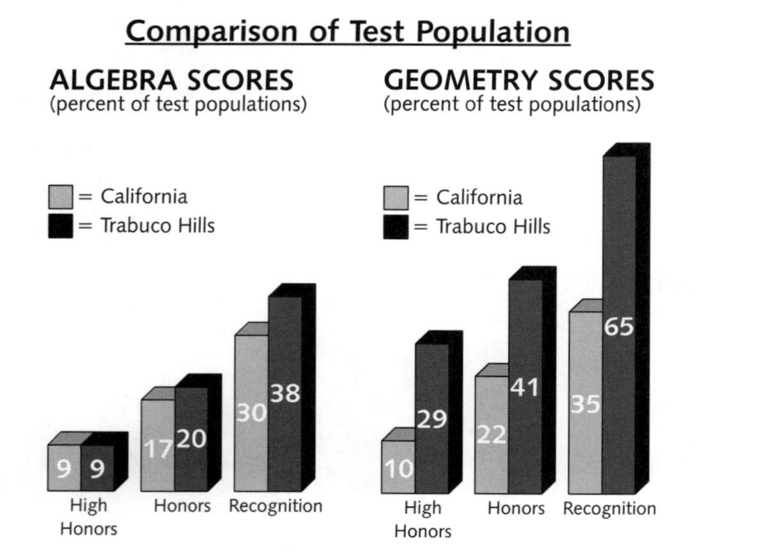

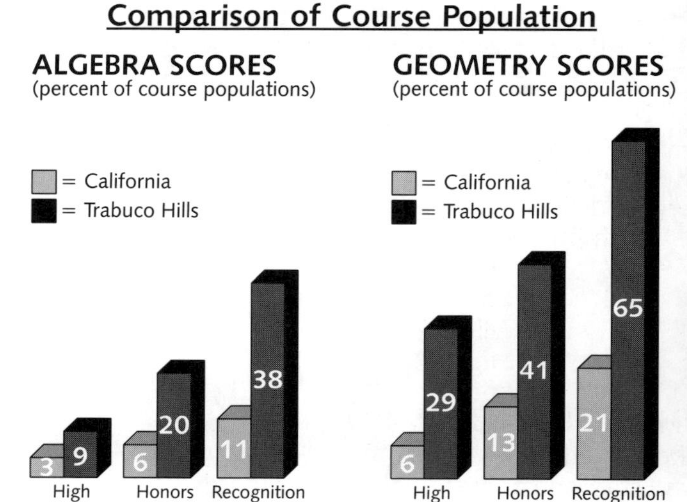

Continued on next page

A Mountain of Vision (continued)

Our Algebra students met or exceeded the state at all levels. The Geometry scores were even better. Two out of every three students taking Geometry at our school received an award for being in the top 35% of the state. Nearly one of every three were in the top 10% as shown below. We were stunned. Furthermore, our school finally climbed from the bottom of the district rankings, surpassing two schools and tying with the third. That third school, however, tested less than half of its students, while we tested all of ours. We were now elated as well as stunned.

Les Axelrod of the Curriculum, Standards and Assessment Division of the California State Department of Education said of our results, "Your instructional program is obviously working very well. You outperformed the state 2 to 1." We investigated further and found that our case of testing all students was a rare exception. Only 35% of all Algebra and 60% of all Geometry students in California that year took the Golden State Exam. When we compared the results of the tests to the entire population of each course (rather than to only those that took the exam) the results were even more outstanding.

Our students outperformed the state by nearly 4 to 1 in Algebra and by over 3 to 1 in Geometry!
The question now being pondered by the members of the team is "What aspects of the instructional program contributed to these results?" An examination of our program and its results offer some insight.

VISION PLANNING

The birth of this program began almost a year prior to the Golden State Exams. As with most educators, the Algebra and Geometry teachers felt that the students could apply themselves better; however, we also believed that we could do a better job teaching.

With this belief, the team began with a vision-planning model that is often used in the business world: Brainstorm, Prioritize, and Categorize. The brainstorming session started with a direct, yet simple question, "What do you expect your students to know by the end of the year?" Rather than trying to remember what is in the textbook, we focused on what we consciously valued and wrote our responses on the board. The teachers of the successive courses were then asked, "What do you expect students to know coming into your class?" Their responses were added to the list.

To prioritize this list, we asked another question, "Which topics would you be most embarrassed if your students didn't know?" We intentionally limited our list to the top ten to focus our attention on what we saw as most important. We then categorized them by common threads and discussed how they might fit into each course.

THE ACTION PLAN

Once all the teachers agreed upon the common vision, we designed the blue print for the year. First, a final exam was written for each course. This gave us a very clear idea of the expectations we held for each concept. Next, a syllabus was created for each course, paying attention to the pace of the topics. In Algebra, we condensed the early lessons dealing with solving basic linear equations. Most students had seen these concepts before, and it seemed silly to us that they received more attention than the tougher, newer concepts of exponential functions and quadratic equations. In Geometry, we chose to weave area and volume throughout many units, rather than springing it on the students all at once near the end of the course.

Once the final exams and the syllabi were in place, we chose and/or created projects and challenge problems for each of the topics. The projects were chosen for their value in raising the level of questioning, engaging student interest, offering context to the topics, and revealing underling mathematical concepts.

We also designed a new testing format. The details of our "progressive testing" are discussed in "Nuts-n-Bolts." The pertinent points of this testing system are its cumulative format and emphasis on problem solving.

Continued on next page

A Mountain of Vision (continued)

IMPLEMENTATION

The vision and action planning all took place over several days during the summer (for which the teachers were paid). The challenge came in making this program work in the classroom for the ten months of the school year. Fortunately, teamwork carried us and the program throughout the year.

Communication was highly regarded among the team. The teachers from each course met once a week at lunch. Over sandwiches and soda, we discussed what worked and what didn't. We adjusted the pace and instruction according to how the students were performing. Although we shared ideas for lessons and projects, teachers were free to use whatever instructional methods they thought best. We stayed on the same track by giving the same tests and quizzes. The content of the tests was reached by consensus so nobody was taken by surprise. This spirit of collaboration, flexibility, and commitment to the vision were the crux of our implementation.

THE ANALYSIS: What Worked

So what components of this experience most promoted the increase in test scores? Opinions may vary, but here is our analysis.

The most important aspect of the program was setting high standards for a limited number of topics. This produced tremendous focus for the teachers. Other contributing factors were: revisiting topics, raising the level of questioning, teaching for conceptual understanding, and emphasizing problem solving. Also, the new pace and sequence of topics in each course allowed time for the more difficult topics to be taught before the state exams were taken. Furthermore, collaboration exposed the teachers to a variety of new methods and philosophies.

So how much influence did the use of math projects have on the increase in test scores? We believe that the projects had tremendous influence on the success of the program, but only because they served as an effective means to the goals stated earlier. That is, projects raised the level of questioning, emphasized problem solving, offered context and engaged student interest. Yet, these things may be accomplished by other means as well. Therefore, even as educators who strongly advocate the use of math projects, we will always be reluctant to claim that they are the best or only way to teach. However, in this case study, we believe projects were an integral component of the success of the program. (Many of the projects published in the first volume of *The Math Projects Journal* were used in this curriculum.)

THE LESSONS: What You Can Do

We understand that not many teachers have the resources available to implement the same type of program detailed in this report. However, we at *The Math Projects Journal* have always offered what we felt "you can use in your classroom today." So here are some ways that you can apply the principles learned from our experience.

1) When beginning a new unit, choose the 3 to 4 topics you believe to be most important. Emphasize these topics throughout the unit, and de-emphasize (or eliminate) the others.
2) Raise the level of questioning. Ask critical thinking and analysis-based questions that will encourage students to look for the underlying concepts.
3) Modify existing tests or write your own. Make an attempt to assess your students' reasoning ability and conceptual understanding of the topics as well as their knowledge of facts and algorithms.
4) Make your tests and lessons cumulative. Keep hammering on the important topics.
5) Use problems and activities that offer context to the topics.
6) Be patient. It will take time for your students to respond and show improvement.
7) And yes, use math projects whenever you feel they will be beneficial.

Hoping to reproduce these results, we at *The Math Projects Journal* and our respective schools will continue implementing the principles that we believe brought our students such great success. We will keep you posted.

The portfolio you made us keep during this year has been a great help in many ways. I've many times to clarify and refresh my memory, but it was of use to my friends w were taking class and were confuse

Portfolios: More than a Glorified Notebook

THE MISTAKE

I remember the first time I used portfolios in my classroom. Having just returned from a conference, hot on the idea of having my students keep a record of their work, I had my students write their names on manila folders and then tucked them inside a cabinet drawer. I taught the rest of the year without opening that drawer again. The reason my first attempt failed was that I did not have a clear focus of what I wanted the portfolios to accomplish in my class. Since my portfolio model had no direction, it went nowhere.

I have since developed a much better system that is very effective for me. I know that it is effective because it meets the very clear objectives that I have established for it. Those objectives come from the answers to three questions that any teacher should ask before designing or implementing a portfolio system: What? Who? Why?

THREE QUESTIONS

What are you going to put in the portfolio? If your students have done nothing more than a mass of textbook problems, then your portfolio is going to be little more than a glorified notebook. A notebook shows what the students have done. Portfolios show what the students can do. The timing of my initial portfolio attempt was premature, because at that time I had no evidence of what my students could do. Now I have many projects, and activities that demonstrate their mathematical competency (or lack there of). I have evidence of my students' abilities, not just their work; and my students have something meaningful to display.

Why do you want to implement portfolios? Your answer should go beyond the doomed response of "Because they are cool." Lastly, who is the audience? With a traditional notebook, the audience consists solely of the teacher, but a portfolio can have multiple audiences and multiple purposes.

In my class, the portfolio audience is me, my students and their parents. By reading student self-reflections (explained herein), I can get inside their heads as to what they will truly take from my class. For my students, their abilities are on display in their portfolios, rather than their point totals in my gradebook. The impact this has on them at the conclusion of the course is best stated by a former student:

> *I wanted you to know that the portfolio you made us keep during the year has been a great help in many ways. I've used it many times to clarify and refresh my memory, but it was also of use to my friends who were taking [that class] this year and were confused themselves. Thank you again.*
> *-Courtney, Hemet High School*

The audience that the portfolios have the greatest influence over is the parents. I have had enormously positive feedback from parents, but the usefulness of the portfolios really shows in those belligerent parent conferences (we all have those once in a while). Parents can argue an abstract percentage in your gradebook, but they cannot argue the evidence that is shown in the portfolio. Laying open the portfolio of a poor student next to that of a model student usually softens even the most abrasive confrontations.

Another audience that is useful is the faculty. Portfolios have great potential for being passed on from teacher to teacher, as an authentic record of the students abilities. Whoever the audience may be, and whatever the purpose may be, keep them both clear in your mind as you develop your portfolio system.

THE STANDARD MODEL

Once you have answered the what, who and why, you still need to understand the basic portfolio process before you begin to design your own system. While portfolio systems vary among teachers, three stages are standard among any good portfolio process: Collect, Select, Reflect. Throughout the year, students collect the evidence that will be displayed in the portfolio. Periodically, students are asked to select various works by certain criteria. It

Continued on next page

Portfolios (continued)

may be that the instructor asks for three pieces that demonstrate growth, or one sample each from four different topics discussed in the course. The criteria is set by the instructor, but the selection is left up to the students. Finally, no portfolio is complete without some kind of self-reflection by the student. This usually involves some writing in which the student addresses the selected evidence in regards to what they have learned in the course.

CUSTOMIZED OPTIONS

The following is a description of the portfolio system that I use in my classroom. It is very customized and based on the objectives that I have set for my classes; however, it still embraces the standard model of collection, selection, and reflection. The actual structure of the portfolio itself is based on what I term "The 4 E's of the Math Mission." These 4 E's are my professional answer to the common question, "Why do we have to learn this?" My response is, in order to ...

Empower you with fundamental math skills.
Enhance your critical thinking and problem solving abilities.
Expose you to a variety of mathematical topics and their related fields
Enrich your life with greater appreciation for the power and beauty of mathematics, and a deeper understanding of its embedded role in the nature of the universe.

These are the first four of seven sections in the students' portfolio. Every time I return a graded project, I have them place it in one of these four sections. They select which one according to what that assignment represented for them. There is no right or wrong choice, but they will be responsible later for explaining their choices. The fifth section is entitled Study Samples, where they keep their Best Test, one sample of their Best Notes, and one sample of their Best Homework.

Notebooks show what students have done. Portfolios show what students can do.

The sixth section is dedicated to Assessment. Here is where they keep their self-assessment and parent evaluation form. At the conclusion of each semester, I have the students sort the evidence in each of the first four sections with the best being first. In their self-assessment, they are responsible for discussing how the best piece in each section exemplifies what they have learned in the class.

Once this is completed, the students take the portfolio home and have their parents complete an evaluation form. The parents are expected to complete three easy tasks. First, they are asked to read their student's self-assessment. Then they are expected to examine the portfolio with the student. Once this parent-student conference is complete, the parent is prompted to write down any of their thoughts and observations.

The portfolios remain in my classroom until the parent evaluation. (They are too valuable to risk losing.) After each parent evaluation, they are returned to the "portcase" — a bookcase in my classroom — until the end of the year when the students leave the course with the portfolios in hand.

MY PROFESSIONAL PORTFOLIO

I keep a portfolio also. My professional portfolio is structured the same as those of my students. The 4-E's, Study Samples, and Assessment sections are bursting with samples of student work from the last six years. I display how my students are learning basic math skills, developing problem solving abilities, experiencing new applications of math, and perceiving math in the natural world. I can also display how they and their parents evaluate the work. My professional portfolio has one extra section though. I call this section Various Works. It includes honors and awards that I have received, thank you notes, samples of writings and brochures of workshops that I have conducted.

I believe my portfolio accurately reflects my style and accomplishments as a teacher. I strongly suggest, whether or not you require your students to keep a portfolio, that you keep one. This will give you a chance to collect, select and reflect your work as a teacher of mathematics.

Cooperative Learning: A Struggle Among Equals

First in a Three-part Series

In our classes, we have run the gamut of cooperative learning techniques. We have assigned groups randomly, by student choice, and by ability level; we have done individual assessment as well as group, peer, and self assessment.

Since using projects implies collaboration among students, we had to determine the best cooperative learning methods. Through our search, we have settled on some standard practices in critical areas such as assigning students to groups, fostering the collaborative spirit, assessing students and arranging the room. Here is what works best for us.

ASSIGNING THE STUDENTS TO GROUPS

Opening day, students choose their own seats, and thus their groups. After each unit exam, every 4-5 weeks, the teacher reassigns. The students are grouped homogeneously by ability — A students with A students, C's with C's, F's with other F's, etc. Research by Dr. Uri Treisman at Berkeley inspired this practice, and it has generated positive results.

The research showed that in heterogeneous groups, where top students are placed with lower students, neither learns much. Contrary to conventional thought, the better student does not teach the lower. (Isn't that our job anyway?) Instead, the A student often takes over most of the work in an attempt to salvage the grade, while the lower student copies in an attempt to pass the class.

The solution is to group students by equal performance and ability in order to form a struggle among equals. In this fashion, students at all levels challenge and support one another. Everyone moves forward.

Contrary to conventional thought... the A student often takes over most of the work in an attempt to salvage the grade, while the lower student copies in an attempt to pass the class.

Of all the things we have tried, homogeneous grouping has made the greatest positive impact. We are still amazed at the quantity and quality of intellectual dialogue generated among our students when they are grouped in this manner. The top students are no longer merely giving away answers; they are defending them. The constant challenge pushes them to probe and analyze the material to the fullest extent of their natural abilities. The mid-range students are engaged and excited, because now everyone has something to offer the group; no one is made to feel inferior. As for the lower-achieving students, the struggling demonstrate self-reliance, the unmotivated show more initiative, and the unruly become tame. In fact, since we have been implementing homogeneous grouping over the last three years, we have had fewer discipline problems. The greatest benefit, however, has been the increase in math competency at all levels.

If your many attempts at cooperative learning have proven futile, try homogeneous grouping. It has worked wonderfully for us. Let us know how it works for you.

Cooperative Learning: Assessing What Our Students Know

Second in a Three-part Series

In our last article, we discussed the struggle among equals. We shared how homogenous grouping—"A" students with "A" students, "F" students with "F;" etc. —has improved the mathematical competency of our students and reduced discipline problems. In this article, you will learn how this grouping technique, coupled with the practice of a weekly group quiz, has enhanced our ability to assess each student.

ASSESSING STUDENTS IN GROUPS

Assessment is a complex topic. Thus, our discussion will focus on three basic points about how we assess students in a project-oriented class. First, our primary concern is how well the students are learning math, not how well they are working in groups. We view group work as a means to an end, not the end itself. Arguments can be made to the contrary, claiming that teachers should be training students for a collaborative work environment. Good point; it still doesn't matter. For us, cooperative learning is a tool, not an objective.

Secondly, therefore, we primarily assess students individually. Although students do the projects and analyze the results in groups, they complete the write-ups on their own. We then evaluate the work on a student-by-student basis. We have our students WRITE about what they thought and did on nearly every major assignment. This is important in a class where discussion and collaboration are free flowing, for we must know that each student has contributed their own work and thinking. Furthermore, student writings offer a wealth of insight into what they truly know and don't know.

Lastly, we have transformed the traditional weekly quiz into a powerful learning tool. At the end of each week, we administer a group quiz. The quiz is comprised of approximately 3-4 questions based on material studied that week, as well as 2-3 questions from previous material and usually a bonus question. The quizzes are the most structured group work they have; during quizzes students may only interact with members of their own group. Keep in mind, the students are grouped by ability. We will grade only one person's quiz in each group at random, and everyone in the group will get that grade. (This is the only assignment in the course for which the students receive a group grade.) This encourages the group to make sure each paper is the same. Most groups will accomplish this the smart way and discuss each problem together. Despite warnings, some groups divide the questions among the members of the group and copy each others' papers. We let them, and after poor performance on the first few quizzes they change tactics.

Our primary concern is how well the students are learning math, not how well they are working in groups.

Eventually, collaborative, academic dialogue develops in each group, including the poor performing groups. In fact, we have seen four students, who otherwise would normally not pass the quiz on their own, achieve at least a "C" by working together. As a group they have performed better than any one of them could singularly. As individuals they have learned more from one another than they would have on their own.

The group quiz solidifies the concepts studied during the week in the students' minds. It is a time for students to pause, discuss and process. From our discussion thus far, it would appear that the group quiz is primarily a teaching tool. It is! The role of the quiz in the assessment program is to leverage the students into an academic dialogue, while our other assessment tools focus on evaluating the math competency of each individual student.

In the next installment, we will share how we foster a collaborative learning environment.

Cooperative Learning: Lighten Up!

Third in a Three-part Series

In previous articles, we have shared how homogeneous grouping has improved the mathematical competency of our students as well as reduced discipline problems, and we have shown how we use this grouping technique with projects and a weekly group quiz to assess each student. In this last article, we will discuss how we foster a cooperative learning environment.

FOSTERING A COOPERATIVE SPIRIT

We give students a lot of freedom when they're working in groups. Although they sit in groups of four, the group structure is enforced only occasionally. Other than the weekly group quiz, which was previously discussed, the only other time that we impose the group structure is during project activities. Many projects require some kind of data collection. This is all done within the structure of the homogeneous groups. In this manner, group members rarely rely on one person to do all the work.

Once the "activity" portion of a project is completed, students are permitted to discuss strategies and calculations with anyone they wish. At this point, the data is so varied that discussion naturally focuses on concepts and procedures rather than simply on answers.

In other words, cooperative learning is allowed, not imposed. It is fostered, not enforced. Reaching out for help and working with others comes naturally for students. We have found that nurturing a spirit of collaboration has more to do with getting out of their way than supplying them with any tools.

Therefore, the best advice we have for fostering the cooperative learning environment is simple: Lighten up! If you are micromanaging the group work, telling students when they may collaborate and walking them through every step — then you are DOING cooperative groups in your classroom, but no one is learning cooperatively.

Lightening up calls for teachers to lead rather than control. This can be a difficult step, since most teachers have been trained to dictate everything that happens in their classroom. However, if students are to truly work together, they must be given latitude to think, share, and explore.

The greatest benefit of giving freedom to students is the freedom it gives to the teacher. This style of group work gives the instructor the opportunity to walk around the room and work individually with students. Any teacher or student will tell you that one-on-one instruction is the most powerful teaching method. In order to achieve that, the instructor must loosen the reins, step away from the comfort of the board and enter into the fray of adolescent learning.

ARRANGING THE DESKS

Desk arrangement is not a critical issue, but it is a practical question for most teachers. In our classrooms, we set the desks in groups of four. The desks in each group face the same direction two by two, as if all four students were riding in a car.

The groups line the perimeter of three walls and face the center of the classroom, so they all form a double-rowed horseshoe. In this configuration, there is a large area in the center of the room for non-board demonstrations and investigative activities. No student is more than two rows away from the teacher, every student can view the board easily and whole-class discussions are more easily facilitated.

In our classrooms, a spirit of cooperation, collaboration, and learning is nurtured by the attitude of the instructor, the desk arrangement, and most importantly, the freedom and respect given to the students.

NUTS & BOLTS

Sample Course Requirements and Policies of a Project-Driven Class

Even the most progressive, project-oriented teacher cannot avoid the need to compute a course grade. Therefore, we offer a synopsis of our grading policies and procedures. We have suggested before that initially, a teacher should simply assign ten percent of the course grade to project-related activities, then increase this weight as the teacher infuses more projects into the curriculum. You will notice that our course requirements go well beyond the simple addition of a project category. This is because prolonged, extended use of projects has changed the very nature by which we teach. Our course requirements and grading policies reflect that change.

REQUIREMENTS

PROJECTS: Projects are group activities and investigations like those that we publish in this journal. The group portion of the project is often done in class, while the individual portion is completed at home. Since the students work in groups, we demand that their individual conjectures and written explanations be thorough and unique.

CONCEPT PROBLEMS: At the conclusion of each unit, our students complete a problem write-up for each of the main ideas in the unit. Typically, there are 3-4 main ideas with 2-4 basic problems assigned for each concept. For each of these concepts the students are to do the 4 S's: 1) STATE the problems, 2) SOLVE the problems, 3) SHOW all work, and 4) SUMMARIZE the topic in writing. Each concept must be presented on a separate sheet of paper for easy inclusion in their portfolio. The students have 1-2 days of classtime before the unit test to complete these problem sets, which are due on the day of the test. We use this classtime to walk around the room in order to check for student understanding and assist where needed. Again, since students will be receiving help from both the teacher and fellow students, it is critical that their individual explanations be thorough and unique.

ORIGINAL WORKS: The Original Work is an assignment in which the students submit a self-generated application or presentation concerning a topic from the course. They may either create or find a problem that involves the topic, or make a connection with something of personal interest. The original works are due at the conclusion of each semester, and are considered a major project assignment.

TEXTBOOK PROBLEMS: Traditional problem sets may be periodically assigned. Since mathematical reasoning and understanding are valued more than merely completing assignments, the weight of the homework category is only five percent. However, forty percent of the overall grade is still comprised of non-exam activity.

QUIZZES: Quizzes are the only portion of the grade that is group-evaluated. (i.e. Group members all get the same grade.) The students take the quiz in their groups, which are assigned according to similar rather than mixed ability level. Group quizzes are given weekly, and serve as evaluation, assessment and instruction.

TESTS: Individual cumulative tests are given each unit and have two parts. Part One focuses on skills and knowledge. Part Two focuses on problem solving and applications and is open resource — that is, the students may use materials such as notes, projects, homework, and their textbook.

THE FINAL EXAM: Prior to the final exam, each student's course grade is calculated and the student is assigned a course letter grade. If the student scores higher on the final exam than the course letter grade, the course grade is raised one letter grade. If the student scores the same as or one grade lower than the course grade, the course grade is maintained. If the student scores two or more grades below, the course grade is lowered one letter grade. Students with a D in the class who score less than 50% on the exam, fail the course. This final exam system encourages all students, not just those with a borderline grade, to study for the final exam. The exam follows the same two-part format as the unit tests.

Continued on next page

NUTS & BOLTS *(continued)*

GRADING

Everything submitted in this class is evaluated on the following rubric (A-I):

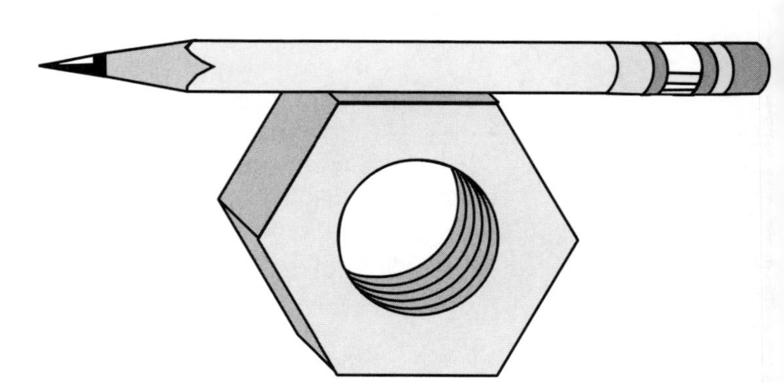

Grade	Criterion
A+	Awesome - extra credit
A	Thorough effort and communication
B	Correct, organized & professional
C	Minimal understanding
D	<70% (tests/quizzes only)
F	<60% (tests/quizzes only)
INC	Incomplete (revise)

The scores are then averaged and weighted as shown below:

Categories & Weights	
Tests*	50%
Projects & Challenge Problems	20%
Quizzes	10%
Concept Problems	10%
Original Work	5%
Homework	5%

Course Scale	
90%	A
80%	B
70%	C
60%	D

PORTFOLIO: Portfolio completion is mandatory, but it does not constitute a portion of the grade, since it is considered a representation of the evaluated work in the class. The details of the portfolio process will be discussed in an upcoming issue.

INCOMPLETE: ***Failure to submit an acceptable product for any assignment will result in an incomplete for the course, which equates to an F on the report card.*** In other words, a student will not pass the course if he/she is missing any graded assignment (homework assignments excluded). If a student turns in an assignment that is not worthy of at least a C, we will return it as an incomplete for revision. So, no matter what percentage grade the student may have in the course, he/she receives an F on all grade reports until the assignments are completed. Only those students that are truly failing will allow this F to remain through their semester grade. This encourages (forces) students to complete every assignment, and in the long run, accomplishes two goals. One, the students at least SEE all the material, and two, the students avoid collecting low scores which would mathematically bury their grade by the semester's end. This last benefit is very popular with parents. However, it is advisable that you clear this policy with your administration before implementation.

LATE WORK: Late work is penalized one letter grade within the first week of the due date and two letter grades during the second week. Any work submitted more than two weeks late will be given a zero, but must still be submitted in order to eliminate the incomplete.

***PROGRESSIVE TESTING:** Since the tests are cumulative, the overall test score is computed by the test average or the most recent test, whichever is greater. Our tests are cumulative and contain approximately 40% review material from previous tests. The cumulative format allows students an extended period of time to develop an understanding of key concepts. It also sends the message that these concepts will not go away and that they must be mastered, instead of encouraging a "study for 2 weeks, forget in 2 days" regime. The "most recent test score" option encourages students to study; and as the course goes on, there is an increasing incentive to do so. Moreover, the industrious but slow are rewarded for their determination and perseverance, because, it is never too late for the students to show us what they know. The results have been an improvement in student performance as well as in their grades.

Kick the Habit!

TEXTBOOK FREE

Kicking the Habit

I kicked the habit! I am no longer a textbook junkie. I no longer rely on my daily fix of some publisher's bloated curriculum. I am free of my addiction without the help of an arm patch, rehabilitation clinic or twelve-step program. I quit cold turkey. Here's how.

At my school, the students are issued a math book that they leave at home and each teacher is issued a class set. I usually keep one underneath each desk. This year, however, the librarian informed me on the first day of school that we were out of Geometry textbooks. Our student population had grown so large that our library ran short. In fact, for two to three weeks many of my students would not have a book at home either.

There was talk of teachers sharing class sets and photocopying pages for students. I decided to try a different strategy. I took this as a professional challenge to see how long I could teach without a textbook. I knew whatever happened would be a growing experience for me as well as my students.

Well, by no fault of the school library, two to three weeks stretched to seven. By that time, I was well into my "textbook free" strategy, so I just kept the ball rolling...for the rest of the year. I used only 12 assignments from the textbook in those 180 days. Here is how that unique experience of being textbook free has changed my teaching, forever.

Firstly, I am now much more focused on standards. Rather than leafing through the textbook, I looked at my state and district standards, and established my curriculum from those. After all, shouldn't they be determining what we teach? From there, I grouped the topics into units, and then scheduled individual lessons. This process naturally pared down the number of topics that I taught and allowed me to allocate a full week of instruction to each concept, rather than one day to each section of the textbook.

The second big change that has occurred is the structure of my lessons. Everything from my homework to my instruction has radically changed. My typical textbook free lesson was comprised of three to six problems of various difficulty. Oftentimes, I began a lesson with one to three review problems from previously learned material which applied to the current lesson. This is similar to a traditional warm-up with the exceptions that the problems are very relevant to the new lesson, and not simply arbitrary review.

Sometimes, I began with THE big problem from the previous night's assignment, and solicited student responses. It is not hard to see that my old practice of dedicating 20 minutes of class time to questions on how to complete the previous homework disappeared. The intent of the class slowly evolved from getting the answers correct to understanding the mathematical principles behind the question.

These introductory problems served as a terrific assessment tool, also. Previously, it was difficult to know how well the students were doing when only a handful of them were asking questions from a truck-load of exercises. However, when the whole class was engaged on the same few problems, it was easy to walk the room and evaluate their performance and understanding.

The introductory questions naturally lead to the main problem or small set of problems that would drive the lesson. The students were engaged in an investigation, project or activity relating to the concept. Each day my students came to class to solve problems, rather than take notes — a huge change from all the previous "textbook years." This

Continued on next page

TEXTBOOK FREE [CONTINUED]

process of problem-solving and investigation consumed the full class period. Gone were the days of having the students start homework in class. I taught the entire class period.

The homework assignments were only one to three problems long and were typically extensions of the day's topic, not just practice exercises. I had learned from the international comparisons that America is one of the few countries that pushes the drill-n-kill regime and yet we are at the bottom of the performance pile. So I tried to limit both the number and size of my assignments, and to make them more challenging and contextual.

By doing that, I firmly settled the argument regarding the quantity and frequency of homework that students need to be successful. For the skeptics that are still reluctant to abandon their practice of assigning 30 homework problems a night, I have some strong evidence. My class averages led the district on the district final. With this in mind, I can at least make a case that this new homework philosophy is not hurting my students in anyway.

Another significant change was my lesson planning. Rather than writing examples of how to complete an algorithm or creating cute acronyms to remember esoteric rules, I actually wrote lesson plans. I started planning each lesson by asking: "What do I want the students to know? What is their common misconception of the topic? How can I best get them to understand the topic? How can I challenge them within the context of the topic?" I would then try to create a story/context/scenario and a small set of problems that would best develop understanding of that topic. It was so much fun. This change in my approach to lesson planning was actually a reflection of my new attitude towards teaching. My job description truly shifted from covering material to uncovering knowledge.

Focused, standards-based curriculum; in-depth, problem-solving instruction; short, conceptually-based homework assignments. This experience was so exhilarating that I am now a junkie all over again. I traded my old addiction to the textbook, for a new one — creative lesson planning. This is one habit, though, that I never intend to kick.

On a Good Day

On numerous occasions, I have been asked what a typical lesson looks like within the context of project-intensive instruction. While I share that typical model elsewhere (see "Textbook Free"), I would rather not laud the merits of a typical lesson because it is not what I personally aspire to.

As with my golf game, I don't want to tell of my 100 typical shots (yes, I am that bad at golf), I want to talk about the one perfect shot. There is always one every round. It's the one that keeps me coming back thinking I can play this game. Let me share one of those rare shots from the classroom — a lesson that I strive to make my regular experience. Rather than a typical lesson, let me share my ultimate lesson.

LESSON OVERVIEW

The lesson actually spans three days, with the main goal being to have students acquire the necessary skills needed to calculate the length of an arc. The first day of the lesson is intended to familiarize students with the concept of pi, and how to use it to calculate the circumference of a circle given its diameter. The second day of the lesson is intended to show students how the diameter can be determined given the circumference. It also perks student interest in circles and develops the principle of writing equations, which is an ancillary theme of all three days. The third day is the climax of the lesson in which the students use the concepts and skills of the previous two days to derive a formula for finding the length of an arc, given the radius and central angle.

DAY ONE

A Slice of Pi

I take the students through "A Slice of Pi", an activity that helps students understand that pi is the ratio of the circumference to diameter (rather than simply approximately 3.14). The students measure several round objects, "How far around? How far across?" They then divide "how far around" by "how far across." In other words, they find the ratio of the circumference to the diameter. Each group writes its results for each object in a chart on the board. The students get spooked. The value that suspiciously appears throughout the chart is approximately equal to 3.14. Now they understand the relationship between circumference and diameter. Part of the activity includes measuring the circumference of a tree and recording the diameter of a basketball hoop (known to be exactly 18 inches). I then have the students use the spooky value in the chart to find the diameter of the tree and the circumference of the hoop. Without any algorithmic training, the students can find these dimensions — no problem.

The Staggered Start

For homework that night the students are given a few traditional problems requiring them to calculate the circumference given diameter and vice-versa. They also had the following problem.

> *A standard high school running track is 440 yards around on the inside lane, with each straight-away being 120 yards long. A runner standing in lane 8 gets what is termed a staggered start. If the race is 440 yards long, and each lane is three feet wide, how much of a lead should the runner in lane 8 receive?*

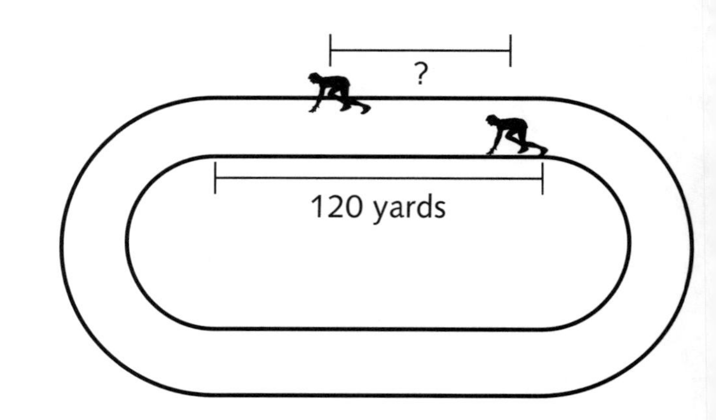

DAY TWO

I begin the day with the following statement, "We have three problems to solve today. They deal with a running track, the globe and pizza and how each relates to the ratio of pi that we investigated yesterday."

Continued on next page

On a Good Day (continued)

We start with the track problem from the previous assignment. The students discuss their solutions (mostly partial solutions) with their group members. After some time, I solicit responses for the board.

A common strategy is to reduce the track diagram to a circle. Since the straight-away distance is the same for both runners, the extra yardage for Lane 8 must be in the curves. So the students *changed the question* to "What is the diameter of a circle with a circumference of 200 yards?" [440 - 2(120)]. The entire class quickly jumps on board with this strategy and accurately calculates 64 yards for the diameter of the circular track.

The class now identifies that the easiest solution is to find the diameter of Lane 8 and use that to calculate the circumference of Lane 8. Then, by subtracting that from the 200 foot circumference of Lane 1, it will yield the correct distance of the staggered start. Their strategy is correct, but their execution is wrong. First they assume, that since each lane is one yard wide, that the runner in Lane 8 is eight yards from the runner Lane 1. Actually, runner 8 is only seven yards away from runner one, just as a runner in Lane 2 is only one yard not two yards from the runner in Lane 1. The students also fail to add this extra width to the diameter at both sides of the track.

With a bit of discussion, they all agree with the correct method, and they accurately add 7 yards twice to the 64. Now the question reads, "Given a diameter of 78, what is the circumference?" The students have previously demonstrated the ability to solve this problem with the basketball hoop, and quickly agree that the circumference of Lane 8 is 245 yards. Therefore, the runner should get a 45 yard lead on the staggered start!

Just when the students are feeling pretty good about solving a tough problem, we extend it with the following:

> *Write an equation that will allow a coach at any high school to calculate the distance of the staggered start given the lane number and width of the lane.*

The class works diligently with a variety of equations, and settles on something like the one shown below:

$$D = C_8 - C_1$$
$$D = [d + 2(L - 1)w]\,\pi - d\pi$$
$$D = 2(L - 1)w\pi$$

At some point in solving the problem, one or more of the students inevitably ask for the dimensions of the new track. Who's to say that all tracks are congruent? I encourage them to substitute d for the diameter of the circular track and recalculate. Amazingly, the students get the same equation as they did the first time. In other words, their formula will work for any oval track of any size, given lane number and width of the lane. The students think that is very cool!

Around the World

This one is more of a demonstration than an activity. It begins with the following question (see additional activity "Around the World"):

> *Place a string around the earth's equator and pull it snug. Add a length of one meter to the string and push it away from the equator so that the string is equidistant from the surface of the earth. Which of the following is the largest animal that could walk under the string: flea, dog, horse, giraffe?*

The solution to the problem above is found by solving the following system of equations:

$$\begin{aligned} C &= 2\pi r_1 \\ C + 1 &= 2\pi r_2 \\ (C + 1) - C &= 2\pi r_2 - 2\pi r_1 \\ 1 &= 2\pi(r_2 - r_1) \\ 1/2\pi &= r_2 - r_1 \\ r_2 - r_1 &\approx 1/6 \end{aligned}$$

Continued on next page

On a Good Day (continued)

The students can write the first equation on their own and, with a little explanation, can see the meaning behind the second one. An answer of approximately $^1/_6$ means that the string will be one-sixth of whatever unit is added onto the string. In this case, one-sixth of a meter is about 6.5 inches. The kids find this unbelievable, especially since we never used the actual value for the radius of the earth. Here is the teachable moment in this activity. Since our equations were written in terms of r, this works for any circular/spherical object of any size.

The demonstration really drives this point home. Take a string around several round objects of various size (racquetball, globe, group a students standing in a circle). For each object, first wrap it with the string, then add a meter more of string and rewrap. Guess what? Each time the string is exactly 6.5 inches away from the object — just as they predicted.

Pizza Slices
The second day is concluded by assigning the homework problem.

I don't ever eat the crust of any pizza. With the following portions of the pizza, in each case, how much crust (length of the arc) will I be wasting?

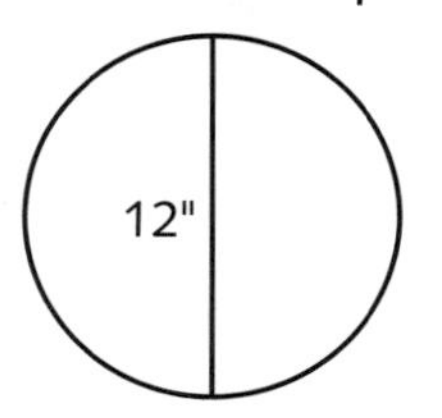

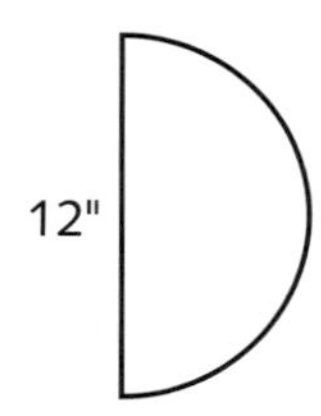

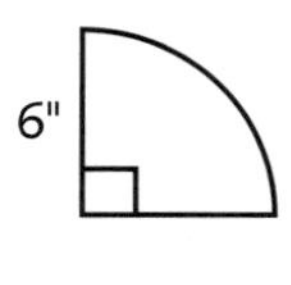

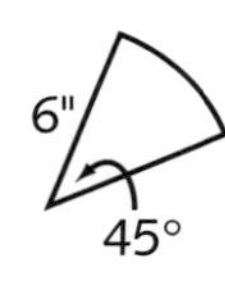

The students are told to answer this for all six situations shown. Six? "While you only see five diagrams, there are actually six problems to be solved. What is the sixth question?"

The students respond, in near unison, "What is the length of the crust for all the infinite pieces of pizza in the world!"

"Given what?" I probe, to which they retort, "Radius and angle."

DAY THREE

Group members share their solution to the pizza crust problem and then share them up on the board. Amazingly, the students all seem to have the same equation: $2\pi r/(^{360}/_A)$. In my excitement, I am tempted to show the students how they are one step away from the conventional formula for arc length $[(^A/_{360})2\pi r]$. I resist this temptation for one very strong reason: this is not the way students think about the problem. In the conventional formula, the angle is used to determine what fraction of the entire circle is represented, which is the way I and probably most mathematicians think about the problem. In the student-generated formula, the angle is used to determine into how many pieces the circle is portioned, and then the circumference is divided into that many parts. Technically this difference is subtle, but conceptually it is very profound. I leave the students with their creation, and the remainder of the day is spent applying it in a variety of situations.

CONCLUSION

What I enjoy most about teaching this lesson is the way it flows. It starts with a concept of π, proceeds to calculations on basic circles, then moves onto seeing circles within a bigger problem and to understanding the relationship of diameter to circumference, and finally arrives at a new take on a very old equation. And the students are with me every step of the journey. Throughout the lesson is an interwoven principle of writing and solving equations. In fact, I asked for feedback from a colleague who has observed this lesson. He claims that the whole lesson boils down to deriving three equations — all involving pi in a different fashion. The finest quality about this lesson, though, is that it is very effective. Students retain a high level of competency regarding circles, arcs and sectors.

The lesson is a keeper. I need more keepers like it. I use this and other similar lessons as models — much like analyzing Tiger Woods so I can improve my own golf swing. I wish all of my lessons were this tight, rich, rigorous — and this fluid. Until they are, I'll keep swinging.

ARTICLE

Sleeping with the Enemy:

Being Project Minded in a Textbook Curriculum

You feel trapped. You see value in teaching with projects, but you feel pressure to cover the material in your textbook. You find yourself teaching section 5-2 simply because yesterday you taught section 5-1. You do not know how to fit in anything new, and you are not sure what you can throw out. You see the gains made by the use of projects, but it is not as simple as homework checks and answer columns.

You are facing the crux of the problem. Projects can greatly enhance teaching and learning, but do not interface well with traditional curricula and grading policies. While you may value the former, the school system and the culture value the latter. So the ugly fact remains. You are relegated to sleeping with the enemy.

We suspect that this is your experience, or else you would not be reading this publication. We know that you are as bored talking to your students as they are bored being talked to. After covering all the material, most students still do poorly on the final exam. The defenders of the old order blame the poor results on unmotivated students, uninvolved parents and an undisciplined culture. We know, as do you, that it is the pedagogy.

The key to using projects effectively within a traditional curriculum is to answer the 6 FITS:

WHAT FITS? This basically asks, "With what types of activities is the teacher comfortable?" Long or short? Simple or complex? Sometimes, teachers create their own projects, but generally, a project is used because the teacher discovers it and feels that it would work in his/her classroom. For our discussion, let us walk through the FITting process with "How Big is Barbie", a project where students predict the dimensions of life-size dolls.

WHERE DOES IT FIT? This is the simplest question to answer, since most projects focus on one or two main concepts. For instance, "Barbie" deals with ratio and proportion, so it fits into a unit on proportion or similarity.

WHEN DOES IT FIT? Should the project be used for the introduction, reinforcement, enrichment, or review? As an introductory activity, "Barbie" could be used to hook the students' attention for a unit on proportions. "Barbie" is also excellent for reinforcement. After using traditional direct instruction to review proportions and teach scale factor, the instructor can use "Barbie" to strengthen those skills. It also lends itself to enrichment, since the students are required to read a chart, analyze data and develop hypotheses. The nature of the project and the discretion of the instructor determine when a project should be used within a unit.

HOW DOES IT FIT? Time is the number one issue when trying to integrate projects into the curriculum. The best solution we have found is to pare down your current curriculum. In order to do this, we must have a clear idea of the skills and knowledge that we wish our students to learn. Besides fitting projects into the curriculum, we also need to fit them into our grading policy. We suggest you start by making the sum of all projects a small portion (10%) of the students' grades. Once you add more project-oriented lessons, you can increase their weight.

WHY DOES IT FIT? This is basically a question of assessment. Does the project facilitate the students' skills and knowledge? One of the worst mistakes a teacher can make is to implement a lesson because it is "cool," when the lesson offers no mathematical substance or educational value. Be sure the lesson is what your students need.

DON'T HAVE A FIT. This warning is for three reasons. One, more projects means more papers to grade; yet by de-emphasizing textbook problem sets, you will grade less often. Two, you will see that your students know less than you previously thought. They do not read, write, think, or compute to your level of expectations. Have faith because through time and multiple projects, they will learn to do so. Three, the more you teach with projects and other alternative learning activities, the more you will be dissatisfied with teaching in a traditional curriculum.

So keep up the good fight, but don't panic. Find ways to make the projects fit, and most importantly, find the best blend of the progressive and traditional methods and materials that works for you and your students.

That's what ***understanding*** *is all about!*

INSIDE and OUT: Multiple Representations

"If you know one problem inside and out, you can do a hundred just like it." This shining bit of insight was given to me by a woman named Seheli, a math teacher who was raised and educated in India, but is now living in the United States. She claims that in India students learn a small set of problems thoroughly so that they may grasp the underlining principles involved. In this model, if a student understands the concepts, they may apply them to a broad range of problems.

The Indian philosophy "knowing one problem in order to do a hundred" is in sharp contrast to our American approach of "doing a hundred in order to know one." This concept is the reverberation of a presentation on fractions that I witnessed by my friend and mentor, Dr. Tom Bennett of California State University, San Marcos. The crux of his presentation went something like this:

Let us see how many ways we can represent $1\frac{1}{3} \div \frac{1}{6}$

He started with a repeated **addition model**:	He then moved to a sample of **repeated subtraction**:	Next, he illustrated a **multiplication model**:	Then there is the **area model**:	It can also be represented with **multi-link cubes**:
$\frac{1}{6} + \frac{1}{6} + \ldots + \frac{1}{6}$ $= 1\frac{1}{3}$.	$1\frac{1}{3} - \frac{1}{6} = 1\frac{1}{6}$; $1\frac{1}{6} - \frac{1}{6} = 1$; ... $\frac{1}{6} - \frac{1}{6} = 0$.	$\dfrac{\frac{4}{3} \bullet \frac{6}{1}}{\frac{1}{6} \bullet \frac{6}{1}} = \dfrac{\frac{24}{3}}{1}$	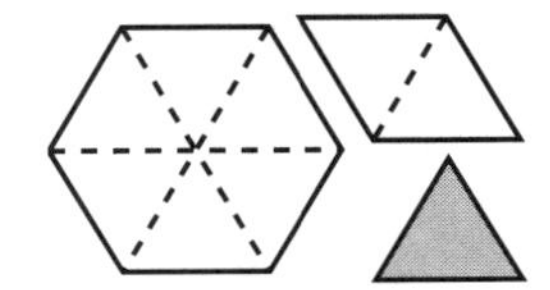	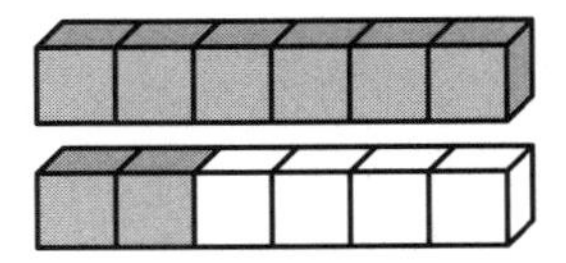
How many times do we need to add $\frac{1}{6}$ to get $1\frac{1}{3}$?	How many times do we need to subtract to get zero.	What is $1\frac{1}{3}$ times 6?	How many small triangles ($\frac{1}{6}$) are needed to cover the hexagon (1) and rhombus ($\frac{1}{3}$)?	How many single cubes ($\frac{1}{6}$)will it take to make the shaded area ($1\frac{1}{3}$)?
Eight times.	Eight times.	Eight.	Eight.	Eight!

The presentation struck me. I said, "Dr. Bennett, you just showed us several ways to do one problem, instead of the traditional method of showing one way to do many problems." His response... "Exactly."

"If you know one problem inside and out, you can do a hundred just like it," echoed in my head again. This idea of multiple representations reminded me of a presentation by renown author and educator, Ruth Parker. Mrs. Parker criticizes the algorithmic approach to teaching arithmetic. We all know that children in America are taught only one way to do addition, multiplication, etc. She shared an anecdote that is very relevant to this point of multiple representations.

At a "Back to School Night" meeting, Ruth Parker was sharing with parents alternative ways to do arithmetic. One of the fathers, who was Korean, supported her efforts with the following story.

> *In Korea, they play a game called "How Else?" in which students share strategies for doing arithmetic. For example, in multiplying 45 times 23, one student may double 45 and add a zero to get 900; then add three 40's and three 5's. Another student might multiply 50 times 23 (1150) and then subtract 5 times 23 (115). The idea is not to know one particular method, but to have enough number sense to know several methods.*

Again the echo, "If you know one problem inside and out, you can do a hundred just like it."

I suggest we all take the advice of Seheli, Dr. Bennett, and Mrs. Parker, and use multiple representations to pave the way to greater understanding for our students.

MEASUREMENT

A Bridge Between the Concrete and the Abstract

The word "geometry" literally means "to measure the earth." Yet, how often are students of geometry given the opportunity to measure? Much of the mathematics that students are called to study in today's classroom was discovered out of practical needs that arose in areas such as agriculture, architecture and navigation. However, much of this pragmatic quality of mathematics is not found in many math classrooms. Most students experience only formal, abstract symbol manipulation.

By international comparisons, American teachers tend to begin too often with the abstract concepts rather than the concrete principles. Ironically, we do not take the formalization of mathematics far enough either. Many of the top performing countries in mathematics education begin the teaching of a topic with concrete instances and incrementally develop the abstract concept.

Measurement is one way of bridging that gap between the concrete and abstract.

For example, measurement can be used to help students understand the illusive concept of irrational numbers — square roots in particular. A sample of how measurement can be used to develop the understanding of irrational square roots, particularly, the square root of two, is as follows.

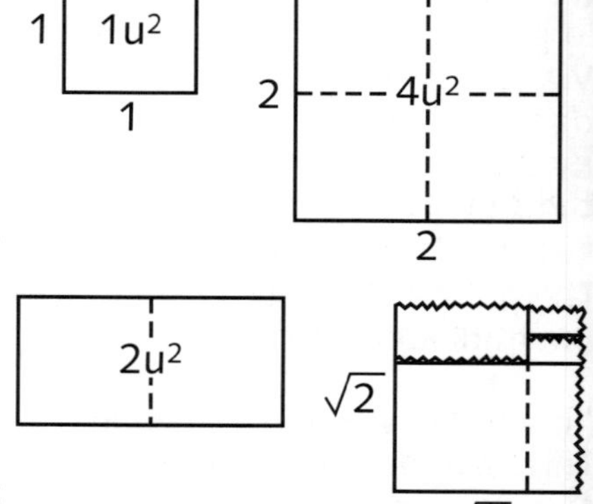

Begin with two squares, one having an area of one square inch and the other of four square inches. The squares should be actual size on paper so that the students can measure the length of the sides of the square. These lengths will be the square root of one and the square root of four respectively. Then offer the students a rectangle created by two adjacent unit squares.

Challenge the students to cut the rectangle into pieces so that they can arrange the pieces to make a perfect square. Encourage the students to start with one unit square, and slice the other into a variety of rectangles as shown.

By having students use different cutting strategies, you will end up with a variety of resultant figures, all of which will only be approximations of a true square. Once the students have their shapes cut, ask them to measure the sides of the nearly-square figures.

Of course, the figure that most closely approximates a square will have a side that is a little less than 1.5 inches. At this stage, the key issue to reinforce with the students is that the length of the sides will always be the square root of the area, just like in the first two squares. Therefore, the side of this square must be the square root of two. Estimating the decimal value of the square root of two manually or with a calculator will make this approximate measurement more meaningful.

The fact that the students can measure this quantity is critical for two reasons. The first is that students often see a number like the square root of two as simply another symbol to manipulate, like a variable. This is partially due to the fact that the rules for simplifying expressions that involve variables are identical to those involving square roots. Second, to further complicate matters, students often have the misconception that irrational numbers "go on forever," since their decimal values do not repeat or terminate. Many students have a difficult time understanding that while the decimal value of a number goes on forever, the quantity does not. This quantity (in this case length) can be measured and has a finite value.

Continued on next page

MEASUREMENT *(continued)*

This principle of being able to measure an irrational length can be further extended using the Pythagorean Theorem.

Ask students to draw two right triangles with legs of exactly one inch each for the first triangle and one and two inches for the legs of the second. Then have the students measure the hypotenuse of each triangle. The hypotenuse of the first triangle should be about 1.4 or approximately the square root of two. Similarly, the hypotenuse of the second triangle should be about 2.25 or approximately the square root of five. Again, the goal of this example is for students to understand that the value of an irrational number can be measured.

Another important topic that can be developed using measurement is trigonometric functions. Many students know that the tangent of an angle is the ratio of the opposite side to the adjacent side, yet they still don't know how that concept relates to the output of the tangent function on their calculator. To help remedy this, the following simple measurement activity will allow students to make the connection between the abstract concept of tangent and the concrete value it represents.

Draw right triangle ABC, with C being the right angle. Draw auxiliary lines B'C' and B"C" as shown. Measure BC, B'C' and B"C", and also AC, AC' and AC". Using angle A as a reference, find the decimal value of the ratios of the opposite side to the adjacent side for each of the corresponding triangles. What do you notice? (They are approximately the same.) Measure the angle and find the tangent of that angle using your calculator. What do you notice? (It matches the ratios that we found in our triangle.)

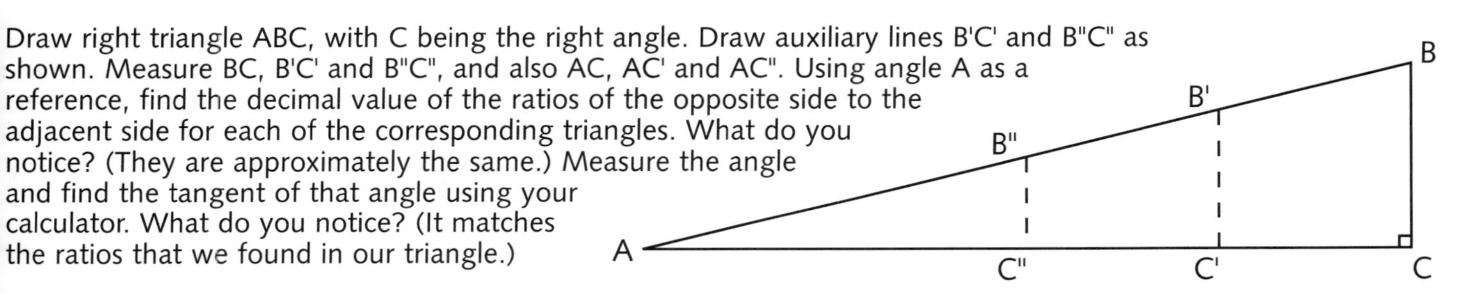

While measurement can help students bridge the gap between concrete and abstract concepts, the measurement does not have to be strictly length, nor does the principle of study have to be something as difficult as irrational numbers or trigonometric ratios. For example, a common theorem taught in geometry is the polygon sum theorem which states that the sum of the interior angels of a polygon is equal the product of 180 degrees and two less than the number of sides: S = (n - 2)180. We can derive this formula by partitioning a variety of polygons into triangles; and many teachers have their students do just that. However, how many students are asked to draw a variety of polygons and actually measure the interior angles to confirm the conjecture?

We teachers often assume that because a student responds with the appropriate algorithm that the student understands the mathematical principle to which the algorithm applies. I believe students can remember to apply the polygon sum formula whenever they see "find the sum of the interior angles," without ever understanding what the theorem really means. By using a protractor to measure the angles of a variety of polygons, the students will see that the formula that was derived intellectually holds true physically.

$$S(n) = (n-2)180$$
$$S(6) = (6-2)180$$
$$= 720$$

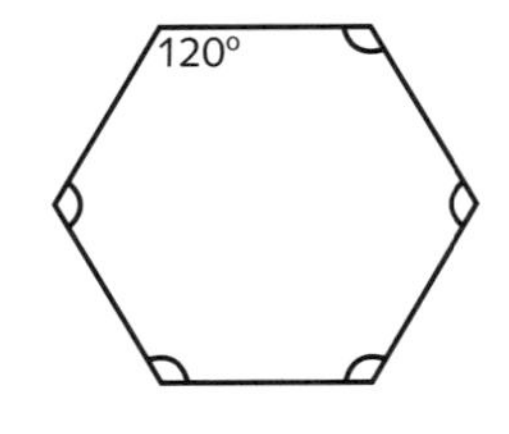

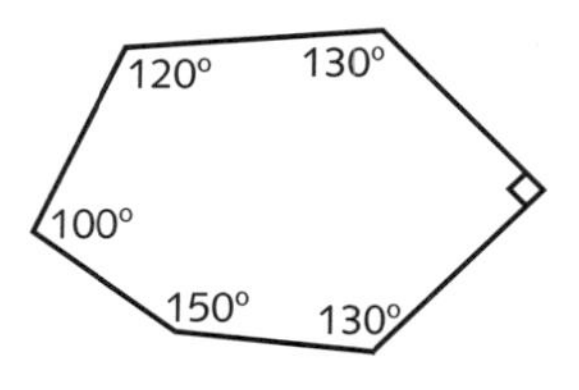

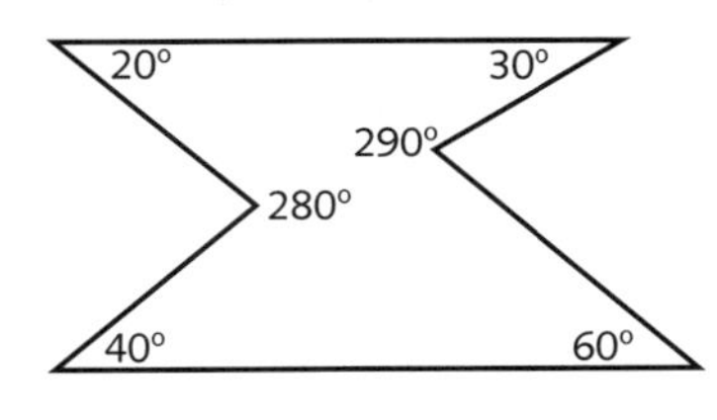

Measuring with a variety tools stretches beyond geometry. Algebra lends itself nicely to measurement, especially with time. Since so many of the equations that are used in a typical algebra course deal with rate (i.e. speed), it may be justified to have a class set of stop watches, as well as the standard rulers and protractors. The list of tools doesn't have to stop there; don't be afraid to borrow equipment from the science department. Thermometers, barometers and balance scales are just a few of the possible tools that may be used in bridging the gap between the concrete and the abstract.

Moving Students FORWARD by Teaching BACKWARDS

When I was a small child, I was always intrigued by ambulances. It was not so much their high speeds and sirens that grabbed my attention, as much as their propensity to spell ambulance backwards. Each letter as well as its order within the word was reversed. Of course, once I began to drive and caught my first glimpse of an ambulance in my rearview mirror, I understood its purpose. It needed to be presented backwards in order for people to understand when it really counted.

That same childhood story has repeated itself in my teaching. For several years I have been posing projects and problems to my students before I actually teach the skills needed to complete or solve them. I then refer back to the problem throughout a unit. By doing so, the students begin to understand the usefulness of a particular technique as well as the mathematical principles underlying the topic.

This practice of presenting the problem first is contrary to the way most adults were taught when they were in high school. All the algorithms were presented for the first 14 chapters, and then all the word problems were thrown in during the last few weeks of the course. In contrast to that method, I now front-load many lessons or units with some kind of word problem, then follow up with the instruction of the techniques and principles to solve it. Thus, I affectionately refer to this method of instruction as teaching backwards.

When a teacher offers students a context in which to think about a topic, or a purpose for which to solve a problem, all the formulas and procedures no longer exist simply as esoteric facts to be memorized. The otherwise meaningless stream of textbook information becomes a comprehensive description of a phenomenon of their world that they are exploring. The previously unrelated techniques become connected tools to accomplishing something worthwhile, and on our best days of teaching — something fun. When students are presented with a project first, the purpose for learning is elevated to solving a problem or understanding our world, not just to prepare for a test and achieve a grade.

This idea of placing context first is nothing new. Over the course of human history, mathematical concepts have developed in order of the most simple and concrete to the complex and abstract. For teachers, this principle goes to the heart of the nature of learning. Psychologist and educational theorist Jean Piaget discovered that children learn to think concretely before abstractly. In the same fashion, most of the great early discoveries in mathematics and science were made due to people's need to solve a particular problem. The ultimate goal in both psychology and education today is to abstract and generalize. In the same vane, we should push our students to these higher levels of thinking. However, we must start where they are — in the concrete and contextual. We need to put the meaning back into the mathematics that we are teaching!

Even though generalization is the ultimate goal in mathematical thinking, once the abstract understanding has been achieved, context is still needed in order to communicate the ideas. The two most famous physicists of this century, Albert Einstein and Richard Feynman, are most regarded for their ability to express very abstract concepts in layman's terms. In fact, Dr. Feynman has insisted that if we cannot reduce a concept to "the freshman level," then we really do not understand it at all. (*Six Easy Pieces*, R. P. Feynman, 1995.)

So perhaps, I am not teaching backwards at all. Could it be that our students do not need us to spell mathematics backwards so that they can read it in their rearview mirrors? Somebody else has already turned it around on them, and they need us to simply flip it back. We can do that for them by offering context before imposing algorithms, and by immersing them in concrete application before abstract generalization. We can truly move our students forward by teaching backwards.

ULTIMATE COSMIC POWER

in an itty-bitty thinking space

ARTICLE

ULTIMATE COSMIC POWER

abstract generalizations of pervasive patterns

"Give me any combination of two numbers that have a sum of seven," I said to my students. One person offered, "Two, five," which I wrote on the board as (2, 5). I asked for a few more and got (5, 2), (1, 6) and (0, 7).

"Good," I praised, "now give me ALL the combinations of two numbers that have a sum of seven." They chuckled. "I want them all, and I want you to write them down." The students were hesitant, because they knew there are an infinite number of pairs that have a sum of seven. So I challenged one of them to a race. "You write them down on your paper, I'll write them on the board. Nobody goes to lunch, until one of us is done. Ready, Go!" I scribbled on the board $x + y = 7$. "Done!"

They didn't buy it. "I have just written ALL the combinations of two numbers that have a sum of seven. Since you don't believe me, I'll do it a different way. In fact, I'll take you all on. All of you write down combinations of numbers, that way you get done in one-thirtieth of the time, and I'll still woop ya. Ready, Go!" I quickly sketched the graph of $x + y = 7$. "Done!"

This goofy little exercise was intended to impart the idea that mathematics gives us the ability to represent an infinite number of elements in a finite time with little effort. I spread my arms wide in front of the class and exclaimed "Ultimate cosmic power..." then brought my hands to rest on a student's head and continued, "...in an itty-bitty thinking space." (A play on the Genie's words from the Disney movie *Aladdin*.)

For example, I used a lesson in geometry that is quite popular among many educators for teaching the Interior Angle Theorem. The theorem states that the sum of the measures of the interior angles of any polygon will be equal to 180 times the quantity of two less than the number of sides: S = (n - 2)180. This is most commonly developed by using a chart similar to the one shown in Figure 1.

	△	□	⬠	⬡	octagon	decagon	circle	
# of sides	3	4	5	6	8	10	100	n
# of diagonals from 1 vertex	0	1	2	3	5	7	97	n-3
# of triangles	1	2	3	4	6	8	98	n-2
Sum of all angles	180	360	540	720	1080	1440	17640	(n-2)180
Single angle measure (regular)	60	90	108	120	135	144	176.4	$\frac{(n-2)180}{n}$

Figure 1. A chart developing the Interior Angle Theorem through a series of instances.

My students had little trouble completing the chart corporately in class. I pointed out to them that the first four columns were the "concrete" side of the chart. Anybody can do this, because anybody can count. Mathematicians don't count, though, they compute. To demonstrate this, I asked them to explore the octagon and decagon. Among the students, the common method in completing these two columns was simply to keep adding repeatedly until they got to the column they needed. This was too cumbersome for the 100-gon, so the students then had to devise a strategy. This is the "abstract" side of the chart. The first seven columns showed seven different instances of the pattern, however, the last column showed us ALL the infinite examples at the same time. Again, ultimate cosmic power in an itty-bitty thinking space.

While this exercise is traditionally used to demonstrate the conceptual underpinnings of the theorem and its formulas, it also offers a great opportunity to allow students to generate abstract generalizations of pervasive patterns. Knowing the formula, S = (n - 2)180, will serve the students for as long as they are in the class. Being able to generate formulas like S = (n - 2)180 will serve them a lifetime.

Continued on next page

ULTIMATE COSMIC POWER (continued)

While the students had a firm command of the theorem, I was curious whether they understood the process behind generating abstract expressions for unfamiliar patterns? In other words, if presented with a series of diagrams and a chart, could the students correctly write a formula or expression for the pattern. My instincts said probably not yet, but I decided to test them. On the next exam, I included the problem in Figure 2 and allowed them to use the chart that they completed in class for the interior angles. The directions were as follows:

Below is a chart showing the number of "perimeter triangles" in polygons that have more than three sides. Perimeter triangles are the triangles that are formed by the diagonals of a polygon, and which share a side or vertex with the polygon itself (shaded below). The chart displays the number of sides of a polygon, the number of diagonals from each vertex, and the number of perimeter triangles. Complete the chart.

# of sides	4	5	6	8	100	n
# of diagonals from each vertex	1	2	3			
# of triangles	4	10	18	40		

Figure 2. A test question similar to the Interior Angle chart (Figure 1).

I was pleased to see a few students accurately complete the chart. In particular, they realized that the number of perimeter triangles would be equal to "n(n - 3)." Interestingly, many students saw a different pattern. They assumed that since the second row of this chart matched the second row of the previous chart, then the third rows must also match. Therefore, many wrote "n - 2" for the number of triangles. I was disappointed that only some of my students possessed this Ultimate Cosmic Power, so I continued to present them with similar lessons.

During the next unit, I conducted a lesson that involved calculating surface area and volume of simple cubes. The pattern and chart are shown in Figure 3. The students were given only the first row and column; then they were expected to complete the rest of the chart during the activity. They were supplied with blocks to construct the first three diagrams, but then had to create strategies to finish the chart. Again, collaboratively, they had little trouble. Again, I pointed out that the last column offered ALL the infinite cubes at the same time.

# of cubes (along the edge)	1	2	3	4	10	n
Surface Area	6	24	54	96	600	$6n^2$
Volume	1	8	27	64	1000	n^3

Figure 3. A chart investigating patterns of surface area and volume with stacking cubes.

Continued on next page

ULTIMATE COSMIC POWER (continued)

After a few more weeks of instruction, it came time to assess their knowledge of surface area and volume, and their ability to generate abstract representations. On the next exam, I posed the problem in Figure 4. The directions for this question were:

Several cubes are lined up to form a long "block" as shown below. Complete the chart below relating the surface area and volume of the "block" to the number cubes that form it.

# of cubes	1	2	3	4	10	n
Surface Area	6 units^2					
Volume	1 unit^3					

Figure 4. A test question based on the surface area and volume investigation chart (Figure 3).

Again, the students were allowed to use the chart from the lesson in class. The number of students that accurately completed this type of chart was much greater than that on the previous test. They were making progress. However, I still had several students who filled in the chart with the identical numbers in the chart from the lesson, even though the diagrams were drastically different. They saw the first columns of the chart were the same, and assumed the rest would be congruent as well.

I interpreted this phenomenon as the persistence of old habits rather than the lack of cognitive ability. The student habits that I speak of specifically are the copying of each other's work, and the copying of information from the teacher. I suspected that the more I persisted with these types of questions, the more those old habits would fade, and the better the students would become at recognizing pervasive patterns and representing them algebraically. My suspicion was correct. On the next exam, I presented the students with the question in Figure. 5. The directions for this question were the same as those for the previous one:

Several cubes are aligned to form a "block" as shown below. Complete the chart below relating the surface area and volume of the "block" to the number cubes that form it.

Width	1	2	3	4	10	w
Surface Area	6 units^2					
Volume	1 unit^3					

Figure 5. Another follow-up test question based on the surface area and volume investigation chart (Figure 3).

From their comments after the test, it was apparent that students found the problem to be difficult, but the great majority of them accurately wrote $2w^2 + 4w$ and w^2 for the surface area and volume, respectively. In other words, the problem was tough, but they conquered it. I was pleased to see that persistence in both problem-solving and abstract reasoning were resulting from these activities. We still have a long way to go in the school year. However, I feel confident that with persistent training my students will make significant strides in developing Ultimate Cosmic Power in their Itty-Bitty Thinking Space.

CANNONBALLS & CONCEPTS

Going beyond algorithms for a deeper understanding of mathematical principles

We American math teachers sure do love our algorithms. That is the contention of the Third International Mathematics and Science Study (TIMSS). So what's wrong with that? Absolutely nothing. It is the lack of attention that we give to developing student understanding of mathematical concepts that contributes to our pathetic performance in international comparisons (see "A Call for Substance").

I stumbled upon a math problem that radically reveals the extent to which we math teachers are focused on algorithms, and suggests that this is due to our own lack of conceptual understanding. This revealing math question comes from a book that I received as a Christmas gift titled, *The IQ Obstacle Course*, by David J. Bodycombe (1997, Barnes & Noble). The question from the book is a follows:

> *A cannonball is fired and after traveling 5 meters it has reached half its maximum height. After what (horizontal) distance will the shot land? 12m, 18m, 20m, 22m or 28m?*

maximum height

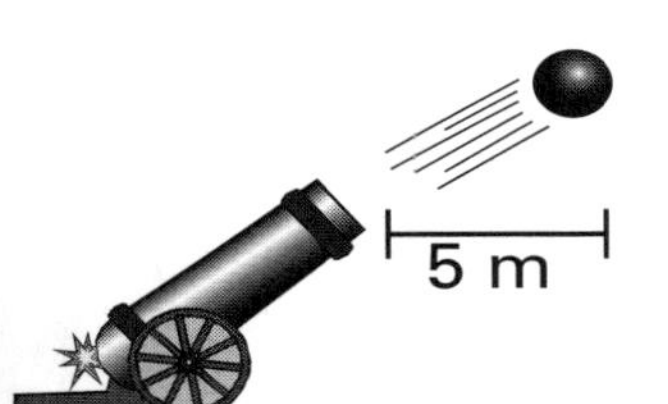

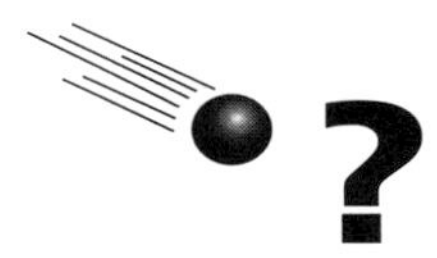

I thought this would be a rather straight forward problem. Recognizing that the trajectory of the cannonball is a parabola, my first instinct was to write and solve some form of quadratic equation. I checked my instincts with a different revelation: this is an IQ book, and so the author of the question is assessing cognitive ability, not knowledge of algebra. I subsequently scrapped my equation and simply "thought about" the problem. For the sake of simplicity, I assumed the landing point and the mouth of the cannon to be at the same height. I mentally divided the parabola's width into four 5-foot wide "quartiles," but knew that it would take the cannonball further than 5 more feet to achieve its maximum height. Therefore, I eliminated the first three answers.

The question was then simplified: Is the answer a little more than 20 feet or a lot more? Picturing that the cannonball's rate of ascent was continually slowing, I opted for a lot more. Yes, twenty-eight was my final answer.

I looked up the answer in the back of the book, and confirmed my assumption. However, the reasoning that the book offered was most interesting. It stated that the greater majority of the path of a projectile is in the upper half of its trajectory.

The point of the exercise is simply to reveal how shackled we are to the teaching of algorithms.

The book was assessing the reader's conceptual understanding of parabolas, or more specifically, trajectories. Yet, for myself and the hundreds of math teachers in workshops around the country to whom I have presented this problem, the response is the same. The first instinct is to write an equation. Apparently, we are all algorithm junkies. It is reasonable to think that we solve that way, because we were taught that way; and thus, we teach that way.

Continued on next page

CANNONBALLS & CONCEPTS (continued)

It should be questioned whether a preference for algorithms over concepts is of any true concern. I believe so due to one glaring fact, the greater majority of the teachers that I present this problem to get the wrong answer. In fact, the most popular answer is 20. They assume that the rate of ascent and descent is constant. Others claim that the rate will change and therefore the parabola's width at height zero will be less than twenty, and they opt for 18 as an answer.

My anecdote is not to imply that American teachers are less intelligent or educated than their international counter parts. (I have not given this question to any teachers from other countries to make a fair comparison.) The point of the exercise with the teachers, and now with you the reader, is simply to reveal how shackled we are to the teaching of algorithms, and how desperately we need to focus on conceptual understanding, for ourselves and our students.

It is found in the TIMSS, that instruction in the top achieving countries, is not void of the teaching of algorithms. The instruction simply begins with the conceptual development that lends to the understanding and long term memorization of the processes. Obviously, algorithms have enormous value, and play a vital role in the study of mathematics at any level. In fact, it is through the use of an algorithm that it can be shown that the book offered an incorrect answer. The cannonball would actually land 34 feet away. The solution is as follows:

Given a vertex of (h, k), the other two known points of the parabola, are (0, 0), (5, k/2). Our answer as to where the shot will land, will be 2h. Using the general form of a quadratic, our first point (0, 0) yields the value for a. Substituting a and our second point (5, k/2), we solve for h as shown:

$$y - k = a(x - h)^2 \qquad y - k = (-k/h^2)(x - h)^2$$
$$0 - k = a(0 - h)^2 \qquad k/2 - k = (-k/h^2)(5 - h)^2$$
$$a = -k/h^2$$

Expanding and using the quadratic formula, the solution for h is $10 \pm 5\sqrt{2}$ or approximately {3, 17}. Doubling h gives the landing point of the shot or {6, 34}.

After first glance, 6 looks like an unreasonable answer. However, it actually would be where the shot lands, if the 5 feet of horizontal distance were on the descending portion of the trajectory. The answer of 34 is where our favorite cannonball will land.

Victory goes to the algorithms! Not entirely. Although the book's answer is wrong, it does not diminish the need for greater conceptual understanding of mathematical topics. This is not an either-or argument. To demonstrate the need for teaching both concepts and algorithms in our daily instruction, let us analyze our solution.

The simplified answer in radical form of $10 \pm 5\sqrt{2}$, can be viewed as $5(2 \pm \sqrt{2})$. Going through our entire solution again replacing the horizontal distance of 5 with a variable, "m," gives a solution of $m(2 + \sqrt{2})$ for any projectile achieving half of its maximum height after m meters of horizontal distance. (Hallelujah for algorithmic prowess.) This means that maximum height will be reached after a little more than three times the distance of m. In fact it will be about 3.4 times more. (Praises for conceptual understanding). However, what does the radical two have to do with the trajectory? (Oh, for the want of more conceptual understanding.)

The solution of 34 tells us that the book's answer to the question is not accurate, but the problem still serves as an effective call to arms. Give this question to your colleagues, and I believe that you will be quickly convinced that we American math teachers are in serious need of developing our conceptual understanding of mathematical principles. You may also agree that unless we do so, we doom our students to our same fate. Who else is going to save them from it? Hopefully, this mental odyssey has also demonstrated that the teaching of algorithms is not inherently evil. Instruction that encompasses teaching to both algorithmic knowledge and conceptual understanding is the best ammunition we have. Fire away.

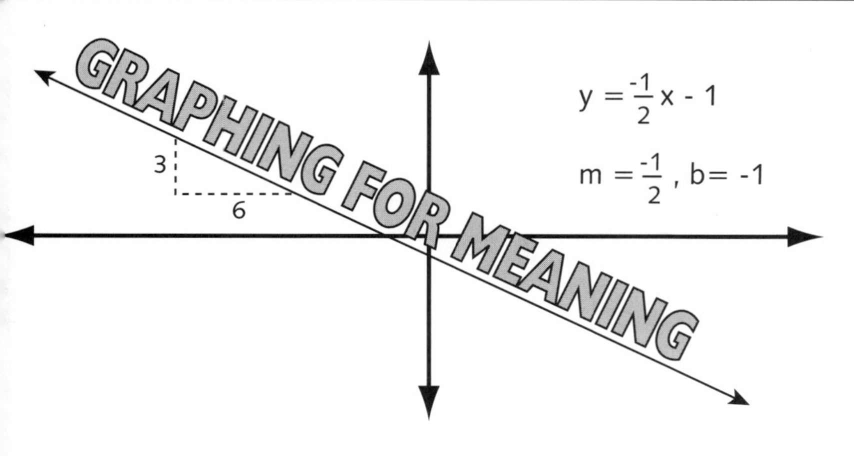

"Give me a pair of numbers such that the second number is two more than the first," I say to the class.

"Three and five," Jennifer offers.
"10 and 12," adds Jake.
The class spouts out another half dozen.

"Very good," I state, "Now write down ALL the combinations of REAL numbers, that means fractions too, such that the second number is two more than the first."

A few jump on it right away only to give up quickly. The rest know immediately that there is an infinite number of possibilities so there is no sense in trying. I insist that it can be done, so I challenge them to a race. "You do it on your paper, while I do it on the board, and I guarantee that I'll get done first, and that it won't take forever. Ready, Go!" I quickly write $y = x + 2$ and with hands raised victoriously in air, I exclaim, "Finished."

Of course, they rebel against the notion that my equation has truly accomplished the task, so I challenge them to another race. "Ready, Go!" I quickly draw a coordinate plane and then graph the line of my equation. "Done!"

They still don't buy it. I defend my case by stating, "Pick any pair of numbers that satisfies our condition, and you will find that it lies on the line. Pick any pair of numbers that doesn't work, and you'll find that it lies off the line." Eventually they give in, mostly because it is almost lunch time.

The point of the exercise is to impress upon my students that the graph of a line is an infinite set of points that describes a relationship between two numbers. My desire is that they understand that the line is more than simply the result of starting somewhere on the y-axis and counting up-and-over. When they look at a graph, I want them to see more than just its y-intercept and slope. I want them to be able to derive meaning from graphs and to communicate information and ideas through the use of graphs.

I also hope to save them from the travesty exemplified in the following story:

A colleague was presenting ideas of the math reform movement to some very skeptical state bureaucrats. She showed the following diagram and question as an example of how to enhance students' understanding of graphs.

"These lines represent three runners in a race. Describe what each runner is doing in the race."

The most skeptical member of the state board retorted, "That is not a fair question, the runners are not even staying in their own lanes!"

The fact that public education policy is made by people who do not understand the subject is not a surprise; however, the point here is that unless we want our students to leave high school with the same misconceptions about graphing that this bureaucrat did, we need to take our students beyond the plot and sketch routine of graphing. We must develop their understanding of graphs and their ability to use them in effective communication.

"That is not a fair question, the runners are not even staying in their own lanes!"

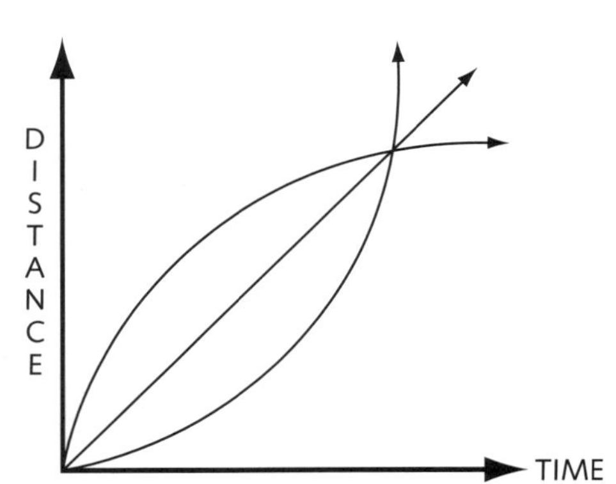

WHO WILL PASS YOUR CLASS?

WHO WILL KNOW MORE MATH?

BEAVIS & Barbie Holding Students Accountable... To What?

There is a bet that I like to make with teachers. The wager is this.

Imagine that on one side of your classroom you have Ken and Barbie, Mattel's version of the typical American teenagers. On the other side, you have Beavis and Butthead, MTV's version of the typical American teenagers. In years past, Beavis and Barbie were placed in different math classes. But now in the political climate of public education, they are enrolled in the same class with the same expectations. "Algebra for all" includes Beavis as well as Barbie.

Here is my bet. I will bet $100 that Barbie passes your class and that Beavis fails. This is no guarantee, but I think that eight out of ten times I will win. In the long run, I am going to make a lot of money.

That is my bet. Now it is your turn. You must wager the same $100 — that Barbie knows more math than Beavis. Are you willing to bet your own cash on that one?

I have yet to have anybody take me up on the bet. I would not take that bet either. That is the point. Who determines whether or not these students pass or fail? We do! Yet, we are not willing to bet that those who pass actually know more than those who fail.

Barbie will pass...

If mathematical knowledge and skill are not the criteria we use to evaluate students, then what is? Theodore Sizer, Brown University professor and author of *Horace's Compromise*, suggests that the number one criteria for getting a high school diploma in America is docility. In other words, the students who show up, sit down and shut up will pass. Those that do not conform, fail.

I believe that Sizer's critique, while harsh, is candid and true. While much of Sizer's criticism is aimed at the current public education system as a whole, my hypothetical wager points out that part of the problem is with us — the teachers.

...and Beavis will fail.

Take another Beavis and Barbie example. Let us imagine that at the end of the course, they both have a 68% in the class, needing a 70% to pass. Barbie of course has done all her homework, attended school consistently and asked numerous questions. Beavis, on the other hand, has ditched a day or two, has many missing assignments and has either slept through class or disrupted it regularly.

They both have 68%. Again, I bet $100 — Barbie passes and Beavis fails. I know this happens, because throughout my career I have heard teachers share this very story. "Oh, she worked so hard and he was such a slacker." They tell it with pride, because they believe they are instilling the great American work ethic and promoting a just and orderly society.

Yet, think about the scenario more carefully. While Barbie completed all her assignments, Beavis did few to none. That means Beavis must be scoring better on tests in order to achieve the same grade. In other words, Beavis actually knows more math, yet Barbie gets the nod and Beavis gets the boot.

Why? Because Beavis is Beavis.

I must admit, that for as critical as I am of this phenomenon, I know that it exists in my own classroom. If I were to rank my students according to the grades they earned in my class, I know the upper portion of the list would be ruled by Barbies, and the bottom dwellers would be predominantly Beavises.

Continued on next page

Beavis & Barbie (continued)

This year I have found data at our school that further supports the hypothesis that the school system is inherently biased against Beavis. As with many schools, ours has extended courses in Algebra and Geometry. These courses are designed to take the "struggling" student through the normal curriculum at a much slower pace. We found that these extended students scored better on the state's standardized test than the "regular" students. In other words, the extended program is effective in improving the mathematical abilities of these students.

Unfortunately, we have no extended program for Algebra Two. So what happens to the extended students when, after two years of a slow but fruitful pace, they enroll in this advanced course? They again become a struggling student. The course moves at a blistering pace that favors the student who is good at taking notes and memorizing for tests — Barbie. We improved the extended pupils' intellectual abilities, but not their work ethic. They are better mathematicians, but not better students. Therefore, after all the previous success, Beavis will still fail.

Why? Because Beavis is still Beavis.

So now that my little but mighty wager has called us on the carpet, and exposed our propensity to favor the Barbies over the Beavises, what do we do? How do we help both Barbie and Beavis to learn more math and, in doing so, help them both to succeed academically?

He are a few thoughts. Confess, focus, question, experiment and commit. First we must *confess* that we teachers are both part of the problem and a key part of the solution. Then we must *focus* on holding students accountable to what truly matters — their mathematical abilities. To do that we need to *question* everything that we do from instruction to classroom management to grading, and evaluate whether or not these procedures are promoting mathematical competency. We will find that many of them do not, so we must *experiment* and find techniques that do. Lastly, we must firmly *commit* to developing the conceptual understanding of students. We must teach them to truly understand what they are doing and not just mimic what we are doing so that they can pass the next test.

We can do these things and help the Beavises of the world; or we can point a finger at them and let them shoulder the blame. If we choose to criticize instead of help, we will continue to perpetuate the development of more Barbies — academically successful, but mathematically incompetent. However, if we do care about Beavis' academic success, and the mathematical competency of both Barbie and Beavis, then we the teachers need to change. Beavis will not succeed unless we do.

Why? Because Beavis will always be Beavis.

STILL ALL FULL OF STUFFIN'

WHAT THE SCARECROW REALLY RECEIVED

**"I would not just be a nuffin'
My head all full of stuffin'
If I only had...
enough unit credits, grade points and course requirements to get... a diploma."**

No, no, no. That's not how the song goes. The Scarecrow really sings, "If I only had a brain." Unfortunately, he never gets one. In *The Wizard of Oz*, the Scarecrow's ultimate quest reaps a mere diploma. This sad revelation is made by Dr. Theodore Sizer in his book, *Horace's Compromise*. Dr. Sizer claims that too often a diploma is awarded for the collection of unit credits and grade points rather than for mastery of material and acquisition of skill.

The obvious retort from the school system would be that although the awarding of the diploma is based on the unit credits, the students earn the credits by passing certain required courses. The implication here is that a prerequisite for passing a course is a minimal amount of knowledge.

Unfortunately, any argument that the Scarecrow's certificate signified some level of intellectual ability is squelched as soon as the Scarecrow opens his mouth. Upon receiving his credential, the Scarecrow attempts to demonstrate his newly found brain by impressing his friends with the following statement:

> "The sum of the square roots of any two sides of an isosceles triangle is equal to the square root of the remaining side."

An analysis shows that this conjecture is false. In fact, these conditions are impossible for any isosceles triangle. The Scarecrow made it evident that while he had his diploma, he still lacked a brain. Furthermore, this scene offers other disturbing parallels to our public education system.

The first is the Scarecrow's false sense of confidence. After stating his faulty conjecture he claims, "Oh joy! Rapture! I've got a brain!" The second is that while the Scarecrow struts his unwarranted confidence, his friends accept his statement as valid evidence of his brain. The Scarecrow, his friends, and the movie audience are all duped. All except the Wizard. He never claims to give the Scarecrow a brain. The Wizard clearly states his point,

> "Why, anyone can have a brain. That's a very mediocre commodity. Back where I come from we have universities, seats of great learning, where men go to become great thinkers. And when they come out they think deep thoughts. And with no more brains than you have. But they have something you haven't got...a diploma."

A brain is a mediocre commodity? You can get through college without a brain? All you need is a diploma? Okay Wizard, how does one get a diploma without a brain? Dr. Sizer answers this question with a quote from Charles Silberman's, *Crisis in the Classroom*. "The most important strategy for survival is docility and conformity." Few teachers would argue that students who sit quietly, take good notes, and turn in their homework will pass, and those that do not will fail. I believe Dr. Sizer's experience says it all, "Almost no school that I've seen gives mastery more prominence than it gives its routines." Therefore, if teachers wish to produce more competent students, then they need to make mathematical competency the number one criterion for passing their classes.

Continued on next page

STILL ALL FULL OF STUFFIN' (CONTINUED)

While teachers may have little power to change the entire public school system, each has the ability and responsibility to guarantee that they do not produce well-papered incompetents. Attendance, note taking and manners are all worthy attributes of a good student; knowledge, problem-solving and intellectual curiosity are all worthy attributes of a good mathematician. Good students, good mathematicians—one should not be sacrificed for the other. Teaching methodologies and reward systems will determine what is produced in the classroom.

I want my own students to know very clearly that they will be held accountable for their competency, not just their compliance. As a constant reminder, I have a poster of the Scarecrow in my classroom, accompanied by Sizer's quote "The Scarecrow didn't need a brain, just a diploma." Encircling the two is a hoola-hoop. Early in the year, I explain to the students that simply "jumping through hoops" during high school may get them a diploma, but not necessarily an education. Very few understand that how well they learn is just as important as how well they behave.

Hopefully, the political wave of accountability that is washing over the nation will make it obsolete for teachers to have to fight this battle for mastery so fiercely. Until then, everyone needs to hold the line and work fervently to give the Scarecrow what he was so earnestly looking for.

THE SCARECROW'S CONJECTURE

"The sum of the square roots of any two sides of an isosceles triangle is equal to the square root of the remaining side."

This is the conjecture that the Scarecrow makes when he receives his diploma from the Wizard of Oz. Have students come up with both an instance and a counterexample for the conjecture. In other words, have them draw and label one isosceles triangle for which the statement is true, and one for which it is false. The counterexamples will be plenty, but there are no valid instances for this conjecture. We can algebraically disprove the conjecture as follows. On the left, it is assumed that the first two sides chosen are the congruent sides. On the right, it is assumed that only one of the two sides chosen is a congruent side.

$$\begin{aligned} \sqrt{x} + \sqrt{x} &= \sqrt{y} \\ 2\sqrt{x} &= \sqrt{y} \\ (2\sqrt{x})^2 &= y \\ 4x &= y \end{aligned} \qquad \begin{aligned} \sqrt{x} + \sqrt{y} &= \sqrt{x} \\ \sqrt{y} &= 0 \\ y &= 0 \end{aligned}$$

Let x = the length of each of the congruent sides of an isosceles triangle.
y = the third side

Allow the students to ponder why these final statements cannot be true for a triangle. They will most readily see that one side cannot be zero. Also, by the Triangle Inequality Theorem, the sum of any two sides must exceed the length of the third. Therefore, if the sum of the two congruent sides is twice x, there is no way they will exceed y, which is shown to be 4 times as long as x.

Variations can be made on the Scarecrow's Conjecture. A good one for teaching instances and counterexamples in Geometry is: "The sum of any two sides of an isosceles triangle is equal to the square of the third side." Students may come up with instances such as 8, 8, 4 or 10, 10, 20 and usually the equilateral triangle of 2, 2, 2. Stress "ANY two sides," and the only true instance left is the equilateral triangle with a side length of two units.

GODZILLA OR Godzuki

Competent vs. Cuddly

Godzilla is one of the most infamous movie monsters of all time. Yet, few people know that Godzilla had a child. The 1969 film "Son of Godzilla" introduced Godzuki to the world. Godzuki is contrary to everything you would expect from the offspring of the king of the monsters. Rather than stomping, roaring and breathing fire like his raging father, Godzuki waddles, squeals and puffs rings of smoke. Godzuki can also speak. In the motion picture, he actually communicates in a child-like voice with the humans.

Godzilla is frightening. Godzuki is cute.

What holds true for the movie, holds true for the classroom. We teachers love the Godzukis and abhor the Godzillas. Given a choice, who among us would not opt for a room full of articulate cuddly creatures over a band of raucous monsters. While the Godzukis of our classrooms get our praise, the Godzillas get our attention. We crown the cute creatures as our heroes, while we send in the armed forces to manage the loud, obnoxious beasts.

However, this is more than just an issue of manners. It is an issue of ability. In regards to our movie monsters, we are speaking of a novice versus a master. Papa can level Tokyo, while his apprentice son cannot hurt a fly.

Godzuki is cuddly. Godzilla is competent.

While we would rather teach the lovably polite, we are paid to produce the intellectually powerful. Yet, these two are not mutually exclusive. The point is that we often demand and reward the passive and compliant, yet we wish to develop the capable and independent. The trouble is that being effective and productive often means being loud, active and messy, too. It is, thus, easier to deal with docility.

However, our job is not to merely process children; it is to educate them. So we must ask ourselves what we truly want for our pupils. Do we train our students to waddle and blow smoke, or do we teach them to roar with confidence and breathe the fire of intellectual prowess as they stomp through any task or challenge?

® Godzilla and Godzuki are registered trademarks of Toho Co, Ltd

WHAT DO YOU DO WHEN YOU DON'T KNOW HOW TO DO SOMETHING?

THINK!

Don't Panic

Make A Chart

Create a Simpler Example

Use Your Resources

Devise a Plan

Write an Equation

Change Your Stategy

Graph

Draw a Picture

Start Over

Find a Ratio

Check if the Answer is Reasonable

ARTICLE

SCULPTING YOUNG THINKERS

What do your students do when they don't know how to do something? If yours are like most students, they quit before they even get started. Upon presentation of a problem, if the solution does not strike their minds like a bolt of lightening, they shut down. A quick look at the headlines of any newspaper will show that the lack of ability or desire to solve problems is pervasive in the adult world as well.

So, what do you want your students to do when they don't know how to do something? Our personal survey of several teachers on this topic has generated a long list. The following is an excerpt from that list:

- Formulate a strategy
- Change your strategy
- Break down the problem
- Create a simpler example
- Use your resources
- Look for a pattern
- Draw a diagram
- Write an equation
- Sketch a graph
- Check your answer
- Find a ratio

Wouldn't our classrooms be just heavenly if our students initiated some of these behaviors, without our prompting? Teachers would be elated to see their students persevering through a problem by changing their strategy several times, even if that meant starting over. Seeing a student write an equation on their own volition would be reason for great celebration.

We know what we want our students to do, but the important question begs to be asked. What do we do in our daily instruction to promote these problem solving behaviors? (Ouch, that one was painful.) We must be honest. Demanding that students take notes and complete 50 textbook questions per night promotes good study habits, but not good problem-solving abilities.

In order to successfully teach students to solve problems, we must present them with problem-solving opportunities. They need assignments that will teach them how to think, not just memorize procedures. Problems and activities that require complex thought should be an integral part of any mathematics curriculum.

Beyond just presenting complex problems, we must also teach our students what to do when they are stuck on a problem. They need to know how to look for a pattern or break down a problem just as much, if not more, than they need to know how to apply the distributive property.

In my own class, I have a picture of Rodin's "The Thinker" on my wall with the word "THINK!" above his head. Throughout the course, whenever my students are working on a complex problem or project, I have them discuss their methods and strategies. I then post their strategies around "The Thinker." As the course continues, my wall becomes full of problem solving techniques. In this way, Rodin helps me teach problem-solving skills and establish reasoning as a top priority in my class.

"Studying math accomplishes two goals: It prepares some kids to think like scientists and it prepares the rest to think period."

-Marilyn Vos Savant

Christmas Carols & DISCO TUNES

The Strengths and Limitations of a Drill-and-Kill Regime

It happens every year. Everyone knows how to start the song "Twelve Days of Christmas," but no one knows how to finish it. With big smiles, everyone begins singing, "On the first day of Christmas my true love gave to me. . ." Everything is fine and dandy through the first four presents of poultry. Then, with the enthusiasm of an operatic tenure, everyone belts out "five golden rings!" and continues on down to that Partridge in a Pear Tree.

After that, the trouble starts. What comes after the Five Golden Rings? Although some people know the Six Geese a Laying and the Seven Swans a Swimming, there is an inverse relationship between the number of gifts and the number of participants who know the song. An asymptote exists somewhere around the tenth gift. It is a rare Christmas party where anyone can sing from Twelve Drummers Drumming all the way through to the Partridge.

The reason for this phenomenon is that we sing about the Partridge twelve times but sing about the Twelve Drummers only once, and there is an exclamation point at the Five Golden Rings.

Interestingly, this Partridge in the Pear Tree Syndrome is a perennial phenomenon in most American high school math classrooms where the drill-and-kill method is not only the primary means of teaching, but often times is the only one. Material from the previous course is practiced the most, one or two topics get the exclamation point, and the last chapter in the course is rushed through. Ironically, they all carry equal weight on the final exam.

For an Algebra course, pre-algebra skills become the Partridge, the Pythagorean Theorem equates to the Five Golden Rings, and the Quadratic Formula shares obscurity with the Twelve Drummers.

Math teachers get miffed that students do not remember the infamous Quadratic Formula from their previous course, as if it would be appropriate to chastise Christmas party-goers for not knowing the Twelve Drummers Drumming. One reason that many students do not obtain and retain the higher level math skills is that most teachers do not properly practice the drill-and-kill method that we traditionally regard as tantamount to quality mathematics instruction. Repetition is the mother of skill, therefore, we should give appropriate practice for those concepts we deem important.

Yet, pure drill-and-kill can lead to repetition without understanding. The power and potential for misuse of repetition can also be seen in the story of ABBA, the '70s disco band. "Dancing Queen" was just one of the many American hit songs that the Swedish band produced despite the fact that the lead singer, Agnetha Fältskog, did not understand English. She was taught the song's lyrics and with repetitive practice, she reproduced them beautifully, although she could not speak the language nor understand the song without translation.

This ABBA Principle also is a regular part of the mathematics experience in American high school classrooms. We teach our students, through repetitive practice, to reproduce algorithms. Often times they still need a translator for their own work, because we rarely teach them to speak the language of mathematics.

In light of the Partridge in the Pear Tree Syndrome and the ABBA Principle, the need for two critical changes is apparent. Firstly, high level math topics should be taught sooner, with the end of the course being reserved for mostly reinforcement and enrichment. Secondly, we need to teach the understanding of mathematical concepts as well as the memorization of algorithms. May all our students know important topics like systems of equations as well as they know the first five gifts of Christmas. Likewise, may all our students perform math as well as ABBA performs music, but may they also speak its language fluently.

MARTIAN MATH
A Quest for Values

A former teacher and colleague of mine, Rowena Hacker of Trabuco Hills High School, once gave this quiz to the faculty. It has quite a bit to teach us about how our professional values shape student performance. To demonstrate, please read through the given lesson, then answer the accompanying questions.

Example of Martian Mathematics Taken from a Text in Spyloctogy

In this chapter, we will be concerned with a study of the pexlomb. A pexlomb is defined as any Zox with pictanumerals which flotate the Zox into five berta Zubs where each Zub is supramatilated to the Rosrey of the Ord. For example, consider the Zox defined by 3 berta Ooz. It is obvious that any pictanumeral which is Blat must necessarily be Cort to the Ord. This follows from our knowledge of the relationship of a dentrex to its voom. However, if the Ord is partivasimous then the Zox must be Zubious. Thus, if we Kizate the dox pictanumeral, our Zox will floated into 5 berta Zubs. But remember, each Zub must be supramatilated to the Rosery of the Ord. If any one of the Zubs is not supramatilatied, we then have a pixalated pexlomb which requires a completely different procedure.

1. *What is pexlomb?*
2. *If the Zub is not supramatilated, what do we have?*
3. *If we have a partivasimous Ord,what do we have?*
4. *If the dox has been Kizated pictanumeral, what has happened to the Zox?*
5. *Something is obvious from our knowledge of the relationship of dentrex to its voom. What is it?*
6. *What must the Zub be supramatilated to?*

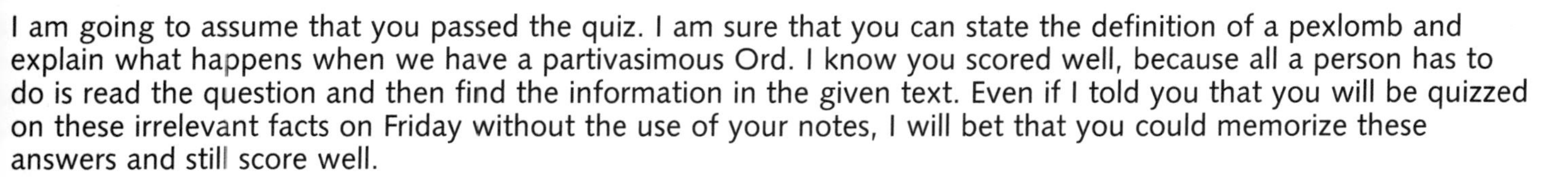

I am going to assume that you passed the quiz. I am sure that you can state the definition of a pexlomb and explain what happens when we have a partivasimous Ord. I know you scored well, because all a person has to do is read the question and then find the information in the given text. Even if I told you that you will be quizzed on these irrelevant facts on Friday without the use of your notes, I will bet that you could memorize these answers and still score well.

Since you are now so well versed in Martian Mathematics, let me ask you this question:

7. *What is a pixalated pexlomb?*

Oh I know — it's not a fair question, because our excerpt doesn't say what a pixalated pexlomb is, only that it requires a different procedure. Since it is so inappropriate for me to ask any question for which I have not directly given you the answer, I will skip that question. I am sure that if I told you what it was and gave you time to study you could memorize that answer, too. The more important question at hand is this:

Are you now good at Martian Mathematics?

Can you solve a martian math problem involving pexlombs and Zubs? Do you truly have knowledge of the relationship of a dentex to its voom? I am very confident that you will get an "A' in the freshman level course titled "Martian Mathematics," and will be passed onto the next course in the curriculum "Jupiterian Geometry." I am also confident that you will turn in all your homework and ace all the quizzes and tests, because you are good at taking notes and studying. The students who are not good at turning in homework, taking notes and

Continued on next page

Martian Math (continued)

studying will flunk and repeat the course during "Mercurian Summer School." (Talk about hot!) The good students will pass and the bad students will flunk, but no student will necessarily know anything about Martian Math. Do you?

The purpose for this activity was to raise yet another more global question: If we taught algebra and other math courses in the same manner that we just conducted this lesson in Martian Math, would we get the same results?

I believe the answer to that question is: We already do and already have.

International studies and personal experience tell us that American mathematics educators are obsessed with the dissemination of information and the memorization of algorithms. The practice is so common that by the time students get to a high school algebra class they already know what the teacher is going to say on the first day of school: Don't speak unless first spoken to, keep an organized notebook, turn in all your homework and whoever collects the most points by the last day wins.

The students' perception of this is apparent in their pervasive questions. "What use is it to answer questions after the test — we don't need to know this stuff anymore? Why am I keeping a notebook if I'm not getting any points for it? How do you expect me to pass, unless you put me in a group with a smart kid? How can I still fail if I've turned in all my homework? I know I only have a 40% in the class with only two weeks left, but isn't there any extra credit I can do?"

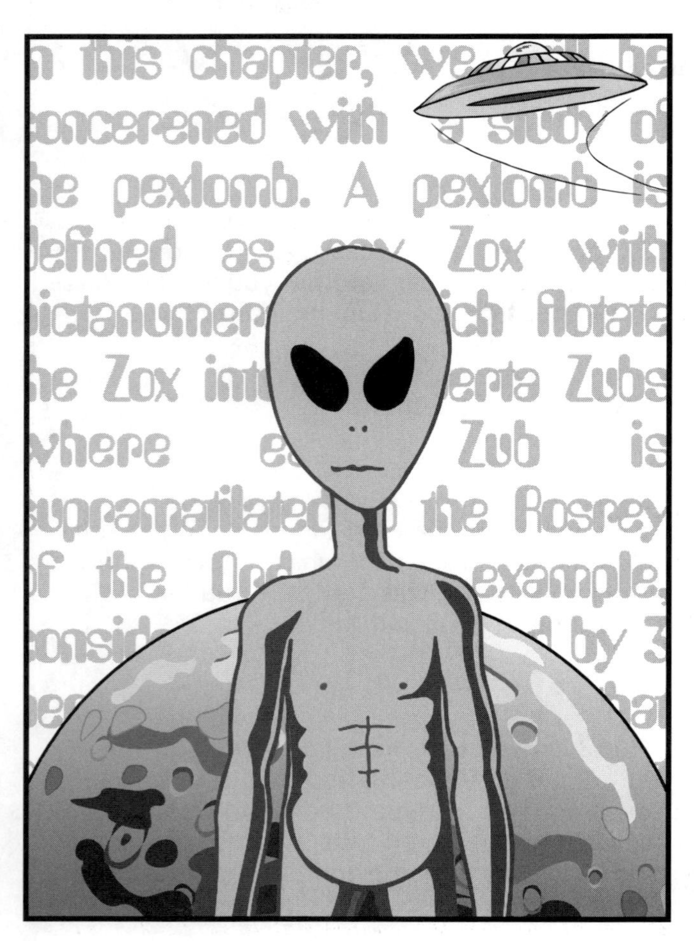

Interestingly enough, my first grader and preschooler never have these conversations with their teachers. I think that is because it is not until middle school that the principle of "Play to Learn" is replaced with the reality of "Work to Earn." Of course, part of becoming an adult is developing a strong work ethic and acknowledging that life is not all fun and games.

However, let us not allow these values to overrun those that are more pertinent to a math class: basic skills, reasoning ability and conceptual understanding. Let us instead create learning environments in which every lecture, discussion, activity, assignment, grading policy, disciplinary action, and vote of encouragement displays these values.

If we want the students to dispense with the belief that learning is irrelevant, then we, the teachers, need to stop sending the message that the name of the game is simply making it through to graduation.

Save time, energy and money!

Mathematics

THE LAZY PERSON'S GUIDE TO EVERYTHING

Learn centuries-old tricks and techniques

- *simplify problems*
- *look for patterns*
- *generalize relationships*
- *compute, don't count*

A Reference for Everyone!

ARTICLE

The Economy of Thought

The Lazy Person's Guide to Everything

"Mathematics was created by lazy people, for lazy people." I preach this to my students constantly, because I think that they do not understand that mathematics makes life easier, not harder. After all, do their textbook problems say "complicate" or "simplify?" Beyond the textbook, math is all about shortcuts. Math was created to save time, energy, and money. This was best stated by Dr. Clavin C. Moore of the University California, Berkeley: "The role of mathematics is to produce economy of thought." (*Sacramento Bee*, Oct 18, 1995)

For instance, the surface of a Rubic's Cube is made of six 3 by 3 arrays of colored squares. How many colored squares are there total? Someone not versed in mathematics merely picks up the cube with one hand and counts colored squares using the forefinger of the other hand.

A mathematician never does this for one simple reason: mathematicians don't count — they compute. A mathematician quickly multiplies 3 times 3 to get 9, and then multiplies 9 by 6 in order to get 54 colored squares. Counting demands too much time and effort for the mathematician. Counting is for the industrious, but computing is for the lazy.

Contrary to the notion of economy of thought, most students find math burdensome, rather than helpful. This phenomenon is a result of the way math is taught rather than the nature of math itself. If all we do as teachers is present mathematics as a set of rules to be memorized and practiced for no practical purpose or pleasure, students will continue to see math as an obstacle, rather than a tool.

Unfortunately, I believe this is the case for too many of my own students. I do my best to counter this by demonstrating economy of thought as often as I can.

For example, take the pattern of dots on the right. Given the first three clusters, ask the students for the next two.

They can all do this by adding two more dots for each successive cluster. However, ask how many dots are in the hundredth cluster, and counting becomes futile. It takes a mathematical strategy to find this answer. However, even stating the strategy is too much work for the lazy mathematician. Why write "Start with four and add two for every cluster," when $D = 2c + 4$ will suffice?

This laziness actually comes in quite handy when proving mathematical principles. Take the distributive property, for example. No mathematician multiplies $3 \cdot 14$ in the traditional manner of carrying values from right to left. That takes too much mental effort. The mathematician's thoughts flow like so: $3(10 + 4) = 30 + 12 = 42$.

Does this work in all cases? The industrious people of the world do several problems until they are satisfied as to the validity of the conjecture. The mathematically inclined take the lazy route, i.e. deductive reasoning. We can show this with the geometric model on the right.

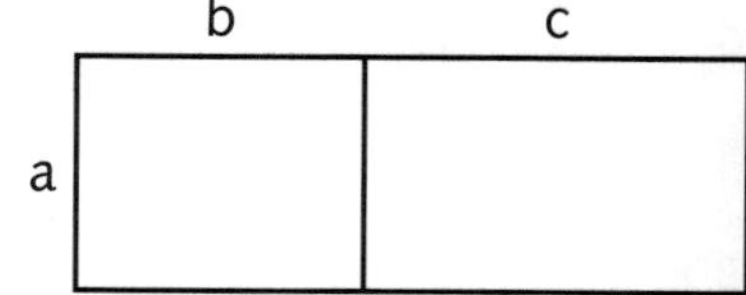

To calculate the area of the figure, we can multiply the width by the length, $a(b + c)$, or add the area of the two smaller rectangles, $ab + ac$. Either way, we should get the same area. Therefore, the following is true: $a(b + c) = ab + ac$.

Following this procedure took a lot less time and energy than generating a multitude of instances (see "Ultimate Cosmic Power"). This economy of thought in proving the conjecture then leads to economy of thought in implementing it as well (e.g. multiplying 3 times 14).

Once we find more ways like this to present mathematical concepts, then our classrooms will be full of lazy mathematicians rather than simply lazy math students.

The Universal Remote

Teaching for Understanding Rather Than Teaching To Memorization

A few years ago, my family and I moved into a new home. Along with the standard utility installations, we ordered cable. The cableman connected the cable box to our TV and then handed us a universal remote control. He said that in order to turn on the system, we were to push the following sequence of buttons: *cable-power-TV-power-cable*. Then he left.

On my own, I punched the appropriate sequence of buttons and watched the TV come to life. The channel and volume buttons allowed us to surf the six dozen channels now at our disposal.

But what about those other buttons on the remote? I was only using about ten of the thirty-plus buttons that existed on this mechanism. The cableman only taught me the specific sequence to turn on the TV. Without an understanding of why this sequence produces the given result, I was ill-equipped to figure out the rest of the remote.

I was lucky, though. By studying the device, this algorithm was easy to figure out. Once you press "cable," everything that you press next influences the cable box. The same was true for pressing "TV." The cableman's instructions were basically telling me, "Turn on the cable box, then turn on the TV, then change the channels at the cablebox." Once I *understood* the sequence, I was able to control the VCR as well. In fact, by understanding the initial 5-button sequence rather than simply *knowing* it, I was able to take advantage of ALL the features and capabilities of my hand-held technology.

This analogy has obvious applications in the teaching of mathematics. International studies show that typical American instruction offers only a demonstration of algorithms. Little is done in the way of conceptual understanding. As with the remote control, the average student is only capable of mimicking the exact sequence of steps that they have been taught. The students who are naturally mathematically inclined can figure out the underlying logic and mathematical principles themselves, and thus can make connections to other realms of understanding, as I did with the VCR commands on the remote. The American model of instruction works for these students, but the rest of the population is left feeling like they "can't do math."

It is easy to forget something that has been memorized, but difficult to misunderstand something once it is understood.

Teaching to understanding would offer students more immediate success (passing the class and standardized tests), but it would also empower students to apply their knowledge to unfamiliar problems or in learning new concepts.

For example, if students understood that slope was a ratio of two units, rather than something determined by simply counting rise over run while plotting a point, then they may more easily understand the tangent function of a right triangle to be a ratio among similar triangles. Besides, giving students immediate success, teaching to understanding also has long term benefits.

For instance, I recently visited the neighborhood and the very house in which I lived while attending fourth grade. I had not been there in 25 years. I could not remember the street address, neither could anyone else in my family. However, I was able to navigate the streets and drive directly to the house.

Continued on next page

The Universal Remote *(continued)*

The things that I had memorized at the time (phone number and street address) had passed from memory long ago, but the things that I understood (how to get home) stayed with me for over two decades. This should not be surprising. It is easy to forget something that has been memorized, but difficult to misunderstand something once it is understood.

In regards to learning mathematics, students who understand what they are learning are much more likely to remember a concept even if they have forgotten the formula or algorithm.

It is not difficult to teach conceptual understanding. It simply involves taking the time to show the students the underlining principles and holding them accountable to this level of understanding. For instance, here is a sample Algebra question that truly reaches for depth in understanding:

> The following equation represents how many candy bars, C, that Steph Yofase gets to eat for every math problem, p, that she accurately solves: $C = (2/9)p + 1$. Describe the $2/9$ in the equation.

Here are some actual student responses.

a) She gets 2 candy bars for every 9 problems that she does.
b) She gets two-ninths of one candy bar for every one problem that she finishes.
c) She gets 9 candy bars for every two problems.
d) It is the slope.

The first two student responses are correct, and their comparison offers a rich opportunity for class discussion. The third is wrong, but demonstrates some general understanding. While the fourth one is technically accurate, it demonstrates little of the student's understanding of slope or its relationship to an equation.

Again, complex investigative activities are good, but not necessary in order to teach to conceptual understanding, and avoid the pitfall of teaching solely to memorization. The key is to focus on your expectations of the students. If you know specifically what you want your students to know, the method will emerge very simply.

ARTICLE

THE RAILROAD AND THE KAYAK

EVALUATION VS. ASSESSMENT

Meet Rod and Brooke, two educators with strikingly different methods of measuring student learning. These differences can be seen by examining their respective choice of vacation activities.

Rod likes to travel across the country by train. His plan is to rate the various sites that he sees. Rod gives every major stop a mark of either good, fair or poor. He is so busy evaluating the trip that he never leaves the train to experience any of the places or people that he sees. Rod enjoys the comfort and security of the train. He also enjoys not having to make any decisions about the trip. The train just moves and stops according to schedule. Rod can even sleep and the train moves along just fine. This leaves Rod free to make his critiques along the trip.

Brooke prefers a much different type of trip. She rides white water rapids in a kayak. She saves her evaluation of the trip until the end. She must do this because all her attention is commanded by the need to steer her craft. While on the river, she must constantly assess her situation by reading the water ahead of her. Where are the currents, the eddies, and the rocks? How deep is the water and how fast is it running? Brooke is very active in determining her course, even though her boat is propelled by the water in which it travels. She makes a plethora of critical decisions as she rides down the river. How should each rapid be approached? Should she paddle hard or back paddle? Riding the big rapids is far from comfortable or secure. It is wet and exhausting, but exhilarating. Brooke loves it.

Back in the classroom, Rod assigns his marks to the performance of each student, but what he sees has little, if any, effect on his teaching methods. Whether student performance is good or poor, tomorrow's lesson plans remain the same.

Although Brooke will eventually give formal marks for her students, she chooses to assess their knowledge and progress instead of evaluate their performance. This difference is both profound and critical. It is a matter of understanding, rather than judging, the abilities of the students. For the same reason that Brooke reads the water while kayaking, she reads her students in order to make crucial decisions regarding her next instructional step.

For Rod, the boundaries between instruction and evaluation are as evident as the starts and stops of the train. For Brooke, the lines between instruction and assessment get blurred. She does both simultaneously. On her best days, even her tests teach her students something new.

The presence of grades and the need for standardized testing makes formal evaluation methods a viable and intelligent option in certain situations. However, very few teachers go beyond these methods. The underlying reason may be that they are easy to implement. Just like the train, it is simple and effortless to sit and ride. However, it takes training and experience to steer the kayak. The more often you ride in the kayak, the better you get to know the river.

Unfortunately, the more you ride the train, the only thing you get to know better is the train itself.

It is time for teachers to expand beyond merely evaluating students. Evaluation has its place, but students need instructors who are willing to wade into the river, gauge the waters, and take the helm.

CERTIFIED
FLUFF

Fluff -n- Stuff

DETERMINING THE VALUE OF A MATH PROJECT

"I used to think that doing math projects meant having my students create mobiles." That was a comment recently made by a school district administrator at one of my "Cool Projects" workshops. This gentleman's statement conjures images of hangers, string and cardboard polygons. They may decorate the classroom nicely and look great for Back-to-School Night, but they are certainly not very rich in substance or mathematical content.

Though he no longer views mobiles as appropriate projects, his previous misconception is not rare among educators. *The Math Projects Journal* has received project requests and submissions involving string art and origami. We, the editors, have made it a practice not to publish those kinds of activities because they do not move students through conventional curricula or state standards. These activities may have subtle mathematical connections, especially with geometry, and may have value in developing spatial reasoning in elementary school students. But how does creating a paper swan make a student better in algebra? It obviously does not; therefore, it does not hold a valid place in a mathematics class.

I used to think that doing math projects meant having my students create mobiles.

These types of activities tend to give math projects a bad name. In fact, a colleague in our math department, had his own name for our use of projects in math instruction. He called it "that touchy-feely stuff." He assumed that we were leaving the wisdom of traditional instruction in order to boost kids' self-esteem, make the class more fun, and produce more passing students with little regard for their math competency.

I cannot blame him for his skepticism. Many of the projects that we do in our classes are considered hands-on or open-ended. Our fellow teacher probably assumed that, under the origami-mentality, we were teaching conic sections by having our students make paper party hats. He believed that we had abandoned true math instruction for the sake of the latest "touchy-feely" trends, that we had replaced the stuff with a bunch of fluff.

That was our colleague then, several years ago, when we first began using math projects. Now he is an integral part of the Trabuco Hills High School C^3 (C-cubed) team that has led students to tremendous improvement on California state achievement tests (see "A Mountain of Vision"). Although he, like us, teaches mostly by direct instruction, something enticed him and the other detractors within the department to embrace the idea of incorporating project-based instruction.

The incentive came when they saw that the projects our students were doing were packed with stuff, not fluff. That is, the projects were rich in mathematical content, intellectually challenging, and developed conceptual understanding of the topics. (see "The Four and a Half Principles of Quality Math Instruction") They also held students to high performance expectations. When examined closely, origami, string art and mobiles do not meet these criteria.

Continued on next page

FLUFF -N- STUFF (continued)

The teachers in our math department saw the power in the types of math projects that we were doing in our classrooms. They saw rich content in projects like the "The Jogging Hare," "How Big is Barbie," "Princess Dido," and the "Shot Put Arc." They saw the enthusiasm and reasoning abilities of our students demonstrated in the "Original Works." The teachers were impressed with the difficulty and complexity of our test questions; they were more impressed with the students' answers.

In other words, we did not sell them on math projects; our students did. They did it, not by showing off a bunch of fluff, but showing their stuff. Our students were not getting good grades for producing pretty projects, neatly organizing notebooks, and politely playing with manipulatives. They were getting good grades for achieving high levels of mathematical competency.

It is important that these expectations of mathematical prowess be communicated to the students as well. The students need to know that their projects will be evaluated according to the stuff they demonstrate, not the fluff they contain. Computer clip art and water color markers should not help them pass the class. Only the demonstration of their mathematical knowledge and ability will help them do so. If both students and teachers stay focused on the purpose of math projects — to learn mathematics — than neither will be tempted to stray into the less rigorous "touchy-feely" realm.

After hearing these ideas at the workshop, the school district administrator who made the comment on the mobiles, had more to say regarding the "stuff" contained in the math projects.

He said, "Thank you for showing me projects that actually have some mathematical substance."

You are welcome, sir.

LESSON STUDIES: A Collaborative Effort

THE MODEL

Last year, I was involved in an extraordinarily powerful professional development experience called Lesson Studies. Through this program, I have expanded my knowledge of how students learn mathematics and have traded some of the limitations of my teaching for new tools and insights to help improve student understanding. The power of the Lesson Studies program is best stated, though, by a colleague of mine who described her first day of Lesson Studies by claiming, "This day has been the most significant day of my professional career."

Modeled after the Japanese practice of professional development, the main characteristic of the program is teacher collaboration. In Lesson Studies, teachers form teams to design, implement, modify and document a particular lesson.

My colleagues and I were first introduced to the concept of Lesson Studies by Dr. Tom Bennett of California State University, San Marcos. During a district in-service, Dr. Bennett shared that Japanese teachers are given one full day a month to conduct Lesson Studies. While we did not think that idea would happen in America, we did think that our school would give us a few days a year to experiment with the collaborative model.

"This day has been the most significant day of my career."

THE PROGRAM

Our site and district administration gave us their enthusiastic support and funded three Lesson Study days, with three corresponding planning days. Each high school site formed two teams. Four Algebra teachers made up one team; four Geometry teachers comprised the other. We were given three full release days to conduct the actual lessons. A few weeks before each Lesson Study, we were given a half-day release to plan the lessons.

During the first period of the implementation day, the teachers convened at one site to share their lesson plans with the other teams. During the second period, each team had one teacher conduct the lesson with his or her students while the other members of the team observed. The third period was used for modifications of the lesson. The team then took the modified lesson back into the classroom, during fourth period. This time, though, teachers "crossed-over" to observe lessons from other teams and schools. The final debriefing of the day was done over lunch.

THE PROCESS

While the opportunity to work with colleagues was a valuable experience, collaboration alone was not the power of Lesson Studies. There were four unique aspects of Lesson Studies that made it a transforming endeavor: misconceptions, context, observation, and external force. ("The America's Cup Sail" lesson will be referred to during the following discussion.)

1) **Misconceptions:** Lesson Studies is not about teacher methodologies; it is about student understanding of mathematics. Therefore, designing the lesson must always start with the conceptual understanding, or often the misunderstanding, of the students. Before "The America's Cup Sail" lesson was designed, the teachers chose to address trapezoids and the principle that the median is half the sum of the bases. From there, one of our teachers expressed his opinion that students have a tough time recognizing trapezoids and their properties. So the decision was made that the goal of the lesson would be to strengthen student recognition of trapezoids and their ability to calculate the length of the median given the length of the bases.

2) **Context:** To help students understand the basic principles being studied and to enhance the teachers' own understanding of those same principles, the teams were challenged to create a context for the topic. This is often a single problem that drives the lesson. In our case, with the goal of the lesson (medians of trapezoids) firmly established, the teachers went through several contexts and diagrams before deciding on the *America's*

Continued on next page

LESSON STUDIES (Continued)

Cup Sail. Here, multiple trapezoids were embedded in the diagram, with the task being to calculate the lengths of the seams and the measure of the angles of the various trapezoids.

3) **Observation:** The aspect of Lesson Studies to which American teachers have the most difficulty adjusting is that the observation should be centered around the lesson not the teacher. We must overcome the belief that the most important actions in the room are those of the teacher. Then we can focus our observations on the actions of the students and on how the lesson, not the teacher, is influencing those actions. During a Lesson Study, the observing teachers walk the room without assisting the students. In the case of "The America's Cup Sail," it turned out that the students came up with all the important principles to solve the problem — medians, corresponding angles, consecutive interior angles, etc. However, several students did not see the horizontal seams as medians. They instead saw three unknown lengths between the two given lengths of 1.8 and 9. Subtracting these two values and dividing by three yields 2.4 as an increment of each successive seam; therefore, the three missing seams were 2.4, 4.8 and 7.2. This pervasive error was easy to correct, but would not have been caught had the observers been focused on the instructor instead of the students.

4) **External Force:** The great potential of Lesson Studies is to get us to think outside the box — to transform the way we think about and conduct math education. Therefore, simply collaborating with colleagues will not bring about that change. There must be an "external force," as Dr. Bennett calls it. This force can come from traditional in-services, conferences, research, the internet or a mentor. For us, it initially came from Dr. Bennett's in-service and our department's exposure to international research — specifically, the TIMSS report. Since then, teachers have been involved in other training programs and have brought their findings back to the department.

THE LASTING EFFECT

The effectiveness of any Lesson Studies program can be determined by how the teachers' practices and attitudes have been changed by the experience. Although not every participant of our program has been sold on the idea of Lesson Studies, last year's experience has already bore fruit. Within our department now, Algebra and Geometry teachers collaborate weekly at lunch to informally design lessons. The content of the professional dialogue is slowly changing from complaints about student failure to inquiry about student understanding. With this gradual yet continual change, the math teachers at our school are growing professionally and, in turn, the students are improving their mathematical ability and understanding.

Inspiration, Perspiration, Desperation

The Forces That Give Birth to a Math Project

Math project enthusiasts borrow and steal material from any and all sources they can find. Sooner or later, the desire will arise for the enthusiasts to develop their own math projects. We, the editors, have frequently been asked where we get the ideas for our projects. Sometimes we find existing material and adapt it to our needs. Other times, we have the thrill of devising our own. Here are a few of our stories about creating math projects.

INSPIRATION

Some projects have emerged serendipitously. "Princess Dido and the Ox Skin" was one such project. Chris found the original story while reading the book *Mathematics and Optimal Form*, by Hilderbrandt & Tromba. It was a small leap of the imagination to go from envisioning Dido's loyal subjects cutting the ox skin, to envisioning math students cutting a bed sheet. The only real question was just how big they could make the encompassing area. It's about the size of a football field. Now, that would be something that the students would remember!

Through the process of deriving the solution, it was evident that the problem was rich with opportunities for the review and reinforcement Geometry concepts. The scope of the project was tremendous and fell together easily.

PERSPIRATION

Other times, the development of a project has required a bit of work. Such was the case with Greg and his "How High" unit, which began as a series of lessons all relating to similar triangles.

The idea from this unit initially came from a textbook problem using 45-45-90 triangles to calculate the height of a tower. Greg quickly expanded this concept to include 30-60-90 triangles as well. Then he remembered that a fellow teacher had given him a project using similar triangles with a device called an inclinometer. He also ran across more textbook problems that used proportions and shadows to estimate heights. While students were working on these problems, they mentioned that their science classes had calculated height using protractors and trigonometry. Upon research for a multicultural math class, Greg came across an ancient Chinese technique called the double difference.

He knew that all these separate techniques were based on common mathematical principles. Therefore, their incorporation could be used to reveal the relationships between trigonometric functions, special right, similarity and proportion. These activities would also provide the students with invaluable lessons in measurement, data collection, and error analysis. With time and sweat, Greg compiled a very cohesive thematic unit.

DESPERATION

Desperation comes when we realize that our teaching has gotten truly boring and ineffective. When instruction needs a fresh and effective twist, again, necessity becomes the mother of invention.

"The Empty Box" is a product of such necessity. Previously, we had used the common version of the shoe box to introduce polynomials, but then we found ourselves going back to the same old drills. Therefore, we sat down and shared ideas for various extensions of the shoe box lesson. From these conversations came the pizza box and donut box problems. Later, we added the graphing questions to give students more practice and help them understand independent and dependent variables. The visual aid and context enriched our students' understanding of quadratic functions.

The solution to our professional crisis was quite fruitful. Our students now understand that polynomials are an abstraction of a pattern or phenomenon. They also see the relationship between the product of sums and the sum of the products, the relationship between the factored form and the quadratic form of an equation, and how factoring may be used to find the roots. In this case, desperation led to powerful instruction.

Additional Activities

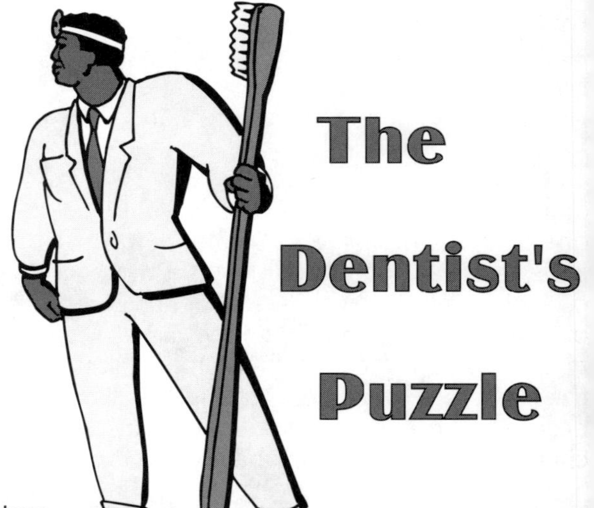

THE MEETING

While on an airplane, I met a man who was going home to Tennessee. He was a recruiter for the Army, and asked me what I did for a living. I shared that I was a high school math teacher, and as is typically the case with people I meet, he shared his childhood horror stories of math. I told him that I would do everything I could to make sure his experience was not wrought upon the next generation. After hearing him out, I turned to reading my book. He then nudged me and handed me a scrap of paper with a math puzzle on it. "My dentist gave me this problem, and I can't solve it. Can you?" he asked. I did and I offer the same puzzle to you to make a very important philosophical point. It also may serve as a great lesson for your students on number sense and variables.

THE PUZZLE

Find the value for E, F, G & H, such that the following statement is true.

$$\begin{array}{r} E\ F\ G\ H \\ \underline{\times\ 4} \\ H\ G\ F\ E \end{array}$$

MY SOLUTION

1) I assumed that all 4 digits must be unique. (The recruiter gave me a blank look when I tried to confirm that, so I was left to my own assumptions.)
2) Since no value is "carried" at the end, E must be 1 or 2. However, E in the product cannot be 1 since we are multiplying by an even number, so it must be 2.
3) The only values of H that will yield a product ending in 2 are 3 and 8. There is no way to get a value of 3 in the product even with a significant carry, so H must be 8.
4) Since the product of F and 4 has no value to carry, F must be 1 or 2. Since 2 has already been used, F must be 1.
5) Now, 4G + 3 must be a number that ends in 1, so 4 times G must be a number that ends in 8. Since 2 has already been chosen, 7 is our only option, which also fits for G in the product.

$$\begin{array}{r} 2\ F\ G\ H \\ \underline{\times\ 4} \\ H\ G\ F\ 2 \end{array} \qquad \begin{array}{r} 2\ F\ G\ 8 \\ \underline{\times\ 4} \\ 8\ G\ F\ 2 \end{array} \qquad \begin{array}{r} 2\ 1\ G\ 8 \\ \underline{\times\ 4} \\ 8\ G\ 1\ 2 \end{array} \qquad \begin{array}{r} 2\ 1\ 7\ 8 \\ \underline{\times\ 4} \\ 8\ 7\ 1\ 2 \end{array}$$

THE LESSON

After about ten minutes I handed my solution to the gentleman. His face was in shock; he had been working on the problem for FOUR YEARS! In explaining my solution, I said that I rephrased the problem by asking "What number when multiplied by four will reverse the order of the digits?" His face contorted. "I thought the letters EFGH meant that the digits had to be consecutive, like 4567." With that thought, there are only a small handful of possible solutions, none of which work. No wonder it took him four years to get nowhere! Then he shared how he took algebra four times, and only graduated because they did not want him to return. Poor guy — four years of algebra and no one taught him the concept of a variable. If he had even a vague notion of the concept, he would have properly interpreted his dentist's puzzle, and not been so haunted by it for so many years.

INTRODUCTION

Psychic Probability is an engaging game that demonstrates the use of the Fundamental Counting Principle. It is basically a fancy, parlor-game version of "Can You Guess the Card That I'm Thinking Of."

THE SET-UP

Start with a deck of 25 index cards. On each card draw one of five symbols: star, square, circle, triangle, and plus sign. The completed deck should have five copies of each symbol, so each symbol has an equal chance of being drawn. At the front of the classroom, place two desks facing each other. One desk is designated as the "Hot Seat," and should be arranged with its back to the class. The other desk, facing the class, is for the teacher and has an open notebook standing upright on the desk. The notebook serves as a blind to keep the students from seeing the cards. On the board, have a simple chart with five columns labeled "1 2 3 4 5."

THE GAME

Tell the students that you are going to play a game that will test their psychic abilities. Call each student one at a time to sit in the Hot Seat. The student is to guess which card the teacher is going to pull from the deck next (star, square, circle, triangle, or plus sign). The game works well if you play it up by looking at the card behind the notebook and saying things like, "Use your psychic abilities to read my mind and tell me the image on the card." If the student guesses correctly, then the student is allowed to guess again. The objective for the student is to guess correctly as many times as possible. Once the student misses, the student is dismissed from the Hot Seat and the initials of the student are written on the chart with a number of checks that corresponds to the correct number of guesses. For example, if Susie guesses three times, then she writes her initials to the left of the chart and places a check in each of the columns 1, 2 & 3. If Jack does not guess correctly at all, he places his initials left of the chart with no checks in the chart. The game concludes once the initials of every student in the class are on the board.

THE LESSON

When the game is complete, most of the chart should be filled with initials that have no checks. Several should be accompanied by one check. A couple of the students should have two checks. One student will possibly even have guessed three correctly, but rarely will a student guess correctly four or more times. Does this experiment imply that the students in the class with the most checks have psychic ability? The point to make for the students is how many checks we should expect in the chart. With five symbols in the deck, each with equal chance of being drawn, we can expect at least one out of every five students will guess correctly at least once. Since the card is returned to the deck each time, the probability of guessing the second card is the same as the first. The probability of guessing correctly twice is the product of the probability of each guess. In other words, $^1/_5$ times $^1/_5$, or $^1/_{25}$. Therefore, we can expect one out of every 25 students to guess correctly twice. In an average California classroom of thirty plus students, we can expect one or two students to accomplish this. The probability of three correct guesses, then, is one out of every 125. I can expect that one student out of all my classes in a day might guess three. The chances of four in a row are 1 in 625. In fact, in ten years, I have had only one student guess four correctly. How long should I expect to wait for someone to guess five?

Submitted by Larry McGehe of Prime Presentations

THE QUESTION

Place a string around the earth's equator and pull it snug. Add a length of one meter to the string and push it away from the equator so that the string is equidistant from the surface of the earth. What is the largest animal, from the following list, that would be able to walk under the string: a Flea, Mouse, Poodle, Calf, Horse, Giraffe?

THE SOLUTION

Circumference around a sphere:	$C = 2\pi r_1$
Add one meter to the circumference:	$C + 1 = 2\pi r_2$
Solve the system by subtracting the two equations:	$(C + 1) - C = 2\pi r_2 - 2\pi r_1$
Simplify:	$1 = 2\pi(r_2 - r_1)$
Solve for the difference of the radii:	$1/2\pi = r_2 - r_1$
The resulting difference is the distance between the sphere and the string:	$1/6 \approx r_2 - r_1$

Therefore, the string would be approx. one-sixth of a meter or about 6.5".

THE PRESENTATION

Now it is time to prove that the string will be approximately 6.5" from the earth, even if we do not know the circumference of the earth. In fact, this will be true no matter how big the earth is. To demonstrate this, have on hand a long string and several spherical objects of varying size. A handball and a globe will be used in this example.

Start by discussing the matter of adding a meter to the length of the string. A meter is a little bit more than a yard, which was traditionally defined as the distance from a man's nose to the tip of his outstretched hand. So place the end of the string at the end of your nose, and then extend the string out to your right. "This here is a yard. We need a little bit more." Turn your head away from your outstretched hand, increasing the length of the string. "Now we have a meter." (This crude estimate will suffice for the purpose of the demonstration.)

Then grab the handball, and wrap the string around its perceived equator. Be sure to keep your fingers pinched on the string to mark the circumference of the handball on the string. Then add the yard by out stretching your arm and bring the string to your nose. "We need to add a yard and a little bit more." Keeping your fingers pinched on the new position on the string, you now have added a meter to the circumference of the handball.

Form a loop with this length. Place the loop equidistant around the handball. This will be challenging, of course, because the loop will sag. You can have a student assist you in holding the loop, or you can place the props on the floor where everyone can view them clearly. What the students will able to see is that the string is approximately 6.5 inches from the handball.

Repeat this process with a globe. Every school has one, and it is the natural representation of the earth. Measure its circumference. With an outstretched arm and turn of the head, "We need to add a yard and little bit more." When you form the loop around the globe, the string will be about 6.5" from the surface of the globe.

For the final demonstration, choose five students from the class. Have them stand shoulder to shoulder in such a way that they form a circle facing outwards. Wrap the string around the group of five, having them hold the string to their navels. "Add a yard and a little bit more." Then wrap the new loop back around the students, and have them move it away from their bellies until the string is equidistant from the group. No surprise — the string is approximately 6.5 inches from each member of the group.

Be sure to conclude the lesson by emphasizing that the one-sixth of a meter applies to any circumference, C. Furthermore, the one-sixth could apply to any units. For instance if we added a foot to the string, the string would be $1/6$ of a foot, or two inches, from the surface of the earth. Have fun!

Begin class with the following three inequalities and corresponding number lines on the whiteboard or overhead.

Assign each student in the class a number. Each student goes to the board and determines if the number assigned makes the statement true or false. The student places a green dot on the assigned number if true, a red dot if false. For example, Jackie is assigned the number 2. On the first inequality, since 2 is less than 4 she places a green dot on the number 2. For the second, she places a red dot on two because 2 + 3 is equal to 5, not less than 5. Finally, on the third, 2 gets a red dot again, since 3 times 2 is not greater than 12. Be sure each student colors the original number assigned, not the product or the sum. For instance, when Jackie decides that 3 times 2 is 6, she may be tempted to place her dot on the number 6 on the number line. She should place it on 2, instead.

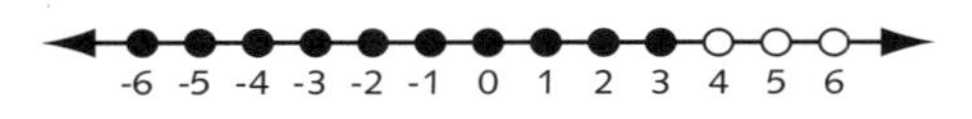

If your class is especially large, you may want to break it down into three groups, one group assigned to each inequality. Once the entire class has placed their red and green dots, it will be very obvious that a pattern exists. Each number line will be divided into two distinct regions of red and green dots. (Be sure to correct any renegade dots). Tell the students that you want ALL the numbers on the number line colored red or green, including the fractions and decimals (real numbers). The students will very quickly tell you to fill in all the gaps with the appropriate color. The only questionable area of shading will be near the boundary point of the colors. For example, in the first inequality, 3 is green, but 4 is red. So the class will vote to shade everything greater than 4 red, and less than 3 green; but what about between 3 and 4? There will likely be disagreement about this so let the class try different numbers between 3 and 4. Soon, everyone will agree that anything less than 4 is shaded green. Then tell the students, that for future problems we do NOT want to test a lot of numbers to know where to shade. Solving the inequality as we do with equations will reveal the boundary point. An open circle at the boundary point represents a red dot (point is false), and a closed circle implies a green dot (point is true). Then we only need to shade to the green side.

This can also be done with inequalities in the Cartesian plane. For example, assume you want your students to graph $y < x + 2$. You can assign students a set of points to test and plot. This is best done in groups. Assign the first group a set of points for which the domain has a value of 2, and the range is the set of all integers from -10 to 10. In other words, they will test and plot the points (2, -10), (2, -9), (2, -8) ... (2, 9), (2, 10). Each point receives a green dot on the coordinate plane on the board if makes the statement true, or a red dot if it makes the statement false. As with the number line, two distinct regions of red and green dots will appear, with an obvious line as the boundary. Show the students how the technique for graphing the line can produce the boundary of the regions (dashed for red dots, solid for green). Shading is to represent the region of green dots — the infinite set of points that satisfy the inequality.

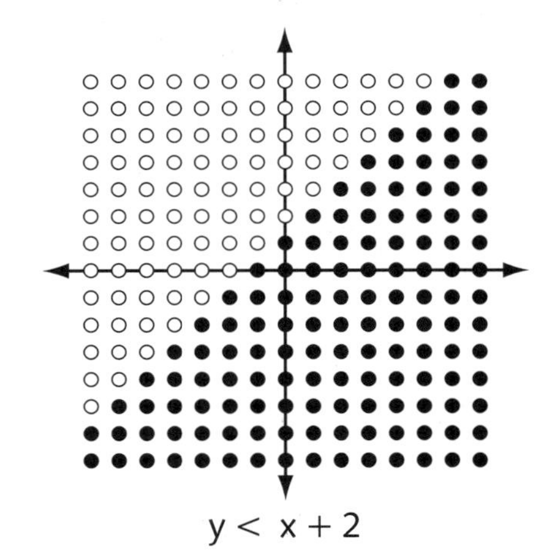

$y < x + 2$

THE TIC-TAC EQUATION

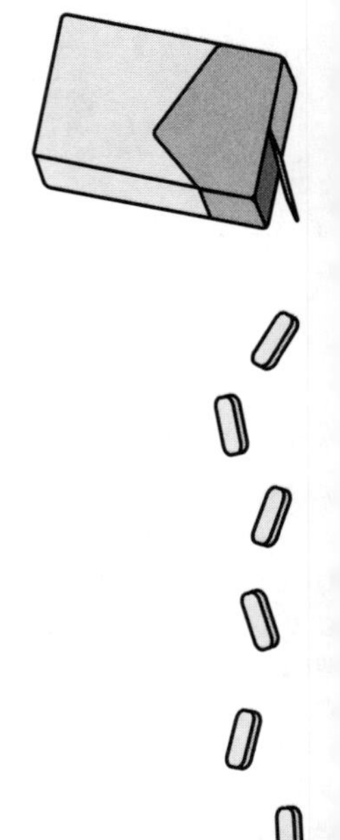

When teaching rates and slopes involving fractions, a useful and fun tool is "The Tic-Tac Equation." The following equation represents the number of kisses, K, that you will receive at the end of a date after eating the number of Tic-Tac candies, t.

$$K = (3/2)t + 1$$

Typical questions to be asked are: how many kisses will you receive after eating a given number of Tic-Tacs, and how many candies will it take to receive a desired number of kisses? Students are usually able to answer these questions as long as the number of Tic-Tacs is an even number, or if the number of desired kisses is one more than a multiple of three. In each case, the answer comes out a whole number. But if the answer is not a whole number, then the students must deal with the issue of getting only half a kiss or eating a third of a Tic-Tac.

When asked, "How do you get only half a kiss?" the students often respond humorously with statements like, "It was only a quick pucker" or "You only get a peck on the cheek" or "You closed your eyes and missed." There is always an optimist in the group that responds, "The half is slipping in your tongue." That's when you stop and move onto the next question, "What does this equation say?"

With a small degree of explanation, the students will see that a person gets 3 kisses for every 2 Tic-Tacs that they eat, not necessarily one and a half kisses for every candy. And let's not forget the obligatory ONE kiss at the end of the date, whether you eat a Tic-Tac or not.

Revisiting this problem in different forms helps students better understand the concept of slope and y-intercept, as well as discrete functions and graphs.

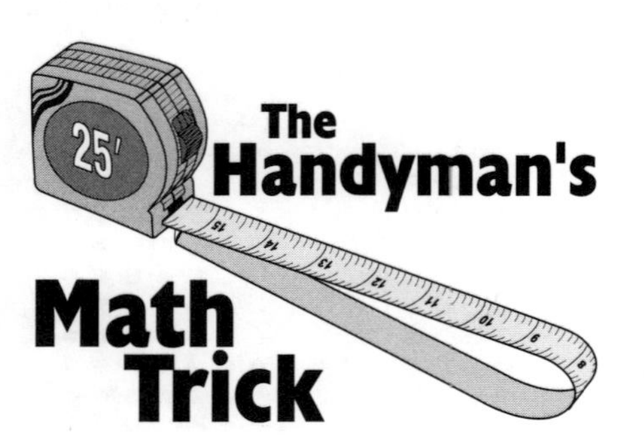

During a workshop I was giving in Little Rock, Arkansas, I had the participants gathering data for the "How High" project. While the teachers were measuring distances along the floor of the lobby, the hotel handyman strolled by. He saw some of the teachers on their hands and knees with tape measures, and paused curiously. "What is this?" he asked, "Invitational carpentry?" He then pulled out his own metal tape measure and offered, "Have you seen this one?" He folded the tape measure back on itself, saying, "Pull the end of the tape measure to align with the current year. Then the age that you will turn this year will align with the year that you were born."

Demonstrating his trick, he folded the tape measure so that the tip aligned with the number 98 (It was December 1998.), and sure enough 64 (I was born in 1964.) aligned with 34 (My 34th birthday was that year.). He then moved the tip of the tape measure one inch so that it aligned with 99. "Next year you will be 35, which now aligns with 64."

The Handyman's Math Trick, as I dub it, is a great demonstration for students to see why the addition property of equality works. This trick can be represented with one of two different equations. The first one, $Y - A = D$, states that if you begin with the current year (Y) and subtract your age (A) you will get the year that you were born (D). The second equation, $Y = D + A$, states that if you start with the year that you were born and add your age that you will get the current year. Both statements are true logically. Students can also see that adding A to both sides of the first equation transforms it into the second.

Brother in the Box

The trouble most students have when first introduced to the concept of the area of a parallelogram is that they want to multiply the base by the "side" of the figure. The purpose for this story is to help them understand why they need to multiply the base by the height.

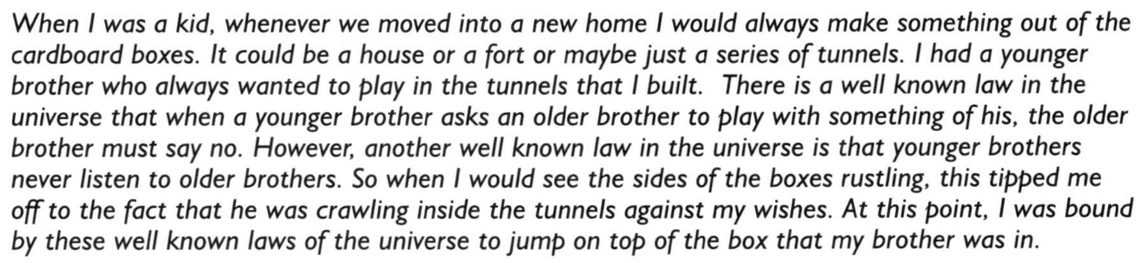

When I was a kid, whenever we moved into a new home I would always make something out of the cardboard boxes. It could be a house or a fort or maybe just a series of tunnels. I had a younger brother who always wanted to play in the tunnels that I built. There is a well known law in the universe that when a younger brother asks an older brother to play with something of his, the older brother must say no. However, another well known law in the universe is that younger brothers never listen to older brothers. So when I would see the sides of the boxes rustling, this tipped me off to the fact that he was crawling inside the tunnels against my wishes. At this point, I was bound by these well known laws of the universe to jump on top of the box that my brother was in.

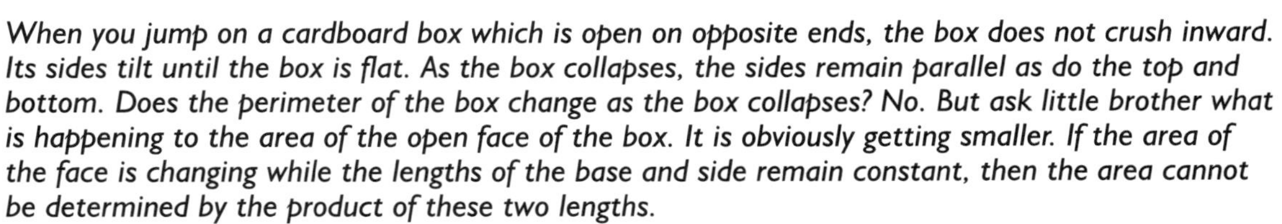

When you jump on a cardboard box which is open on opposite ends, the box does not crush inward. Its sides tilt until the box is flat. As the box collapses, the sides remain parallel as do the top and bottom. Does the perimeter of the box change as the box collapses? No. But ask little brother what is happening to the area of the open face of the box. It is obviously getting smaller. If the area of the face is changing while the lengths of the base and side remain constant, then the area cannot be determined by the product of these two lengths.

What else is changing then as the area is changing? Ask me while I am standing on top of the box and I'll tell you it's the height, measured from the top of the box, perpendicular to the ground. As my feet get closer and closer to the ground (or my brother's head), the area of the face of the box gets closer and closer to zero, until finally my brother is squashed like a bug and I claim victory. Therefore, the area of the parallelogram is the product of the base and the height.

MEGA TRANSLATION

Here is a terrific demonstration of the principle that the composite of two reflections through parallel lines translates the image twice the distance between the lines. In both figures below, the parallel lines are six inches apart. The right triangle is reflected through line a, then line b. In which diagram will the triangle be translated further?

Figure 1 **Figure 2**

Of course, the triangle will be translated the same distance of twelve inches (twice the distance between the lines), however, this is counter intuitive for the students. The second diagram produces such a long first reflection, that it must move the triangle the furthest. This can be demonstrated on the board by drawing the parallel lines exactly six inches apart and having a paper triangle taped to the board. Then, have a few students actually measure the length of each reflection, and use other paper triangles, congruent to the first, to mark the intermediate and final images. The students can see that the triangles have indeed moved twice the distance between the lines. Measure to confirm.

ADDITIONAL ACTIVITIES

Age Trick

Number tricks like this have been around for a long time, getting updated every year. Despite their longevity, people are still always amazed at how well they work. Devise an algebraic expression to represent this trick, then analyze the expression to explain why it works.

Step	Expression
1. Pick any single-digit number (1 to 9).	1. Let x be your original number. Let y be the 2-digit year in which you were born.
2. Multiply this number by 2.	2. $2x$
3. Add 5.	3. $2x + 5$
4. Multiply it by 50.	4. $50(2x + 5)$ which equals $100x + 250$
5. If you have already had your birthday this year add 1753. If you not, add 1752.	5. $100x + 250 + 1753$ or $100x + 2003$
6. Now subtract the four-digit year in which you were born.	6. $100x + 2003 - (1900 + y)$ or $100x + 103 - y$

You should have a three digit number . . .
The first digit is your original number. The next two numbers are your age.

So how does this algebraic expression generate the impressive three digit answer? The 100x guarantees that your single-digit original number will be in the first of the three digits. The 103 - y generates your age. (Note: this exact problem will only work in 2003. Each year, you need to add 1 to the 1753/1752 value.)

Here are some questions for you and your students to ponder: How would you adjust the trick so that it will work next year? What will next year's expression look like? Why is there a choice of 1753 or 1752?

TAKE A MATH WALK

The Math Walk is a simple, fun activity that can be used at the beginning of a new school year in most math classes, but it really lends itself well to Geometry. Have your students grab a pencil and paper and stroll the campus together with the basic task of writing down where they see math. Tell them to "look near and far, high and low, in front and behind." Examples range from the simple geometric patterns involved in tile work or the fractal patterns of plant leaves, to the volume of cement in a building, or the speed of a car driving by.

The real power of the Math Walk isn't at the beginning of the year, though. It comes much later, after you repeat the activity at the end of the first semester and then again at the year's end. If you want to know how your course has opened the eyes and the minds of your students, try this exercise. Here are a few suggestions to help students get the most out of their Math Walk.

1) Encourage your students to be specific. For example, they should write details like "finding the diameter of a tree by measuring its circumference" and not just "a tree."
2) Have your students keep their Math Walk notes from the first walk. Then, on the second, have them draw a line after their first set of notes and begin a second set. At the end of the year, they draw another line and write a third set.
3) Treat their writing as notes; do not make grammar an issue, but be sure they write legibly.
4) Collect the notes "as is" — fresh, informal and unedited. Do not require additional re-writing or explaining

You'll be amazed at the quantity and detail of math that they will see and write. This has been a fantastic assessment tool for us and a great way to start off the year.

GAMES

Twenty-One: A Review Game

Submitted by Randy Hoffman, Trabuco Hills High School, Mission Viejo, California

The purpose of this review game is to prepare for the upcoming test and provide students with a fun setting to display their understanding of the concluding chapter.

This game is an adaptation from a television game show in which playing cards were used to determine a winner. The game show was loosely based on the card game "Blackjack". The object of the game is to have the highest hand without going bust (over 21). In the game show, trivia questions were asked and a correct response allowed the player to gain control of the playing card. The person who got the question correct could either take the card or give it to his opponent. If the player who answered correctly takes the card, that player could freeze his hand or continue playing. If the player chose to freeze, he could not answer any more questions or receive any more cards from his opponents. The onus then fell to the other player to beat the hand or go bust.

In the classroom, I break the class up into groups of four. A question is posed and a correct answer eventually comes up. If a group answers incorrectly they cannot answer again until the next question. The group that answers correctly gets control of the card and usually takes the card, but they have the option of giving the card to another group, possibly forcing the group to go bust. Only the group that answers correctly gets the opportunity to freeze for that round of playing. If a group freezes, I allow them to continue to play to put other groups over 21, but they may not receive any more cards from opposing groups. If a group ever reaches 21, they are declared the automatic winner and another game is started.

The playing cards I use are approximately 12 inches by 18 inches and show up really well in front of the class. A basic search for "oversized playing cards" on the Internet will provide a variety of sources from which you could order them. The groups of students are numbered and placed on my dry erase board. During the game, I use the clips that are above the board to hold the cards so that the whole class can see them. The cards may also be placed upright on the pen/chalk tray at the base of the board. If you do not have oversized cards, you may also do this with a normal deck of cards, and simply write the results for each group on the board.

Example: For the sake of simplicity, let's assume there are three groups A, B & C. Group A answers the first question correctly. A King is drawn from the deck, and group A, which is in control of the card, keeps it. Group C then answers correctly and takes the next card, an 8. On the third question, Group A gains a 9, and freezes — now they may no longer take any cards, but they may still answer questions and give cards to other groups. Group B finally answers a question, and draws a Jack. They give the Jack to Group C forcing them to bust (over 21). Group B then answers the next two questions correctly drawing and keeping a Queen and a 10. Group B's score of twenty beats Group A's score of 19 without busting, therefore, Group B wins!

Question 1	Question 2	Question 3	Question 4
A: K	A: K	A: K	A: K 9
B:	B:	B:	B:
C:	C: 8	C: 8	C: 8

Question 5	Question 6	Question 7	Question 8
A: K 9	A: K 9	A: K 9	A: K 9
B:	B:	B: Q	B: Q 10 (win)
C: 8 7	C: 8 7 J (bust)	C: 8 7 J (bust)	C: 8 7 J (bust)

WIPEOUT

by Randy Hoffman
Trabuco Hills High School
Mission Viejo, California

I love game shows. I have adapted many existing and "has been" game shows as review games for my math classes. One of my particular favorites is a variation of the game show called "WIPEOUT."

WIPEOUT is played as a game were all students are involved in an elimination style tournament. (See elimination ladder insert.) Extra Credit can be given to the winners of each round. Subsequent rounds are worth more as the elimination rounds proceed.

Before playing the game, the teacher needs to prepare several sets of ten index cards (one topic card, six correct response cards, and three incorrect response cards). For example, in one set, the topic card may read "supplementary angles." Then one correct response card could have a diagram of a linear pair, while another lists two angles for which the sum is 180°. The degree of difficulty of the remaining correct response cards can vary in difficulty. One of the incorrect response cards could have a diagram of complementary angles instead. On the backs of the correct cards is a smiley face, while on the backs of incorrect cards is the phrase "WIPEOUT." (Both should be written in yellow marker so that the students can't see through the front of the card!)

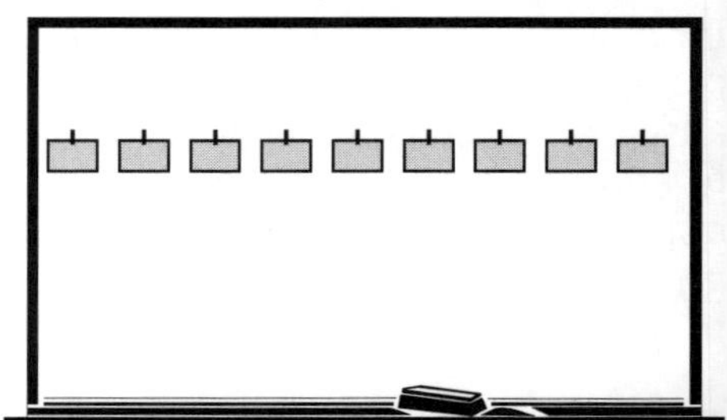

To begin the game, choose two students who take turns bidding on how many correct response cards they can identify in a row. Eventually, one student says to the other "go ahead." The higher bidder must consecutively choose the number of correct response cards that matches the bid in order to advance to the next round. If an incorrect response card— wipeout—is chosen, then the other student advances.

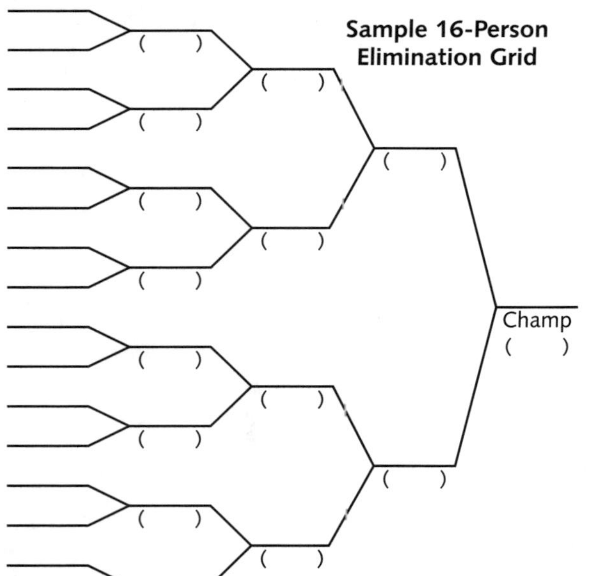

After a winner has been established, two new students are chosen, a new set of cards is placed on the board, and bidding begins again.

Sure, there is an element of luck — if a student only bids three cards, it is possible that the student could randomly match the bid. However, I have found that ego is the great equalizer! Oftentimes, a student with greater ability will bid too high and the less talented student will advance.

This game is usually done during the last 10 minutes of class over a period of 7 to 10 days. I have also added a couple of classroom management rules: students lose points for helping other students (they will get their turn!), and for talking while the game is in progress. Cheering or groaning after a card is selected is definitely allowed!

Index

** Topic is addressed in the lesson, but is not the focus of the lesson*
† Article that contains a lesson excerpt addressing the topic

MATHEMATICAL TOPICS

Index *(continued)*

MATHEMATICAL TOPICS *(continued)*

Index *(continued)*

MATHEMATICAL TOPICS *(continued)*

EDUCATION TOPICS

Index *(continued)*

EDUCATION TOPICS *(continued)*

Join the MPJ Network!

The Math Projects Journal periodically offers new lessons to the members of the *MPJ* network. These lessons are posted on a special page of our web site, www.mathprojects.com. Members are notified by email when a new lesson posting is available. Your purchase of *MPJ's Ultimate Math Lessons* serves as the membership fee. All that is needed in order for you to receive past and future lesson postings on our web site is for you to submit your name and email address to journal@mathprojects.com. If you purchased this book online or by phone, and have already submitted this information...you're in!

Professional Development Opportunities

Training for many of the lessons found in this book is offered in the following two-day teacher training workshops available through Prime Presentations.

Cool Projects = Hot Results in Raising Algebra & Geometry Test Scores
by Chris Shore

Ultimate Cosmic Power in an Itty Bitty Thinking Space
by Chris Shore

Algebra for All
by Carol McGehe

All Prime Presentations' professional development experiences are designed with pride and quality to energize and excite participants with classroom-tested ideas and cutting-edge technology. Teachers appreciate learning rich activities that address their specific curriculum and the quality black-line masters that can be reproduced for immediate classroom use. All of the 31 Prime Presentations staff development workshops are aligned with the National Council of Teachers of Mathematics Principles and Standards for School Mathematics or the National Science Education Standards, as well as individual state benchmarks, standards and high stakes tests.

If you are serious about providing the most effective professional development for your K-12 mathematics and science teachers, please call Prime Presentations today, at (888) 917-3950, or visit www.primepresentations.com.

Join the MPJ Network!

The Math Projects Journal periodically offers new lessons to the members of the *MPJ* network. These lessons are posted on a special page of our web site, www.mathprojects.com. Members are notified by email when a new lesson posting is available. Your purchase of *MPJ's Ultimate Math Lessons* serves as the membership fee. All that is needed in order for you to receive past and future lesson postings on our web site is for you to submit your name and email address to journal@mathprojects.com. If you purchased this book online or by phone, and have already submitted this information...you're in!

Professional Development Opportunities

Teacher training is offered in the following workshops conducted by the author, Chris Shore.

Cool Projects = Hot Results in Raising Algebra & Geometry Test Scores

Ultimate Cosmic Power in an Itty Bitty Thinking Space

Extensive professional development programs are also available. The goal of these customized programs is to improve your students' competency in mathematics by the only means possible: improving the mathematics instruction that they receive from their teachers.

For over seven years, Chris has conducted training for teachers across the nation. For more information on how to bring Chris Shore and these lessons to your school, district or region, call today, at

877-MATH-123

or visit

www.mathprojects.com